Lecture Notes in Mechanical Engineering

Lecture Notes in Mechanical Engineering (LNME) publishes the latest developments in Mechanical Engineering—quickly, informally and with high quality. Original research reported in proceedings and post-proceedings represents the core of LNME. Volumes published in LNME embrace all aspects, subfields and new challenges of mechanical engineering. Topics in the series include:

- Engineering Design
- Machinery and Machine Elements
- Mechanical Structures and Stress Analysis
- Automotive Engineering
- Engine Technology
- Aerospace Technology and Astronautics
- Nanotechnology and Microengineering
- Control, Robotics, Mechatronics
- MEMS
- Theoretical and Applied Mechanics
- Dynamical Systems, Control
- Fluid Mechanics
- Engineering Thermodynamics, Heat and Mass Transfer
- Manufacturing
- Precision Engineering, Instrumentation, Measurement
- Materials Engineering
- Tribology and Surface Technology

More information about this series at http://www.springer.com/series/11236

Jun Ni · Vidosav D. Majstorovic
Dragan Djurdjanovic
Editors

Proceedings of 3rd International Conference on the Industry 4.0 Model for Advanced Manufacturing

AMP 2018

Springer

Editors
Jun Ni
Department of Mechanical Engineering
University of Michigan
Ann Arbor, MI
USA

Dragan Djurdjanovic
Department of Mechanical Engineering
University of Texas
Austin, TX
USA

Vidosav D. Majstorovic
University of Belgrade,
Faculty of Mechanical Engineering,
Laboratory for Production Metrology
and TQM
Belgrade
Serbia

ISSN 2195-4356 ISSN 2195-4364 (electronic)
Lecture Notes in Mechanical Engineering
ISBN 978-3-319-89562-8 ISBN 978-3-319-89563-5 (eBook)
https://doi.org/10.1007/978-3-319-89563-5

Library of Congress Control Number: 2018941836

Printed on acid-free paper

This Springer imprint is published by the registered company Springer International Publishing AG part of Springer Nature
The registered company address is: Gewerbestrasse 11, 6330 Cham, Switzerland

Preface

The 3rd International Conference on the Industry 4.0 Model for Advanced Manufacturing, AMP 2018, will be held at University of Belgrade, Faculty of Mechanical Engineering, Belgrade, Serbia, from 5 to 7 June 2018. It is organized by the University of Belgrade, Faculty of Mechanical Engineering and Belgrade Chamber of Commerce and Industry. This year's conference attracted more than 250 participants, including academics, practitioners and scientists from 21 countries, who contributed 32 keynotes on plenary and workshop sessions.

The previous conferences on the Industry 4.0 Model for Advanced Manufacturing, AMP, were:

1. First International Conference USA-EU-Japan-Serbia Manufacturing Summit, Belgrade, 31 May 2016–2 June 2016, Serbia (AMP Conference 2016), with main topic: Advanced Manufacturing Program—Industry 4.0 model for Serbia.
2. Second International Conference USA-EU-Japan-Serbia Manufacturing Summit, Belgrade, 7–9 June 2017, Serbia—Smart and Intelligent Products (AMP Conference 2017).

The main objective of these conferences is to bring together leading world experts to discuss the challenges and opportunities of the new Industry 4.0 model of manufacturing. Our hope is that such an event will assist in the development and growth of new innovative manufacturing industries in Serbia, producing smart products with intelligent characteristics, and relying on modern, new manufacturing processes and systems.

Conference is hosted by the Faculty of Mechanical Engineering of the University of Belgrade. Belgrade is the capital city of Serbia, located at the scenic confluence of two major European rivers, with a uniquely remarkable and turbulent history, and a vibrant cultural and entertainment scene. University of Belgrade has a long tradition of academic excellence, where great minds from Nikola Tesla and Mihajlo (Michael) Pupin to Milutin Milankovic held lectures or were faculty. Its engineering still remains exceptionally respected in Europe, with its alumni scattered in top universities around the globe. Faculty of Mechanical Engineering in

Belgrade is the largest such school in south-eastern Europe and one of the largest in Europe.

Main topics of interest for this conference include:

- Industry 4.0 model framework,
- Design of smart and intelligent products,
- Innovative design and development of intelligent products,
- Big data challenges, data integrity, accuracy and authenticity,
- Cloud computing, cloud-based products,
- Cyber-Physical Manufacturing,
- Manufacturing automation in the Industry 4.0 model,
- Manufacturing systems and enterprise models for Industry 4.0,
- Advanced manufacturing,
- Engineering education for Industry 4.0.

We acknowledge the outstanding contributions of the following colleagues and friends for Conference establishing and development as following:

Conference Founder and Chair

Prof. Dr. Jun Ni, Department of Mechanical Engineering, University of Michigan, Ann Arbor, MI, USA.

Conference Co-chairs

1. Prof. Dr. Dragan Djurdjanovic (Conference founder), Department of Mechanical Engineering, University of Texas, Austin, TX, USA;
2. Prof. Dr. Vidosav D. Majstorovic (Conference founder), University of Belgrade, Faculty of Mechanical Engineering, Belgrade, Serbia;
3. Prof. Dr. Kornel Ehmann, Department of Mechanical Engineering, Northwestern University, Evanston, IL, USA—for North America region;
4. Prof. Dr. Wilfried Shin, TU Vienna, Wien—for EU region;
5. Prof. Dr. Yashiro Takaya, Osaka University, Osaka, Japan—for Far East region.

International Programme Committee

Dr. A. Archenti, KTH Royal Institute of Technology, Stockholm, Sweden; Prof. Dr. Emilia Assenova, TU Sofia, Bulgaria; Prof. Dr. P. J. Bartolo, University of Manchester, Manchester, UK; Prof. Dr. J. Beaman, University of Texas at Austin, USA; Prof. Dr. D. Bourell, University of Texas at Austin, USA; Prof. Dr. E. Budak, Sabanci University, Istanbul, Turkey; Prof. Dr. E. Carpanzano, Institute of Systems and Technologies for Sustainable Production, Switzerland; Prof. Dr. G. Constatntin, TU Bucharest, Romania; Prof. Dr. S. Coritrou, Technical University of Bucharest,

Romania; Prof. Dr. N. Dragulanesku, University of Politechnica Bucharest, Romania; Prof. Dr. N. Durakbasa, TU Vienna, Austria; B. Eyenon, Samsung USA, USA; Prof. Dr. L. M. Galantucci, Politecnico di Bari, Bari, Italy; Dr. S. Heieck, GIZ Belgrade; Prof. Dr. Zivana Jakovljevic, FME, University of Belgrade, Serbia; Prof. H. Hohonoki, Japan; Dr. L. Jalba, Microelectronica S.A., Romania; Dr. M. Janakiram, Intel USA, USA; Prof. Dr. J. Jedrzejewski, Wroclaw University of Science and Technology, Poland; Prof. Dr. F. Jovane, Politechnico de Milano, Italy; Prof. Dr. A. Jovovic, University of Belgrade, Belgrade; J. Kappenmann, GIZ, Belgrade; Prof. Dr. D. Kochan, TU Dresden, Germany; Prof. Dr. D. Kramar, University of Ljubljana, Slovenia; Prof. Dr. B. Lalic, Faculty of Technical Science, University at Novi Sad, Novi Sad; Prof. J. Lee, University of Cincinnati, Cincinnati, USA; Prof. Dr. Lin Ma, Queensland University of Technology, Brisbane, Australia; Prof. Dr. M. Macchi, Politecnico di Milano, Italy; Prof. Dr. J. Mathew, Queensland University of Technology, Brisbane, Australia; Prof. Dr. R. Mitrovic, University Belgrade; Prof. Dr. P. Monka, TU Presov, Slovakia; Prof. Dr. G. Moroni, Politecnico di Milano, Milano, Italy; Prof. Dr. D. Mourtzis, University of Patras, Rio-Patras, Greece; Katarina Obradović-Jovanović, Assistant Minister, Ministry of Economy, Belgrade; Dr. Augusta Paci, CNR, Rome, Italy; Prof. Dr. M. Piska, TU Brno, Czech Republic; Prof. Dr. K. Rajurkar, University of Nebraska, Lincoln, USA; Prof. Dr. G. Seliger, Production Center, Berlin Institute of Machine Tools and Factory Operation, Assembly & Assembly Plant, Berlin, Germany; Prof. Dr. M. Sokovic, University of Ljubljana, Slovenia; Prof. Dr. Vesna Spasojević-Brkić, University of Belgrade, Serbia; Prof. Dr. Miroslav Trajanović, MEF, Nis; Prof. Dr. J. Vancza, TU Budapest, Hungary; Prof. Dr. Birgit Vogel-Heuser, Technical University of Munich, Germany; Prof. Dr. Sanja Vranes, University of Belgrade; Prof. Dr. L. Wang, KTH Stockholm, Sweden.

Organizing Committee

Prof. Dr. Zivana Jakovljevic, FME, University of Belgrade, Serbia; Chair; Assist. Prof. Dr. Slavenko Stojadinovic, FME, University of Belgrade, Serbia; Mirjana Velimirović, FME, Belgrade; Marija Glišić, FME, Belgrade; Julija Kostić, FME, Belgrade; Elena Đura, FME, Belgrade; Milan Rančić, FME, Belgrade; Nemanja Gligorijević, FME, Belgrade.

AMP 2018 Conference can be regarded as a leading global conference in the area of modern manufacturing to several of its special dimensions: (i) it presented a spectrum of scientific and practical advancements in the field of advanced manufacturing (Cyber-Physical Manufacturing, Industry 4.0), and (ii) it offered practical applications and solutions for various problems in the world of modern manufacturing.

Conference planning, preparation and realization required engagement of a number of persons and organizations. We express our gratitude to all of them, and especially to:

- Founder, Chair, Co-chair and Conference International Programme Committee Members,
- All authors, and especially to the authors that prepared keynote papers thus contributing to the high scientific and professional level of the Conference,
- All members of International Programme Committee for the review of the papers and chairing the Conference Sessions,
- Springer and Mr. Pierpaolo Riva for publishing AMP Conference Proceedings within the edition Lecture Notes in Mechanical Engineering,
- Ministry of Education, Science and Technological Development of the Republic of Serbia for the support of the Conference, and
- Belgrade Chamber of Commerce and Industry, Conference co-organizer.

We wish to express a special gratitude to the students at the University of Belgrade, Faculty of Mechanical Engineering: Ms. Mirjana Velimirović, Ms. Marija Glišić, Ms. Julija Kostić, Ms. Elena Đura, Mr. Milan Rančić and Mr. Nemanja Gligorijević for their invested efforts that enabled preparation and realization of the AMP Conference in the possible best manner.

March 2018

Jun Ni
Conference Chair
Vidosav D. Majstorovic
Conference Co-chair
Dragan Djurdjanovic
Conference Co-chair

Contents

Hybrid Machining Processes 1
Davorin Kramar

Intelligent Sensing Systems – Status of Research at KaProm 18
Zivana Jakovljevic, Milica Petrovic, Stefan Mitrovic, and Zoran Miljkovic

Industry 4.0 and Paradigm Change in Economics and Business Management 37
Dragan Đuričin and Iva Vuksanović Herceg

Photogrammetry Applied to Small and Micro Scaled Objects: A Review 57
Luigi Maria Galantucci, Maria Grazia Guerra, and Fulvio Lavecchia

Artificial Neural Networks in Advanced Thermoset Matrix Composite Manufacturing 78
Pierpaolo Carlone, Dragan Aleksendrić, Felice Rubino, and Velimir Ćirović

Recent Streams of Digital Manufacturing, Its Emerging Trend and Future Directions for SME in Japan 89
Hideaki Hohnoki

Intelligent Integrated Management and Advanced Metrology for Quality Toward the Factory of the Future 98
Numan M. Durakbasa, J. Martin Bauer, Lukas Kräuter, Gokcen Bas, and Günther Poszvek

Hybrid Machining: An Industrial Case-Study Comparing Inconel718 Reaming and Drilling with Abrasive Waterjet Technology 109
B. Wanner, A. Archenti, and C. M. Nicolescu

Dynamic Definition of Machine Tool Feed Drive Models in Advanced Machine Tools 115
George Constantin

Strategic Outsourcing of SMEs in the Context of Industry 4.0: Evidence from Serbia 139
Bojan Lalic, Tanja Todorovic, Nela Cvetkovic, Nemanja Tasic, and Ugljesa Marjanovic

Fundamental Requirements for CAPP Software Design Focusing on Industry 4.0 Specific Features 146
Peter Monka, Katarina Monkova, Milan Edl, Helena Zidkova, and Vladimir Duchek

Strategic Note for a Digital Innovation Policy 156
Augusta Maria Paci

Machining Process Improvement Through Process Twins 164
Erhan Budak

Connecting Humans to the Loop of Digitized Factories' Automation Systems 180
Emanuele Carpanzano, Andrea Bettoni, Simon Julier, Joao C. Costa, and Manuel Oliveira

Development of Skills and Competences in Manufacturing Towards Education 4.0: A Teaching Factory Approach 194
Dimitris Mourtzis

Opportunistic Maintenance for Wind Turbines Considering External Opportunities – A Case Study 211
H. Truong Ba, M. E. Cholette, P. Borghesani, and L. Ma

Cyber-Physical Manufacturing in Context of Industry 4.0 Model 227
Vidosav D. Majstorović, Mirjana Velimirović, Marija Glišić, Julija Kostić, Elena Đura, Milan Rančić, and Radivoje Mitrović

An Automated Laser Control Technique for Improving Powder Bed Temperature Uniformity in Selective Laser Sintering 239
T. Phillips, L. Zhang, S. Fish, and J. Beaman

Gaussian Process Regression for Virtual Metrology of Microchip Quality and the Resulting Selective Sampling Scheme 250
Tyler Darwin, Roman Garnett, and Dragan Djurdjanovic

Cloud Enabled CPS and Big Data in Manufacturing 265
Lihui Wang and Wei Ji

Geometric Inspection Planning as a Key Element in Industry 4.0 293
Giovanni Moroni and Stefano Petrò

Hybrid Machining Processes

Davorin Kramar(✉)

Faculty of Mechanical Engineering, University of Ljubljana, Ljubljana, Slovenia
davorin.kramar@fs.uni-lj.si

Abstract. In this paper a definition and classification of hybrid processes is given with the focus on assisted hybrid machining processes developed at the Faculty of Mechanical Engineering in Ljubljana. These processes include high-pressure jet assisted machining, cryogenic machining, laser assisted machining and ultrasonic assisted machining. The principles of each individual process are described, followed by the presentations of the results of machinability and productivity improvements. The results were obtained during the last eight years of research on machinability of different hard-to-machine materials at the Department for Management of Manufacturing Technologies.

Keywords: High-pressure jet assisted machining · Cryogenic machining
Laser assisted machining · Ultrasonic assisted machining

1 Introduction

Modern products, especially in the aerospace, automotive, space and power industry, require high-performance materials that retain desired mechanical properties even at very high temperatures. Machinability of materials, which meet these requirements, is generally poor or economically unacceptable with the use of conventional machining processes. For the purpose of improving machinability and reducing machining costs, newer, hybrid approaches to the machining of such materials are being developed. Hybrid machining processes (HMPs) are based on the simultaneous and controlled interaction of process mechanisms and/or energy sources/tools having a significant effect on the process performance and material machinability [1, 2].

Several definitions and descriptions of the term 'hybrid processes' can be found in literature. As stated by Rajukar et al. [3] hybrid process is defined as the combination of two or more processes of material separation. Hybrid processes increase the advantages and minimize the potential disadvantages found in individual process. According to Kozak et al. [4] the performance of HMPs must differ considerably from the performance of the combination of each individual process. In general, there are two types of hybrid processes: firstly, processes where all participating processes are directly involved in material removal and secondly, processes where only one of the participating processes directly removes material while the other only executes a supporting function by positively changing the machining conditions. The main objective of a HMP, as stated in [5] is to obtain the '1 + 1 = 3' effect, i.e. the positive effect of a HMP is larger than the simple sum of the individual advantages of a single process and may also simultaneously

J. Ni et al. (Eds.): AMP 2018, LNME, pp. 1–17, 2018.
https://doi.org/10.1007/978-3-319-89563-5_1

eliminate or minimize adverse effects that a constituent processes may have contributed when applied individually. As highlighted in [6], in HMPs different forms of energy are applied simultaneously in the same active zone, e.g. laser assisted machining and HPMs are also defined as a combination of effects during a process that can be achieved conventionally in separate but successive processes, e.g. grind hardening.

In 2010, CIRP, namely the International Academy for Production Engineering, refined these definitions and proposed an open and a narrow definition [7, 8]: (i) Open definition: a hybrid manufacturing process combines two or more established manufacturing processes into a new combined set-up whereby the advantages of each discrete process can be exploited synergistically; (ii) Narrow definition: Hybrid processes comprise a simultaneous acting of different (chemical, physical, controlled) processing principles on the same processing zone.

In this paper, a classification of hybrid processes is detailed, with a focus on the assisted machining hybrid processes, followed by the description of the hybrid processes developed at the University of Ljubljana, Faculty of Mechanical Engineering (FME-LJ). The results presented in the following chapters were obtained during the last eight years of research on machinability of different hard-to-machine materials at the Department for Management of Manufacturing Technologies.

2 Classification of Hybrid Processes

According to the literature [2, 4, 7, 8] the HMPs can be generally classified into the following two categories:

1. Assisted-type processes in which only one of the participating processes directly removes material, while the other process only assists in the material removal by having positive effects on the machining conditions.
2. Combined or mixed-type processes in which all constituent processes are directly involved in the material removal;

Figure 1 illustrates a classification of HMPs proposed by Lauwers et al. [8] adapted by the author. The first group (I) consists of HPMs where two or more energy sources/tools/mechanisms are combined and have a synergetic effect on the material removal process in the processing zone. A further classification is made in 'Assisted-type Processes' (I.A) and 'Combined or Mixed-typed Processes' (I.B), as explained above. In assisting processes, a main machining process (turning, milling, drilling …) is defined as the primary process. The secondary process provides only assistance, by utilizing either medium/fluid (I.A. 1), heat (I.A. 2), high-frequency vibrations (I.A. 3), abrasives (I.A. 4), or a magnetic field (I.A. 5). In Combined or Mixed-type processes, two or more processes are present, which according to the definitions above should occur more or less at the same time. The second group (II) of hybrid processes contains processes where a controlled combination of effects occurs which would conventionally be caused by separated processes. For example, in grind-hardening, material removal is combined with controlled hardening due to the grindings' process induced heat. The following section describe the most important group of hybrid processes, namely Assisted Hybrid Machining Processes.

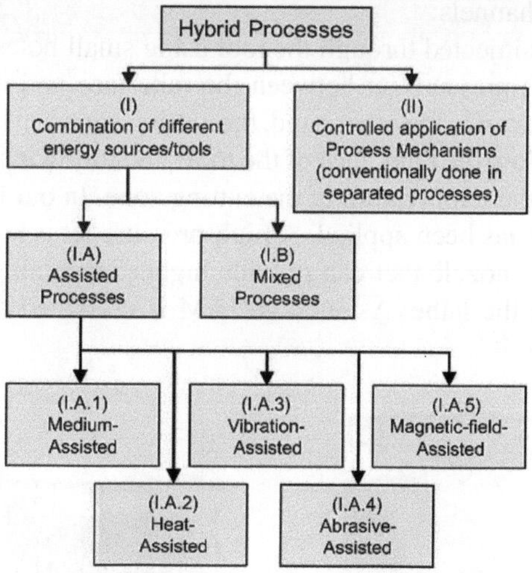

Fig. 1. Classification of the hybrid processes; adapted from [8].

3 Assisted Hybrid Machining Processes

In this paper only Assisted Hybrid Machining Processes are considered, since only this type of hybrid processes were used in our investigations. As stated above this type of processes consist of a primary machining process, which defines the process and a secondary process, which only assists the primary one. The amount of energy applied for the assisted process could be in some applications (jet, laser) relatively high compared to the primary process. As seen from Fig. 1, these assisted-type HMPs can be further classified based on the energy source and/or mechanism of material removal, and/or the source of assistance such as medium-assisted, heat-assisted vibration-assisted, abrasive-assisted, and magnetic-field-assisted HMPs. In all cases these assistances intensify the material removal mechanism and improve machinability of hard-to-machine materials.

3.1 High-Pressure Jet Assisted Machining (HPJAM)

In machining, the chip formation is largely influenced by the heat and friction generated in the contact zone between the rake face of the tool and the machined surface material. Conventional flooding is not efficient enough to prevent extreme thermal conditions in the cutting zone, especially when hard-to-machine materials are processed. Compared to the conventional flooding, the idea of HPJAM is to inject a high-pressure jet of cooling-lubrication fluid into the cutting zone. The jet can be applied in two ways:

- With an external nozzle:
  The jet is injected directly in between the rake face and the chip or can be directed to the gap between the flank face and the workpiece.

- Through internal channels:
  The cutting fluid is injected through the tool using small holes in the insert.
  The supply of high-pressure jet between the rake face and the chip decreases the length of their contact. On the other hand, the cutting zone can be reached by injecting the cutting fluid below the flank face of the tool. The following procedure is the most efficient to reduce the temperature in the cutting zone. In our investigations, the set-up shown in Fig. 2 has been applied. A high-pressure jet is applied on the rake face through an external nozzle that can provide higher machining performances and is easier to be set on the lathe. As such HPJAM is classified as a Medium-Assisted Machining Process.

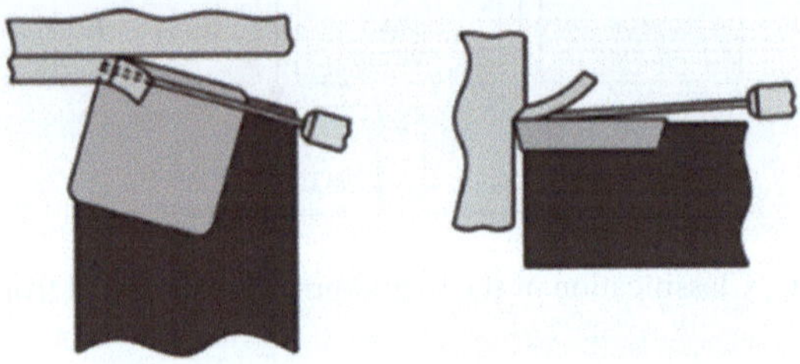

Fig. 2. Direction of high-pressure cooling lubricant application; between the rake face and the chip.

The early research on the HPJAM at FME-LJ included its application to the machining of hardened chrome-plated steel [9]. The aim of this investigation was to compare the capabilities of dry, conventional and HPJAM in turning of the surface

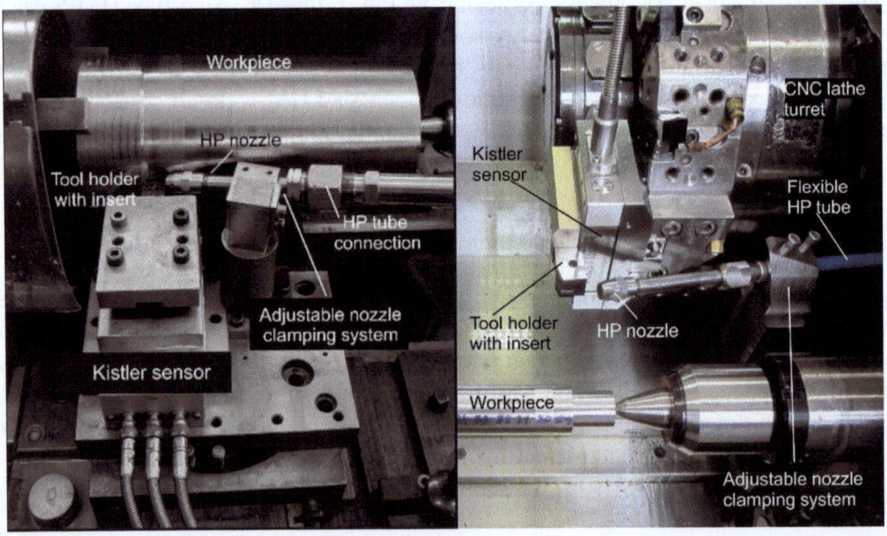

Fig. 3. Machine tool setup for HPJAM used in experimental study; left: conventional lathe, right: CNC lathe.

hardened piston rods. All investigations were performed with conventional cutting parameters ($v_c = 100$ mm/min, $f = 0.25$ mm/rev, $a_p = 2$ mm) using coated carbide tools with no chip breaking geometry on the rake face. The first tests were conducted in longitudinal turning on a conventional lathe of high stiffness, equipped with a high pressure plunger pump (Fig. 3, left) and a nozzle diameter of 0.3 mm.

The experimental study has proved that the hard turning of steels with coated carbide tools is not possible in dry cutting conditions with a flat faced tool. The process capabilities of conventional and HPJAM methods were compared with respect to the chips' breakability and the tool life. Both processes are characterised with their region of operability. The major concluding remarks related to the HPJAM precedence over conventional flooding in hard steel turning with flat faced coated carbide tools are as follows. A great extension of the region of operability for a given tool-material is attained with HPJAM.

More specifically, a 45% increase in the maximum achievable cutting speed and a 27% increase in the maximum achievable feed rate were revealed. With the application of HPJAM a significant improvement in chip breakability was achieved. At a pressure of 30 MPa a relatively good breakability of chips was observed. However, the lack of cooling was noticed because the chips were significantly blackened, hence burned as can be seen in Fig. 4. Insufficient cooling is related to the low coolant lubricant flow rate at such a pressure, with the amounts of 0.7 l/min at 30 MPa. At pressures higher than 70 MPa (flow rate approx. 1.1 l/min and more), good breakability of the chips as well as a suitable cooling was observed.

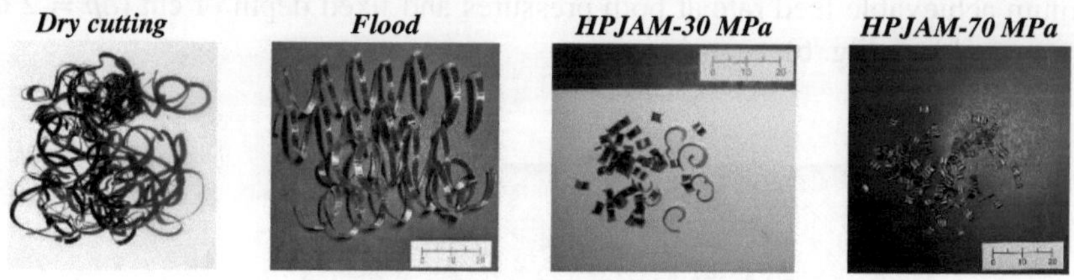

Fig. 4. Chip forms regarding the cooling conditions and cooling lubricant pressure in hard steel turning.

Figure 5 shows the tool flank wear trend in conventional flooding and HPJAM. For the selected criteria VB = 0.1 mm, tool life in the case of HPJAM was about ten minutes, which is approximately five times longer than in the case of flooding.

Later on a high-pressure system was installed on the CNC lathe (see Fig. 3 right). Experiments with this equipment were mostly conducted on the Inconel 718. An extensive experimental research on this Nickel alloy machinability has been performed, and is reported in [10].

The influence of HPJAM process parameters on the region of operability, chip formation, cutting forces, contact length, surface finish and heat generation has been investigated. The extension of the operational area for a given tool-material pair strongly depends on the nozzle diameter and the applied pressure. Or more specific, with a smaller nozzle ($dn = 0.25$ mm) and a lower pressure ($p = 50$ MPa) no significant improvement

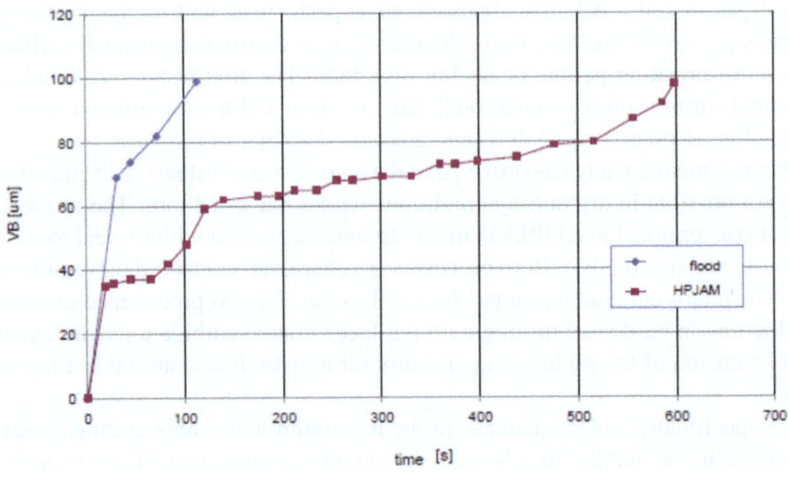

Fig. 5. Tool flank wear development for conventional flooding and HPJAM.

comparing to conventional cooling lubrication was achieved. With the same nozzle, but at the higher pressure ($p = 130$ MPa) the region of operability was enlarged on the side of lower cutting speeds and lower feed rates. A considerably higher performance is achieved with a larger nozzle ($dn = 0.4$ mm). Compared to flooding, more than a 20% increase in the maximum achievable cutting speed and almost a 30% increase in the maximum achievable feed rate at both pressures and fixed depth of cut ($ap = 2$ mm), were revealed (see Fig. 6).

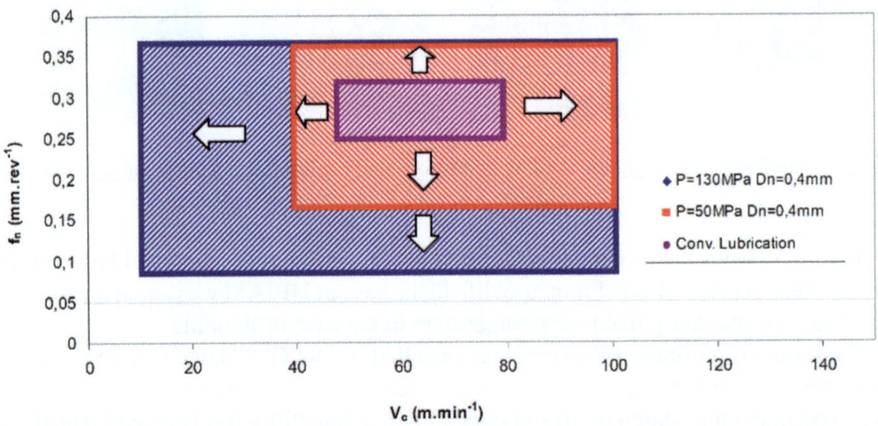

Fig. 6. Region of operability for HPJAM and conventional flooding of Inconel 718.

In HPJAM of Inconel 718 chips were perfectly fragmented. As expected, the pressure was the most influent parameter. An increase in pressure resulted in a considerable decrease in chips' size. From these results [10], it can be concluded that changing the jet parameters does not affect the tool temperature in the considered ranges of pressures and nozzle diameters. It has been revealed that at any speed condition, the interface temperature initially decreases with an increase in jet pressure, up to a critical pressure, above which the temperature is almost constant. Nevertheless, the efficiency of HPJAM in limiting the heat transmitted to the tool is illustrated by the online measurements shown in Fig. 7.

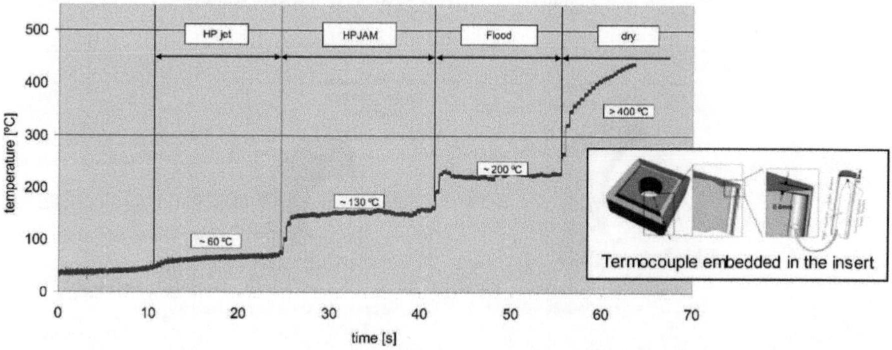

Fig. 7. Comparison of on-line temperature measurements between three lubrication conditions.

It can be stated that a 30% decrease in tool temperature compared to conventional lubrication is achievable with a 450% reduction in coolant flow. The thermal advance is substantial compared to dry cutting with an approximate decrease of tool temperature by 70%.

In the next study [11] a finite element (FE) model for orthogonal machining was developed in order to reach additional data on HPJAM, compare trends with the experimental ones and understand the influence of the high-pressure jet on the cutting process. Mechanical and thermal loads induced by the jet were considered in a four steps simulation (see Fig. 8). The consequences on the primary and secondary shear zones, chip formation and tool rake face have been studied within the whole pressure range ($p = 30 –$ 130 MPa) and compared to dry cutting. It was shown that the jet is able to decrease the cutting forces, chip radius and tool–chip contact length. The contact pressure and temperature fields on the cutting tool are also reduced as well as the tool-chip contact length. It was confirmed that the effects of convection are able to change and even amplify the influence of the pure mechanical load induced by the high-pressure jet.

Some further studies have been conducted at FEM-LJ in order to improve machinability of the Inconel 718 with introduction of HPJAM [12], studies to optimise the process parameters using Taguchi and DOE approach [13], and studies to predict the machining outcomes with the use of a fuzzy expert system optimized with bioinspired algorithms [14]. Numerous benefits of HPJAM in the machining of hard-to-machine materials have been presented within the aforementioned research studies. HPJAM is probably the most promised and commonly used HMP in praxis, and several cutting tool producers are selling high-pressure jet systems (up to 10 MPa). Thought some improvements in jet delivery at the cutting tool with nozzle optimization still need to be

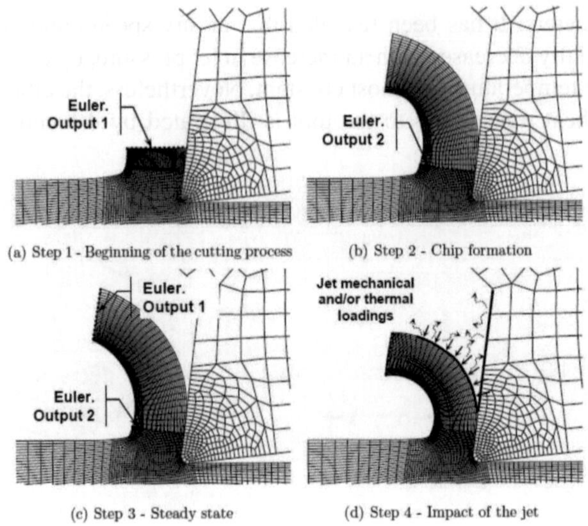

Fig. 8. Simulation procedure – v_c = 50 m/min; f = 0.20 mm/rev; a_p = 2 mm. (a) Step 1 – beginning of the cutting process; (b) Step 2 – chip formation, (c) Step 3 – steady state, and (d) Step 4 – impact of the jet [11].

performed. The cooling fluid closed-loop system requires further development from the filtering and cooling point of view.

3.2 Cryogenic Machining

Cryogenic machining (Cryo) is another media assisted HMP involved in FEM-LJ R&D. It represents an innovative delivery method of the cryogenic cooling media, liquid nitrogen (LN) or CO_2 to cool either a cutting tool, which experiences the highest temperature during the machining process or a workpiece in order to change the material characteristics and to improve machining performance. Cryogenic fluid delivered to the cutting zone at a very low temperature, e.g. −197 °C for LN, quickly evaporates and leaves no residue.

In the study of Pusavec et al. [15] a sustainability evaluation of Cryo and HPJAM in comparison to conventional machining was examined. Sustainability performance measures refer to environmental impact, energy consumption, safety, personal health, waste management, and costs as individual sustainable principles. An experimental study of the Inconel 718 turning was performed. The tool-wear of the carbide cutting tool was measured and used to determine the tool life for each cooling condition. The tool life was normally highly influenced by the cutting speed, and therefore presented for minimum cost criteria. The results revealed that production costs are highly correlated with the cost of cutting tools through the tool life characteristics. In machining of the Inconel 718, the cost of cutting tools amount to 20–30% of the overall production costs. It was shown that both Cryo and HPJAM decrease overall production costs for up to 30% in comparison to conventional machining. The optimum cutting speed for

Cryo and HPJAM, with the respect of lowest overall machining cost is $v_c = 90$ m/min, while in the conventional flooding this has to be decreased to $v_c = 30$ m/min, resulting in lower productivity. Furthermore, HPJAM can be the most cost efficient, however Cryo-machining is completely clean and has the highest sustainability potential. In conventional or HPJAM using oil-based CLFs, the overall technology is completely different in comparison to Cryo-machining, see Fig. 9.

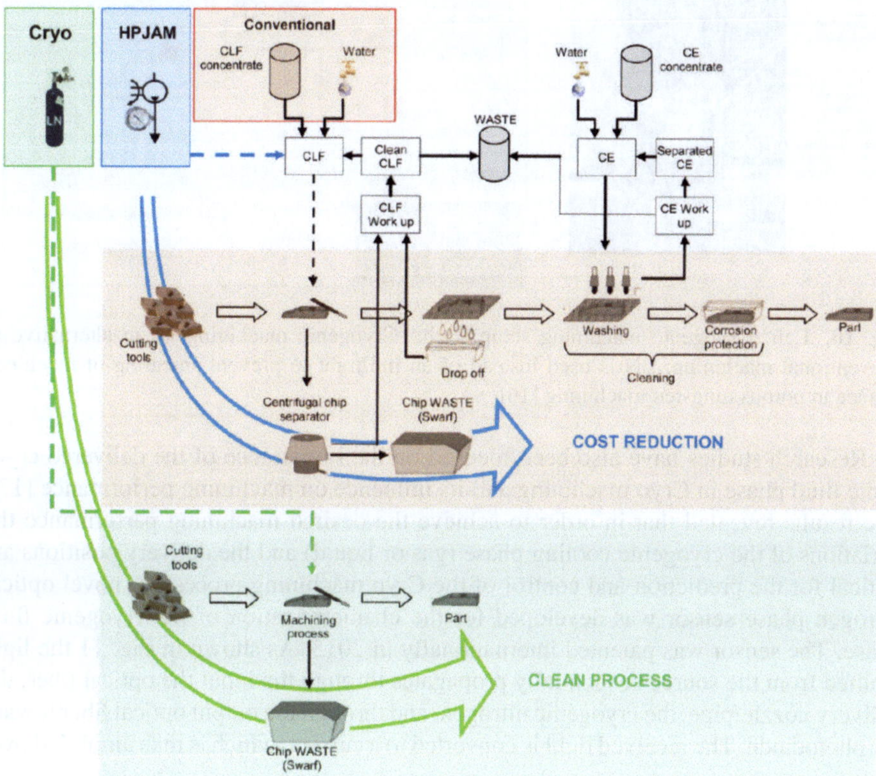

Fig. 9. A comparison of different machining technologies and relations in part production processes [15].

The technology, where CLFs are applied, has to take care of cleaning the parts after machining, drying them, and protecting them against corrosion. This presents additional costs, which are eliminated in cryo-machining. Besides additional equipment costs, the main disadvantage of these two machining alternatives, is the relatively high price of cryogenic fluid (LN), which is not reusable, when the Cryo method is applied, while in the HPJAM method, there is the fact that oil-based CLFs are still in use.

The research in [16] focuses on sustainable machining of hard-to-machine porous tungsten. Investigation of the combined effects of the process technique (cryogenic machining vs. conventional machining with plastic infiltrant), cutting parameters (f, v_c, a_p) and cutting tool grade (carbide, PCD, CBN, and ceramic) on the porosity of the machined surface was performed (see Fig. 10). For the smearing evaluation, pores on the machined surface are individually analysed from SEM pictures. A genetic algorithm-based process optimization was developed for the selection of optimal cutting parameters and the tool grade to obtain the specified machining surface quality with the open pores required for the final product.

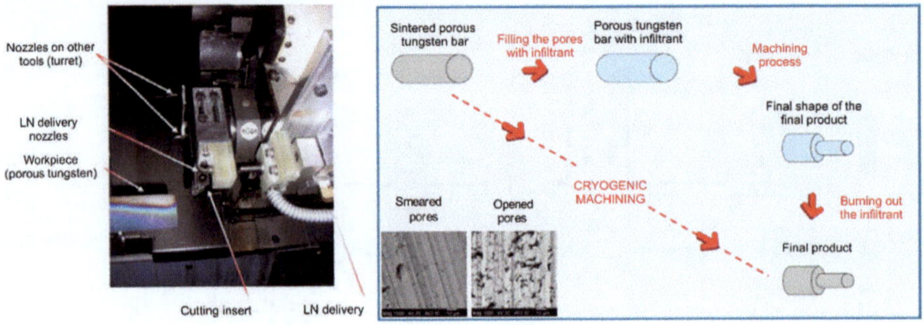

Fig. 10. Left: Cryogenic machining setup; right: Cryogenic machining as an alternative to conventional machining; LN is used instead of an infiltrant to prevent smearing of machined surface in porous tungsten machining [16].

Research studies have also been focused on the importance of the delivered cryogenic fluid phase in Cryo machining and its influence on machining performance [17]. The results revealed that in order to achieve the desired machining performance the variations of the cryogenic cooling phase (gas or liquid) and the delivery positions are critical for the prediction and control of the Cryo machining process. A novel optical nitrogen phase sensor was developed for the characterization of the cryogenic fluid phase. The sensor was patented internationally in 2015. As shown in Fig. 11 the light emitted from the source sequentially propagates through the input the optical fiber, the delivery nozzle/pipe, the cryogenic nitrogen, and through the output optical fiber toward the photodiode. The received light is converted to a current, which is than amplified over

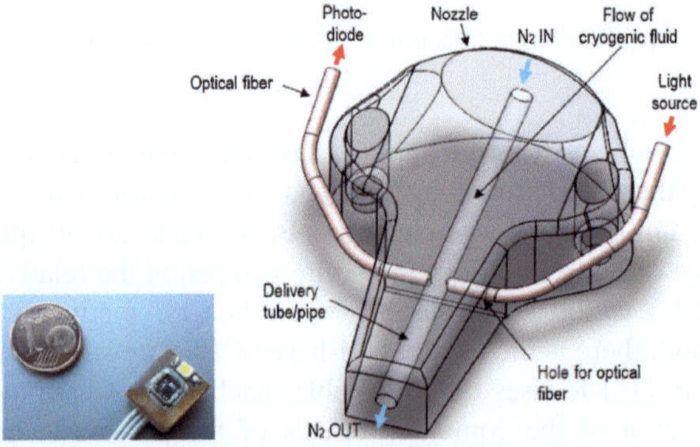

Fig. 11. Innovative cryogenic fluid phase sensor; left: electronic circuit, right: the schematic sensor arrangement integrated into the delivery nozzle [17].

the voltage signal. The sensor does not provide just the possibility to distinguish between the liquid and the gas phase, but also to determine the percentage of liquid in the gas phase or vice versa. The results proved short response times and high robustness of the developed phase sensor.

3.3 Laser Assisted Machining

Laser Assisted Machining (LAM) is a heat-assisted HMP, which consists of a conventional cutting tool and a laser beam (or more laser beams) that heats the to-be-machined material right in front of the cutting tool. Due to the laser heating, the material is softened, which makes the cutting process easier and in some cases even the material removal mechanism changes. The changes in the temperature-dependent material properties significantly affect the machinability of some hard-to-machine materials in terms of cutting forces, tool wear, chip formation and machined surface integrity.

Work material classes that are well suited for application of LAM technology include: (i) hard and brittle materials such as engineering ceramics that can otherwise be subject only to cost-prohibitive abrasive processing, (ii) heat resistant materials like nickel alloys, (iii) materials with abrasive constituents such as high silicon content aluminium alloys, and (iv) materials with a propensity to significant strain hardening like austenitic stainless steels. For each unique workpiece material a different wavelength of the laser source is optimal in terms of absorptivity, but also other factors have to be considered when developing LAM systems. There are three types of laser sources used for LAM, which differ in the active lasing medium: CO_2, Nd:YAG and High Power Diode Lasers (HPDL). In Table 1, the main characteristics of these types of lasers, including the absorptance for a metallic and a non-metallic material, are summarized. In terms of the implementation of laser sources in an industrial LAM process, the most important characteristic of the laser source is efficiency, combined with the absorptance for the machined material. For example, if we take a CO_2 laser, which has a 10% efficiency in transforming electrical power into laser light, and use it for a LAM process on carbon steel (0.1 absorptance), the final efficiency of the laser heating is only about 1%. Very important characteristics in terms of implementation in an industrial process are also the size, rigidity and the possibility of using fibers for transmitting the laser light to the workpiece. The latter constitutes a major advance in terms of flexibility. As seen in Table 1, HPDLs have the highest efficiency while CO_2 laser systems cannot use fibers for transmitting the laser light to the workpiece. In terms of size a 1 kW CO_2, Nd:YAG and HPDL laser systems would have the volumes of about 1 m^3, 100 dm^3 and 100 cm^3, respectively. Another advantage of HPDL-s is that the system needs almost no maintenance. Many LAM studies were conducted and numerous authors had demonstrated the beneficial effects of laser assistance on the conventional cutting, however very few systems were implemented within industrial processes. In most of the studied cases CO_2 and Nd:YAG lasers were used.

With the recent development of HPDLs and their unique advantages, which enable the use of such systems not only for research purposes, but also in industrial environments, the possibility of mass implementation of LAM systems within industrial environments opens up. On the other hand, the laser industry has been evolving quickly in

Table 1. Specifications and properties of different laser sources used for LAM [18]

	CO_2	Nd:YAG	HPDL
Wavelength [μm]	10.6	1.06	0.8–0.95
Output power [kW]	Up to 30	Up to 3	Up to 3
Efficiency [%]	5–10	1–3 (F) 15–30 (D)	30–60
Spec. volume [l/kW]	1000	100	0.1
Using fibers	Impossible	Possible	Possible
Operating voltage [kV]	Up to 20	Up to 1	Up to 0.1
Service intervals [h]	1000	500 (F)	No maintenance
Spec. price [€/W]	75–250	90–400	40–250
Absorptance of Si_3N_4	0.9	0.6–0.8	0.7
Absorptance of carbon steel	0.1	0.35	0.4

the last years and new types of laser sources, which would be suited for LAM, are being developed, for example fiber lasers.

In one of our studies we implemented an experimental system for LAM development, but the system is still in the early developmental stages. The concept of an experimental system developed in 2015 at FEM-LJ is illustrated in Fig. 12. The 400 W YLR-400-AC continuous wave fiber laser from IPG Photonics was applied as the laser source with a wavelength of 1070 nm with a laser beam diameter of 5 mm. The X40CrMoV5-1 tool steel workpiece with a hardness of 59 HRc was bolted onto a Kistler multi-component dynamometer that was clamped onto the working table of the Mori Seiki Frontier M1 vertical machining center. The ZR5220602 carbide tools from WIDIN were applied in the experiments and in order to obtain more accurate cutting force measurements, one of the two cutting edges was grinded off, resulting in a single cutting edge tool. In the experiments, the cutting tool and the laser spot moved relative to the workpiece in the x direction at a constant feed rate v_f (see Fig. 12, right), keeping a constant distance l between the cutting tool and the laser spot. The relations between the measured forces (F_x, F_y and F_z) and the cutting force F_c, the feed force F_f and the back force F_p were determined by calculating the angle ϕ as illustrated in Fig. 12 at every point in time.

The study revealed a decrease in the cutting forces of up to 40% when a laser power of 400 W was applied, in comparison with the dry cutting without laser assistance. Furthermore, two empirical modelling techniques were implemented and both yielded good results in the cutting forces predictions [19]. As stated in the published experiments, these techniques will be used for predictions of the process performance measures in future experiments. The current experiments are directed towards the development of an analytical heat transfer model as the study revealed that the current model was too simplified. In LAM, the control of the heat input is a key component of the process as the deviations from the ideal temperature field in the to-be-cut material can even lead to lowering the process performance in comparison to the conventional process. The first limitation is that the to-be-cut material has to be heated above a certain temperature at which a drastic decrease in workpiece material strength and hardness occurs, through all the depth of the cut. Below the depth of the cut, the temperature should never exceed

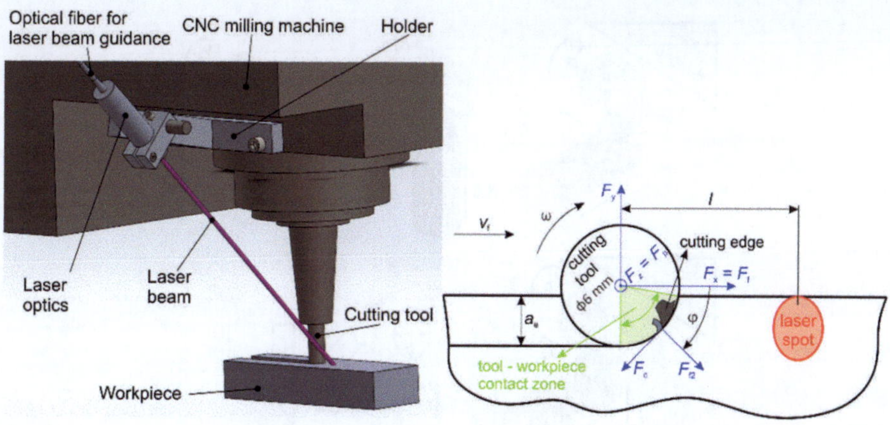

Fig. 12. Left: Concept of the experimental system for laser assisted milling; right: top-view scheme of the LAM process

a transition temperature, as that would prove detrimental to the surface quality, e.g. the tempering effect of heating quenched steel above the critical temperature where the martensitic microstructure begins to deteriorate. The third limitation is linked to the cutting tool. The temperature at the workpiece-tool interface should never exceed the maximum tool working temperature, as that would lead to increased tool wear or even catastrophic tool failure.

3.4 Ultrasonic Assisted Machining

The next group of assisted HMP is the vibration assisted machining. Some first attempts were made at FME-LJ in the implementation of ultrasonic assisted turning (USAT), which belongs to this group. In USAT the cutting tool is induced with small amplitude vibrations, usually couple of microns, and high frequency, around 20 kHz. The results of these movements are the intermittent contacts between the cutting tool and the workpiece. This leads to the generation of a cutting force only during one half-period of the vibration. The cutting force is in this case usually higher than those during conventional turning, but the average forces in USAT are smaller. USAT is a potential technology, suitable for the machining of hard-to-machine metallic and non-metallic materials. It leads to a better surface of the machined workpiece, a longer tool life and a more favourable microstructure of the surface layer. The ultrasonic vibrations can be applied along any of the three cutting directions or in two directions generating an elliptical movement (Fig. 13). The most common configuration is the vibration in the cutting direction. The main condition for vibration assisted machining which has to be met to make an effect, is the cutting speed to be less than $v_c < v_{c,\,max} = 2\pi A\upsilon$, where $v_{c,max}$ is the critical cutting speed, A is the amplitude and υ is the frequency of vibrations [8].

In our research study the sonotrode, which serves also as the toolholder, is made from steel SPM10 and its dimensions were calculated with respect to the acoustic properties for a given working frequency. The modelling of the acoustic behaviour of the

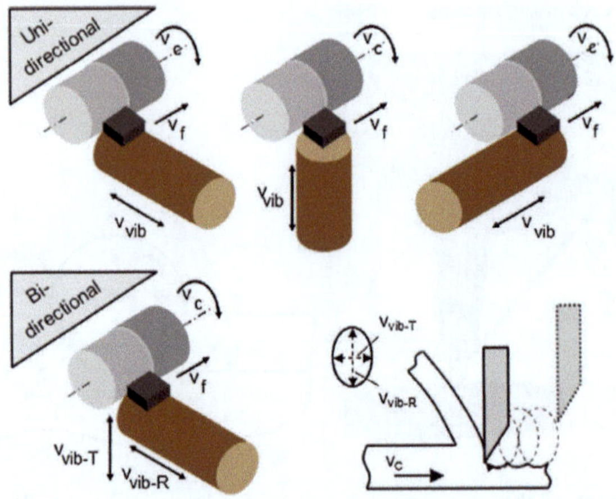

Fig. 13. Variants of vibration assisted turning [8].

sonotrode has been performed and the snapshot of the final result is illustrated in Fig. 14, left. The applied coated carbide tool VCGT 11 03 04-UM 1125, was mounted on both sides of the sonotrode provided the symmetric balance during oscillations. The sonotrode was fixed through the brass ring on a steel holder, customized to fit on the dynamometer and onto the conventional lathe.

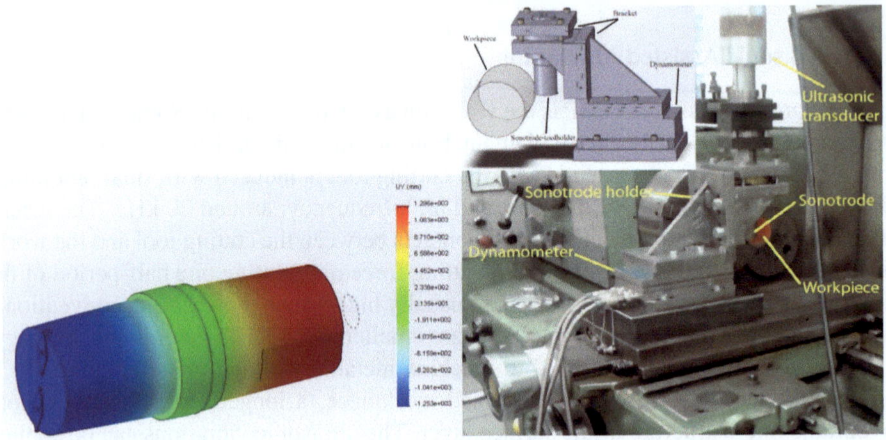

Fig. 14. Left: Sonotrode acoustic analyses; Right: Experimental setup for USAT [20].

The position of the ultrasonically assisted turning device mounted on the lathe is illustrated in Fig. 14, right.

The influences of the cutting parameters and the amplitude of oscillation on the cutting force components were investigated and the comparison with conventional turning was performed. At the beginning of the experiments, only conventional turning was employed. This phase lasted about 30 s and then the ultrasonic vibration was turned on for about the same time. With this kind of approach a direct comparison of the cutting forces can be conducted. An example of the cutting forces variation is illustrated in the graph in Fig. 15, left. Smaller values of the cutting forces are generated during USAT in all combinations of the parameter's setting. The measured average force components during conventional and USAT are illustrated in Fig. 15, right. The average force was computed by calculating the numerical average of the measured force component over time. In USAT the average tangential force was reduced significantly, for more than 40%, as well as the radial force, which was decreased for approximately 47% and the axial force was reduced for about 40%.

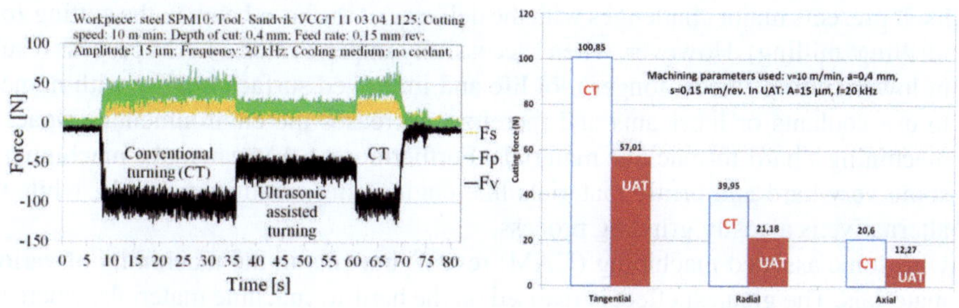

Fig. 15. Cutting forces in conventional and USAT; Left: obtained signal; Right: average values.

4 Conclusions

An overview of the advanced manufacturing processes, i.e. hybrid machining processes (HMPs) is given with the focus on assisted processes, where the primary machining process is combined with the secondary process as an additional external energy source or fluid and/or abrasive media. The process hybridisations are used to enhance the material removal rates as for the improvement of machinability of especially hard-to-machine materials. Hybrid processes are continuously in the process of further development. In the next paragraphs the individual HMP benefits and drawbacks are summarised and the current stage of development is provided.

- High-pressure jet assisted machining (HPJAM): This process provides enormous benefits in machinability improvement over a wide range of metals, from heat-resisted alloys to hard and stainless steels. Excellent chip breakability results in increased process reliability, especially when automation is required. Due to the increased cooling and lubrication capabilities the cutting temperatures are decreased, which leads to longer tool life and/or higher cutting speeds and better surface finishes. The HPJAM is probably the HMP which achieves the highest industrial implementation. The main drawback of this HMP is due to the requirement of additional equipment for high-pressure application. Such additional equipment (pump, filtering and cooling system, HP tubing and nozzle) result in the increase of hourly rates, power consumption, noise level and space requirements.
- Cryogenic machining is the most promising sustainable manufacturing process, which results in cleaner parts, machine tools and chips. Since cryogenic media

evaporates immediately after application, no further washing is needed. In cryogenic cooling cutting temperatures are drastically reduced. This resulted in longer tool life, higher productivity, but sometimes also in higher cutting forces. Since no lubrication effect was identified, further developments in Cryo-MQL (minimum quantity lubrication) combination are under investigation.

- Laser assisted machining (LAM) is certainly the HMP which achieves the lowest industrial implementation, due to the fact that the laser systems used in LAM systems in the last decades were either costly, space consuming or not suitable for an industrial environment (e.g. CO_2 and Nd:YAG), and because the delivery of the laser beam to the cutting zone still poses a major problem (CO_2 and HPDL), or the cutting process itself presents major challenges with the delivery of the laser beam to the cutting zone (drilling, milling). However, when successfully implemented a LAM process results in lower cutting forces, longer tool life and improved surface quality with no need to use coolants or lubricants and therefore decreases the environmental impact of machining a hard-to-machine materials. Furthermore, LAM can make machining of some very hard and brittle materials like engineering ceramics possible, while the alternative is a costly grinding process.
- Ultrasonic assisted machining (UAM) reveals benefits in the machining of various materials. The greatest effect is realised on the hard-to-machine materials, where the cutting forces are reduced and the tool life is being prolonged while resulting in minimal changes on the workpiece materials' top layer microstructure. Furthermore the chip breakability is improved, resulting in favourable chip shapes. Beside, in UAM of some hard materials an increased surface quality can be expected and even different microstructures or textures can be manufactured. Its use is also justified on soft materials, where it enables the use of higher machining parameters without deforming the workpiece, and in composite laminated materials, where it prevents the formation of burrs and the tearing of the material in the cutting zone.

References

1. Janssen, A., et al.: Hybrid production systems. In: Brecher, C. (ed.) Integrative Production Technology for High-Wage Countries. Springer, Heidelberg (2012)
2. Gupta, K., Jain, N.K., Laubscher, R.F.: Hybrid Machining Processes: Perspectives on Machining and Finishing. Springer Briefs in Applied Sciences and Technology. Springer, Cham (2016)
3. Rajurkar, K.P., Zhu, D., McGeough, J.A., Kozak, J., De Silva, A.: New developments in electro-chemical machining. Ann. CIRP **48**(2), 567–579 (1999)
4. Kozak, J., Rajurkar, K.P.: Hybrid machining process evaluation and development. In: Proceedings of 2nd International Conference on Machining and Measurements of Sculptured Surfaces, Keynote Paper, Krakow, pp. 501–536 (2000)
5. Schuh, G., Kreysa, J., Orilski, S.: Roadmap ''HybrideProduktion''. Zeitschrift fur Wirtschftlichen Fabrikbetrieb **104**(5), 385–391 (2009). https://doi.org/10.3139/104.110072
6. Nau, B., Roderburg, A., Klocke, F.: Ramp-up of hybrid manufacturing technologies. CIRP J. Manufact. Sci. Technol. **4**(3), 313–316 (2011). https://doi.org/10.1016/j.cirpj.2011.04.003

7. Zhu, Z., Dhokia, V.G., Nassehi, A., Newman, S.T.: A review of hybrid manufacturing processes – state of the art and future perspectives. Int. J. Comput. Integr. Manuf. **26**(7), 596–615 (2013). https://doi.org/10.1080/0951192x.2012.749530
8. Lauwers, B., Klocke, F., Klink, A., Tekkaya, A.E., Neugebauer, R., Mcintosh, D.: Hybrid processes in manufacturing. Ann. CIRP **63**, 561–583 (2014). https://doi.org/10.1016/j.cirp.2014.05.003
9. Kramar, D., Krajnik, P., Kopac, J.: Capability of high pressure cooling in the turning of surface hardened piston rods. J. Mater. Process. Technol. **210**(2), 212–218 (2010). https://doi.org/10.1016/j.jmatprotec.2009.09.002
10. Courbon, C., Kramar, D., Krajnik, P., Pusavec, F., Rech, J., Kopac, J.: Investigation of machining performance in high-pressure jet assisted turning of Inconel 718: an experimental study. Int. J. Mach. Tools Manuf. **49**(14), 1114–1125 (2009). https://doi.org/10.1016/j.ijmachtools.2009.07.010
11. Courbon, C., Sajn, V., Kramar, D., Rech, J., Kosel, F., Kopac, J.: Investigation of machining performance in high pressure jet assisted turning of Inconel 718: a numerical model. J. Mater. Process. Technol. **211**(11), 1834–1851 (2011). https://doi.org/10.1016/j.jmatprotec.2011.06.006
12. Kramar, D., Sekulic, M., Jurkovic, Z., Kopac, J.: The machinability of nickel-based alloys in high-pressure jet assisted (HPJA) turning. Metalurgija **52**(4), 512–514 (2013). http://hrcak.srce.hr/100829
13. Sekulic, M., Kovac, P., Gostimirovic, M., Kramar, D.: Optimization of high-pressure jet assisted turning process by Taguchi method. Adv. Prod. Eng. Manag. **8**(1), 5–12 (2013). https://doi.org/10.14743/apem2013.1.148
14. Kramar, D., Cica, D., Sredanovic, B., Kopac, J.: Design of fuzzy expert system for predicting of surface roughness in high-pressure jet assisted turning using bioinspired algorithms. Artif. Intell. Eng. Des. Anal. Manuf. **30**(1), 96–106 (2016). https://doi.org/10.1017/s0890060415000189
15. Pusavec, F., Kramar, D., Krajnik, P., Kopac, J.: Transitioning to sustainable production. Part 2, evaluation of sustainable machining technologies. J. Clean. Prod. **18**(12), 1211–1221 (2010). https://doi.org/10.1016/j.jclepro.2010.01.015
16. Pusavec, F.: Porous tungsten machining under cryogenic conditions. Int. J. Refract. Metals Hard Mater. **35**, 84–89 (2012). https://doi.org/10.1016/j.ijrmhm.2012.04.009
17. Pusavec, F., Lu, T., Courbon, C., Rech, J., Aljancic, U., Kopac, J., Jawahir, I.S.: Analysis of the influence of nitrogen phase and surface heat transfer coefficient on cryogenic machining performance. J. Mater. Process. Technol. **233**, 19–28 (2016). https://doi.org/10.1016/j.jmatprotec.2016.02.003
18. Muzenic, D., Kramar, D., Sekulic, M., Gostimirovic, M., Kopac, J.: Hybrid manufacturing: cutting processes with laser assistance. In: Proceedings of 12th International Scientific Conference MMA 2015 - Flexible Technologies (2015)
19. Muzenic, D., Potocnik, P., Govekar, E., Kramar, D.: Optimisation of a laser assisted milling process by using response surface methodology and artificial neural networks. In: Proceedings, International Conference on Innovative Technologies, TECH 2017 (2017)
20. Pucovski, V., Sekulic, M., Kramar, D., Gostimirovic, M., Kopac, J.: Ultrasonic assisted turning. In: Proceedings of 12th International Scientific Conference MMA 2015 - Flexible Technologies (2015)

Intelligent Sensing Systems – Status of Research at KaProm

Zivana Jakovljevic[1(✉)], Milica Petrovic[1], Stefan Mitrovic[2], and Zoran Miljkovic[1]

[1] Faculty of Mechanical Engineering, University of Belgrade, KraljiceMarije 16, 11000 Belgrade, Serbia
zjakovljevic@mas.bg.ac.rs

[2] Lola Institute, KnezaViseslava 70a, 11000 Belgrade, Serbia

Abstract. Within Industrie 4.0 intelligent sensing systems represent an indispensable asset with significant role in enabling shifting from automated to intelligent manufacturing. Instead of being simple transducers, intelligent sensors are able to retrieve useful information from raw signal. They represent systems with integrated computation and communication capabilities, that run sophisticated and real time applicable algorithms and communicate the necessary information to the other elements of the manufacturing facility.

In this paper we present the recent research results in the field of intelligent sensing systems that were accomplished at Laboratory for Manufacturing Automation and Laboratory for Robotics and Artificial Intelligence at Department for Production Engineering (KaProm) at Faculty of Mechanical Engineering in Belgrade. Presented systems are intended for application in various manufacturing processes, such as machining, assembly, manipulation, material transport, rubber processing lines. They are based on application of different non-stationary signal processing (Discrete Wavelet Transform, Huang-Hilbert transform) and machine learning and artificial intelligence techniques (Support Vector Machines, Artificial Neural Networks, bio-inspired algorithms, clustering methods, fuzzy inference mechanisms). The most of developed systems are implemented in embedded devices and their real-world applicability is demonstrated.

Keywords: Intelligent sensing systems · Cyber physical systems
Industrie 4.0

1 Introduction

Nowadays, market conditions are forcing production companies to manufacture diversified products in small lot sizes which can contain even one-off products. Such product variety requires extremely high level of flexibility in manufacturing systems that can only be achieved through their fast and reliable reconfiguration [1]. Nevertheless, effective reconfiguration of manufacturing resources is possible only if they are capable of self-configuration and fast adaptation to new manufacturing conditions and uncertainties introduced by changeable working environment. These requirements can be met only if manufacturing resources are characterized by certain level of intelligence and if intelligent manufacturing automation concept is embraced. Cyber Physical Systems (CPS)

J. Ni et al. (Eds.): AMP 2018, LNME, pp. 18–36, 2018.
https://doi.org/10.1007/978-3-319-89563-5_2

and Internet of Things (IoT) represent the most prominent enabling technologies for shifting from automated manufacturing to intelligent manufacturing within Industrie 4.0 concept [2].

Indisputably, the sensors represent essential assets for any kind of automation. However, their role in intelligent automation is even more emphasized. In Industrie 4.0 framework, the sensor is not just a transducer responsible for acquisition of signals, it is rather a system able to retrieve suitable information regarding system performance [3]. Ability of sensors to provide meaningful information instead of raw signal is crucial for fast and easy reconfiguration of manufacturing assets, and especially in terms of control system adaptation. Since some processing tasks are distributed to the sensors, they become a part of control system, thus leveraging its modularity and reconfigurability [4, 5]. Industry 4.0 sensor is intelligent/smart; it is a cyber-physical system with incorporated computing and communication capabilities.

Embedded signal processing and information retrieval at the sensor bring about additional advantages that are very important for IoT application in industry. Namely, smart sensors for industrial applications are expected to operate autonomously and to be able to perform wired or wireless communication with other devices in the network [6]. Communication inevitably leads to latency in control of manufacturing resources. Nevertheless, transfer of retrieved information instead of raw signal can significantly decrease data transfer burden, eventually having positive influence on the latency. This is of crucial importance for wireless communication [7] which represents one of the pillars for CPS implementation within Industrie 4.0 framework [8].

Information retrieval from raw sensing data requires embedding complex data processing algorithms into intelligent sensing system. These algorithms incorporate not only sophisticated non-stationary signal processing techniques, but also the techniques based on machine learning and artificial intelligence that are necessary for pattern recognition, model fitting and decision making. One of the most important properties of applied procedures is their real time applicability.

Over the last decades an extremely high number of research works has been carried out in application of intelligent sensing techniques in different areas of manufacturing. Machining operations monitoring aiming tool condition (tool wear and breakage) diagnosis, chatter prediction, and part accuracy prediction (surface roughness and dimensional accuracy) was a subject of extensive research [9]. In these purposes, different transducers (force, current and power, accelerometers, acoustic emission sensors, etc.) were employed along with advanced signal processing and machine learning techniques [10]. Similar systems were developed for condition monitoring of other elements of machining systems [11].

Zero defect manufacturing and high product variety impose the utilization of inline data gathering and automatic inspection systems for product quality control [12]. Intelligent systems based on contact and non-contact sensing with integrated data analysis and classification represent the most important enabling technology for product quality increase.

The development of intelligent automation goes hand in hand with advances in robotic technology which is regarded as the backbone of manufacturing automation [13]. Main research topics in robotics refer to the improvement of robot abilities and, in particular, their adaptability, perception ability, decisional autonomy and cognitive ability. The research in the field of robotics represents the cornerstone of the development of intelligent sensing systems. For manufacturing domain, SPARC[1] Strategic Research Agenda (SRA) aims the development of smart robotic systems for low volume production [14] that are characterized by mentioned abilities. In addition, one of the mainstream research directions in robotics for manufacturing is in the field of Human Robot Collaboration (HRC) [15], aiming the combination of superior human perception and cognitive capabilities with abilities of robots to effectively carry out repetitive tasks. Intelligent sensing in HRC is necessary, not only for effective Human Robot Interaction, but also for the development of appropriate safety systems.

In this paper we present an overview of the recent research efforts in the field of intelligent sensing systems conducted within laboratories for Manufacturing Automation and for Robotics and Artificial Intelligence at Department for Production Engineering (KaProm) at University of Belgrade – Faculty of Mechanical Engineering. The reminder of the paper is structured as follows. In Sect. 2 we present the structure of intelligent sensing system along with a brief overview of the techniques employed in its elements. Section 3 reviews the research work that we have done in the field of intelligent sensing systems for machining process monitoring, while Sect. 4 presents our research results in 2D and 3D vision systems. Over last ten years we have invested significant efforts in the development of sensing systems for robotized assembly and mobile robotics, and these results are presented in Sects. 5 and 6, respectively. Finally, Sect. 7 provides some concluding remarks and future work guidelines.

2 The Structure of Intelligent Sensing System

The role of classical sensor is detection/measurement of a physical quantity and its transformation into the suitable, usually electrical quantity. As a rule, it consists of transducer that performs transformation between desired quantities (signal acquisition), and signal conditioning subsystem that amplifies the signal and transforms it into analog or digital form convenient for control system input.

On the other hand, the role of intelligent sensor is not just mere data acquisition and transfer to higher level control. Intelligent sensors represent systems that have integrated computation and communication capabilities, enabling processing and retrieval of useful information from acquired signal. As a rule, these capabilities are introduced by implementing microcontroller based computational devices.

Structure of intelligent sensing system is presented in Fig. 1. In addition to signal acquisition and conditioning, intelligent sensor can have integrated capabilities for signal preprocessing, extraction of relevant features from signals, as well as information retrieval capabilities, depending on the particular application.

For signal preprocessing, different finite and infinite impulse response digital filters or non-stationary signal processing techniques can be employed to remove the noise from the signal or to limit signal band. In addition, various transformations targeting

[1] The partnership for robotics in Europe.

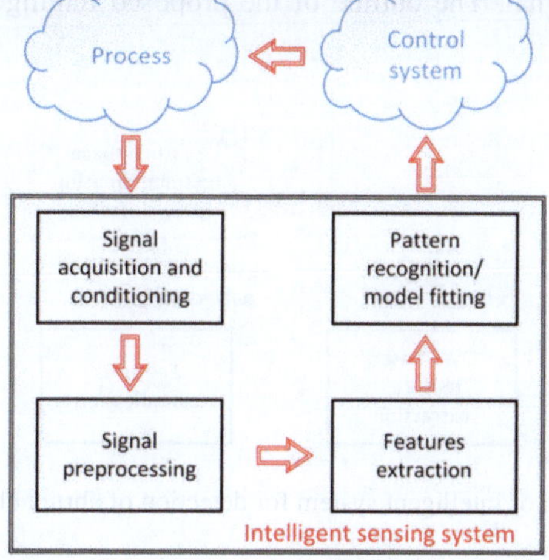

Fig. 1. The structure of intelligent sensing system

enhancement and adjustment of signals, such as normalization or clipping, can be performed.

Extraction of representative and discriminative features represents the key element for the retrieval of information from raw signal. It can be based on different techniques applied in time or frequency domain. For extraction of representative features from non-stationary signals, a number of techniques providing signal representation in time-frequency domain exist, such as Short Time Fourier Transform (STFT), Continuous and Discrete Wavelet Transforms (CWT and DWT), and Hilbert-Huang Transform (HHT). Starting from signal or its time-frequency decomposition, representative features based on signal energy (e.g. signal extrema, range, root mean square) or statistical descriptors (standard deviation, mean value, skewness, kurtosis, etc.) can be efficiently extracted.

Finally, necessary information is retrieved directly from preprocessed signal or from relevant features by using pattern recognition based on supervised or unsupervised classification (clustering) and/or by fitting of predefined or new regression models. Different artificial intelligence and machine learning techniques (Artificial Neural Networks, Support Vector Machines, fuzzy logic, Hidden Markov Models, genetic algorithms, Bayesian networks, to name a few), but also simple thresholding methods can be employed.

The most important requirements that are posed to the methods for signal preprocessing, features extraction, pattern recognition and data fitting are reliability, robustness, and real time applicability.

3 Machining Operations Monitoring

In the field of machining process monitoring, we have carried out the research on the design of intelligent machine for the detection of abrupt changes in machining operations based on vibration signal. The outline of the proposed intelligent sensing system is presented in Fig. 2.

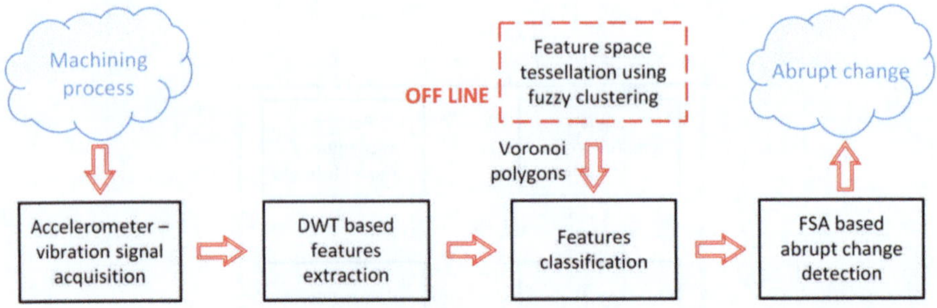

Fig. 2. The structure of intelligent system for detection of abrupt changes in machining

Since abrupt changes in machining are reflected in the non-stationarities in acquired signals, we have explored two non-stationary signal processing techniques – DWT and HHT for feature extraction [16]. Both techniques are able to appropriately localize abrupt change in time. However, HHT is capable of determining instantaneous frequency in the signal with high precision, while DWT can only detect the relative change in signal's frequency content. On the other hand, HHT is computationally expensive with respect to DWT which can be implemented using existing fast hierarchical one-pass algorithm. Since online detection of abrupt changes requires low latency, we have chosen DWT for extraction of features, based on the thorough analysis from [16].

By extraction of discriminative features, raw signal is mapped into the feature space that can be divided into the regions that correspond to stationary cutting and regions that correspond to abrupt changes in the process. To identify these regions, feature space is tessellated offline using experimental results and fuzzy clustering based on feature space deformation algorithm. Starting from the trajectory of features in feature space for regular process and process with abrupt changes, we have developed finite state automaton (FSA) for context sensitive abrupt change detection [17]. Namely, each class corresponds to a symbol in FSA language, and entering of FSA into the certain state refers to the abrupt change. Since each cluster in feature space defines a Voronoi polygon, these polygons are used for fast online classification, and, finally, abrupt change is detected using the developed FSA.

For the experiments, we have chosen intermittent turning that inherently leads to abrupt changes in acquired signal. The grooved workpiece was designed in such a way to simulate tool breakage, and vibration signal was obtained using accelerometers installed at tool holder (Fig. 3). Extracted features were based on the energy of third level DWT signal approximation using db2 wavelet.

We have designed a wireless sensor node (Fig. 4) [18] for in situ monitoring of machining operations. The node is based on low power Atmel Atmega16 microcontroller and Microchip MRF24J40 IEEE 802.15.4 transceiver. Signal is acquired using

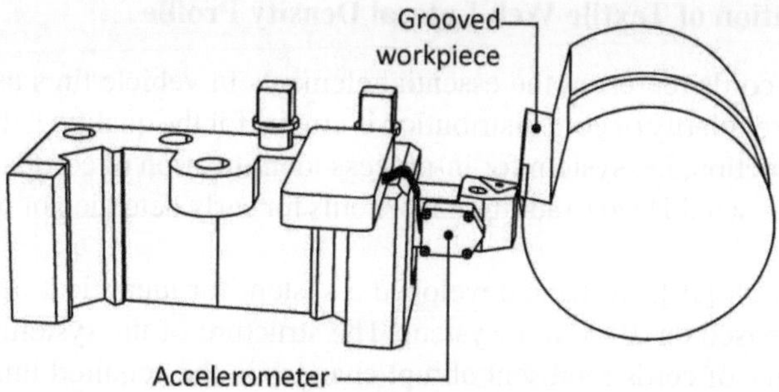

Fig. 3. Experimental setup

ADXL311 Analog Devices dual axis accelerometer. The designed node demonstrated appropriate reliability and latency in data transfer [18].

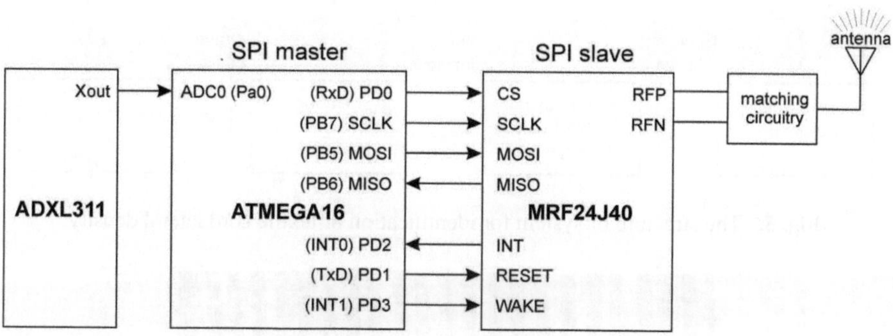

Fig. 4. KaProm wireless sensor node

4 2D and 3D Vision Systems

2D vision systems traditionally provide the most useful feedback information in a large number of control systems in different manufacturing applications, and especially in the robotics. On the other hand, recent developments of 3D scanning devices have leveraged the research in the field of application of 3D vision systems in manufacturing processes control.

In this section we provide some of the results that we have achieved in the area of intelligent sensing systems based on 2D and 3D vision. First two examples refer to the application of 2D vision in identification of textile web at calendering lines and in image based visual servo control of robot manipulator. Our recent research efforts are directed to the development of algorithms for fast and automatic extraction of geometric primitives from 3D point clouds obtained from scanning devices, and in this paper we give the outline of the method for extraction of planar surfaces.

4.1 Identification of Textile Web Lateral Density Profile

Steel or textile cords represent the essential elements in vehicle tires as they carry the main load. The regularity of cord distribution is crucial for the quality of the final product. Automatic inspection, i.e. system for in-process identification of cord lateral density on calendering lines would lead to additional benefits for early detection of possible product non-compliance.

In our research [19], we have developed a system for identification of textile cord lateral density based on 2D vision system. The structure of the system is presented in Fig. 5. The edges of cords represent abrupt changes in the acquired image (2D signal) thus making 2D DWT using 'haar' wavelet suitable technique for their extraction. Pixels

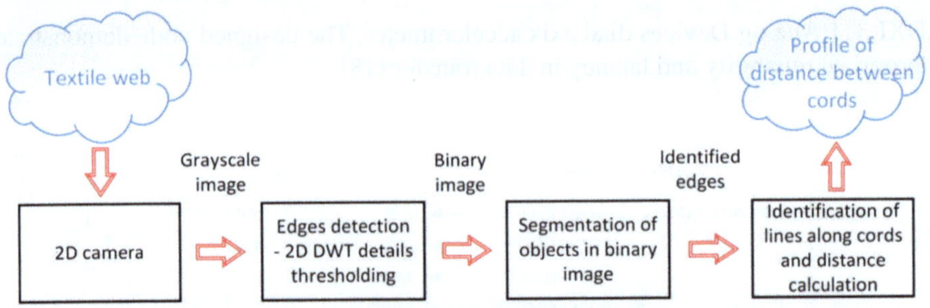

Fig. 5. The structure of system for identification of textile cord lateral density

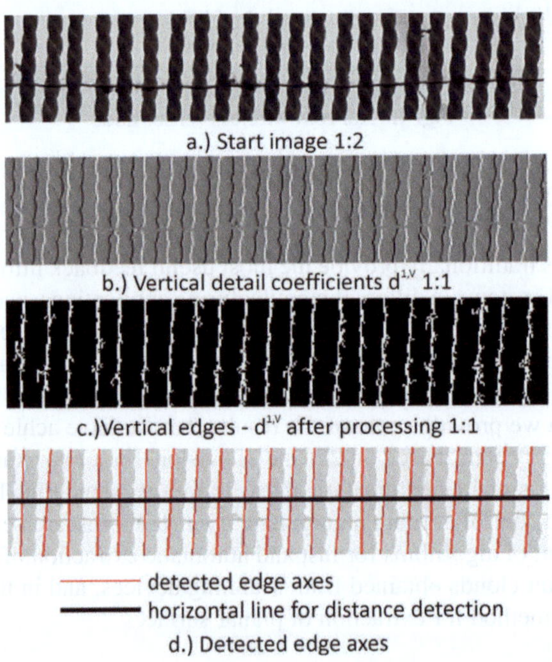

Fig. 6. Edges detection method

on the edges are detected using simple thresholding of vertical detail coefficients. In thus obtained binary image, edges represent distinct objects that can be easily segmented. By calculating the centroid and angle between horizontal line and main axis of the ellipse of inertia of identified objects, a straight line along edge is generated thus making the identification of distance between cords straightforward. Figure 6 presents the application of the method on an image.

4.2 Image-Based Visual Servo Control of Robot Manipulator

Visual servo control or visual servoing represents a solution for motion control of the mechanical systems in structured or unstructured environments. Usually, the mechanical systems of interest are robot manipulators, mobile robots or unmanned aerial vehicles. Aforementioned control approach integrates various techniques from different research fields such as image processing, computer vision, control theory and system mechanics. Here, we summarize results on the robot motion control in which the information from one camera placed directly onto the robot system (eye-in-hand configuration) is used within the control loop in order to ensure the desired position of the manipulator.

Image-based visual servoing (IBVS) has been mostly known through feature-based approaches. The main challenges facing systems for visual servo control in real world applications are the following (i) maintenance of visual features within the field of view of the camera, (ii) incorrect camera calibration parameters, (iii) white image noise, and (iv) modelling error. By implementing machine learning techniques e.g. neural network Reinforcement Learning into the vision based control scheme, the robot is enabled to improve its performance online and to adapt to the changing real time conditions in the environment.

The structure of the intelligent sensing system is presented in Fig. 7. Raw sensory signal is obtained from low cost camera. The SURF (speed up robust features) method, which represents robust image interest point detector and descriptor invariant to scale and rotation, is used to extract and match the image feature points. From extracted

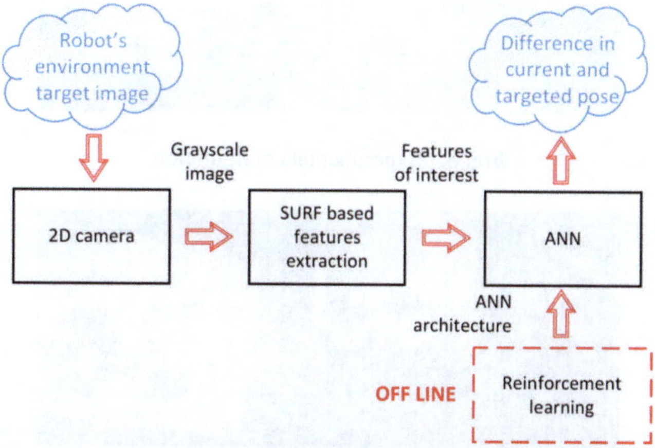

Fig. 7. The structure of intelligent sensing system for pose difference estimation

features, reinforcement learning algorithm based on neural network provides information about the difference in current and targeted image pose to robot controller which carries out corresponding control of robot motion.

In order to prove the effectiveness of the proposed intelligent system developed in [20], the experiments on a real robot manipulator are carried out. The robotic system used for experimental validation consists of a 6 DOF robot NeuroArm Manipulator System from NeuroRobotics and a low cost camera, Fig. 8. The control and data processing are conducted with an AMD Athlon II X4 630 2.8 GHz processor desktop computer with 3 GBs RAM. The scene is composed of three black blob patterns in order to simplify and speed up the camera performance in real time. Captured low resolution greyscale images are in the size of 177 × 144 pixels. In the experiments, SURF algorithm proved out to be fairly accurate regarding feature detection in a low resolution image. Successful examples giving correct detection and matching of the image features from two experimental runs are presented in Fig. 9.

The image at the start of the experiment is presented in the Fig. 10(a). The desired and the image at the end of the experimental run are presented in the Figs. 10(b) and (c),

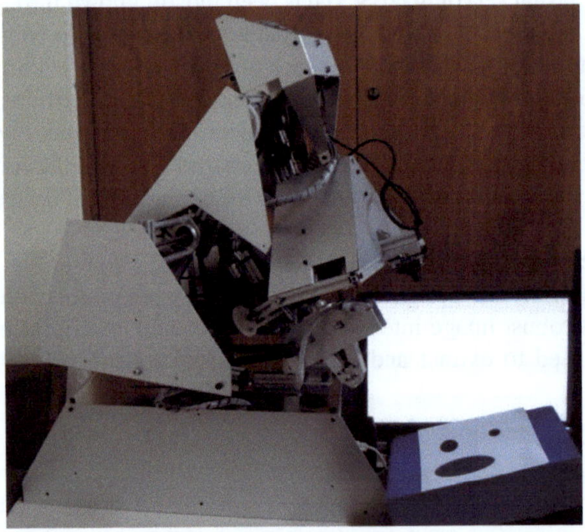

Fig. 8. Experimental configuration

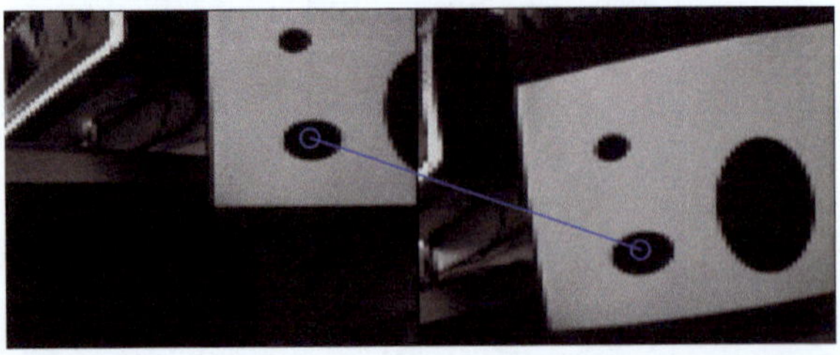

Fig. 9. One correctly matched SURF feature at the start of the experiment

respectively. This last figure indicates the obvious difference of the wanted and actual robot pose, as a result of employed parameter uncertainties.

(a) (b) (c)

Fig. 10. (a) Initial image at the starting robot pose; (b) Target image; (c) Final image

The experimental results demonstrate that neural network Reinforcement Learning controller is able to select an optimal action despite challenging conditions such as the presence of the calibration error, modelling error, and image noise.

4.3 Recognition of Planar Segments from 3D Point Clouds

State of the art 3D scanning devices are characterized by high speed and accuracy. However, their implementation in online control systems in industrial applications is sporadic, if any. One of the main reasons is the lack of automatic real-time applicable algorithms for detection of geometric primitives from point cloud. As a rule, the processing of point clouds requires user intervention at some point, thus making the application of 3D vision systems in control hardly feasible.

To overcome this issue, we have proposed a real time applicable method for detection of planar segments from structured 3D point clouds [21]. The structure of the system is presented in Fig. 11. The method exploits orthogonality of 'db2' wavelets to linear function. This property implies that the DWT detail coefficients of linear signal will be close to zero. In our approach, we have applied 2D DWT with db2 wavelet to the structured 3D point cloud. Afterwards, by simple thresholding, DWT details close to zero

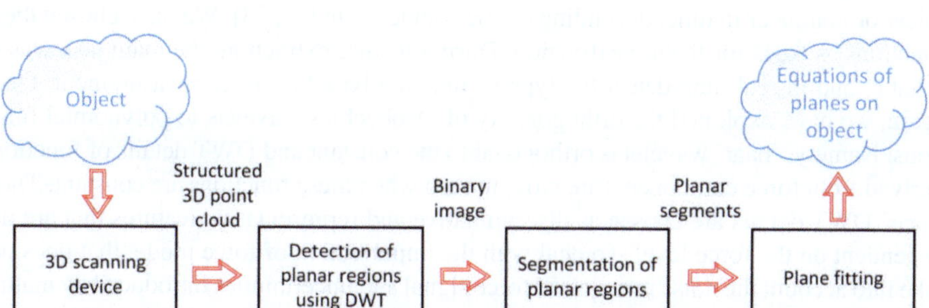

Fig. 11. The structure of real time applicable system for identification of planes from 3D point cloud

were identified and a binary image containing connected regions for each planar segment was obtained. In this way, simultaneous segmentation and recognition of planar segments from point cloud is readily carried out. Once the regions containing points on particular planar segments are detected, plane equations can be easily obtained using ANOVA based regression fitting.

Application example of the proposed algorithm for detection of the seam pose in continuous welding operation is presented in Fig. 12.

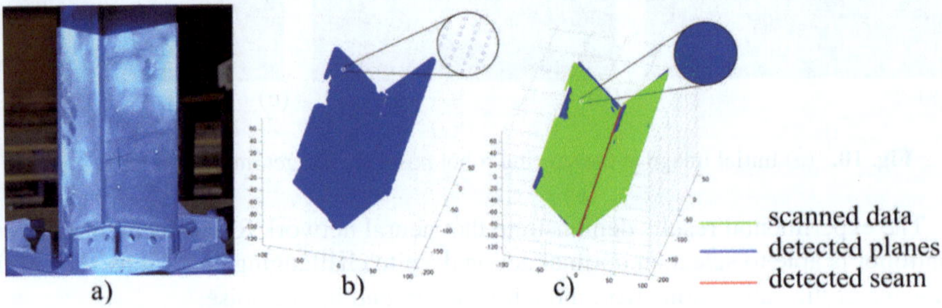

Fig. 12. Welding seam detection from scanned point cloud: (a) scanned objects; (b) 3D point cloud; (c) detected planes and seam position

5 Contact States Recognition in Assembly Processes

Another research field within the Laboratory for Manufacturing Automation referred to the detection of contact states in peg-in-hole assembly tasks. Namely, within peg-in-hole assembly with chamfer, during fine motions, six different contact states can be identified: CS0 - no contact, CSc - contact at chamfer, CS1+ - one point contact at hole rim, CS1 - one point contact at hole cylinder, CS2 - two point contact, and CSl - line contact. Timely detection and avoidance of CS2 is crucial for process success since this contact can lead to wedging or jamming and damage of mating parts.

Contact states are reflected in the generalized contact forces and there exists a planar quasi-static analytical force model for each of these states. Based on this model, we have shown that there exist certain functions derived from force components that are constant, linear or nonlinear in time, depending on the contact state [22, 23]. We have chosen these functions as basis for feature extraction. During feature extraction, the main task was to reliably and in real time detect the type of function based on force measurement. Once again, we have exploited the orthogonality of Daubechies wavelets to polynomial functions. Namely, 'haar' wavelet is orthogonal to the constant and DWT details of functions derived from force components are close to zero when these functions are constant. Thus, 'haar' DWT details are chosen as discriminative and representative features that are not dependent on the force level. To deal with the imperfection of force model that does not take into account the noise in acquired force signal and uncertainties introduced by mating parts real shape and dimensions, as well as the robot that performs mating task, and to achieve appropriate robustness of the system, we have developed a fuzzy inference mechanism (FIM) for contact states detection [24]. For FIM generation, the equivalence of support vector machines and Takagi-Sugeno fuzzy rules based system is exploited. Namely, we have carried out the granulation of feature space using support vector machines, and based on this tessellation, fuzzy knowledge base was generated. One of the most important findings of this research is that it has shown how a priori knowledge regarding the process (represented through quasi static force model in this case) can be incorporated in FIM. The FIM was designed in mathematically formally consequent manner and not heuristically, intuitively or empirically, which is a common practice.

Structure of the system for contact states recognition is presented in Fig. 13, and experimental setup along with an example of the system performance in Fig. 14.

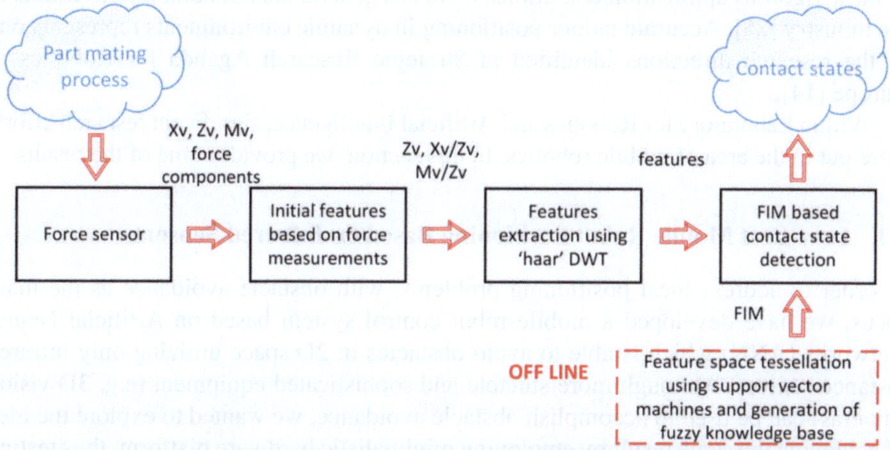

Fig. 13. The structure of real time applicable system for detection of contact states in part mating

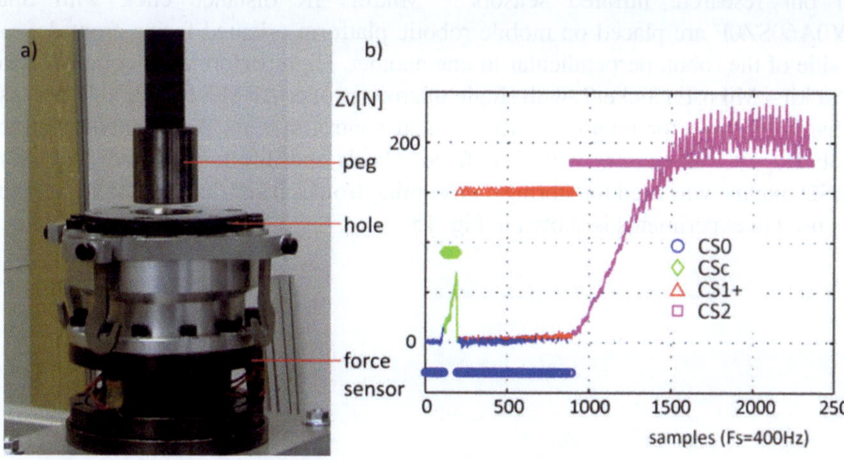

Fig. 14. Experiment: (a) setup; (b) result

6 Sensing Systems for Mobile Robots

Mobile robotics plays an important role in the world of today, allowing implementation of mobile robots of various kinds in a plethora of human activities, such as home and office applications, logistics and industrial applications, military and space applications. Mobile autonomous robots are especially important in flexible production systems, where efficiency, punctuality and accuracy are paramount. Such robots must work in unpredictive environment, with people, other robots and machines, presenting serious design challenges to engineers and researchers, especially in the safety domain. Because of these rigorous application conditions, there is a general lack of autonomous robots in the industry [25]. Accurate indoor positioning in dynamic environments represents one of the research directions identified in Strategic Research Agenda for Robotics in Europe [14].

Within Laboratory for Robotics and Artificial Intelligence, significant research efforts were put in the area of mobile robotics. In this section, we provide some of the results.

6.1 Low-Cost Mobile Robot Positioning Based on Infrared Sensors

In order to address local positioning problems, with obstacle avoidance as the main focus, we have developed a mobile robot control system based on Artificial Neural Networks (ANN) which is able to avoid obstacles in 2D space utilizing only infrared distance sensors. Although more suitable and sophisticated equipment (e.g. 3D vision cameras) can be used to accomplish obstacle avoidance, we wanted to explore the idea of achieving the same result by employing minimalistic hardware platform, thus testing the plausibility of a low-cost solution.

In our research, infrared sensors - MikroE IR distance click with Sharp GP2Y0A60SZ0F are placed on mobile robotic platform oriented facing front, left and right side of the robot, perpendicular to one another. High performance compact development kit – MikroE Clicker2, with single microcontroller ARM STM32F407, was used as a hardware base for intelligent sensing and control system. As a mobile platform, MikroE Buggy was chosen, with four fixed wheels and differential drive. Additional microSD module was used for storing data coming from sensors during training sessions. Robot used in experiments is shown in Fig. 15.

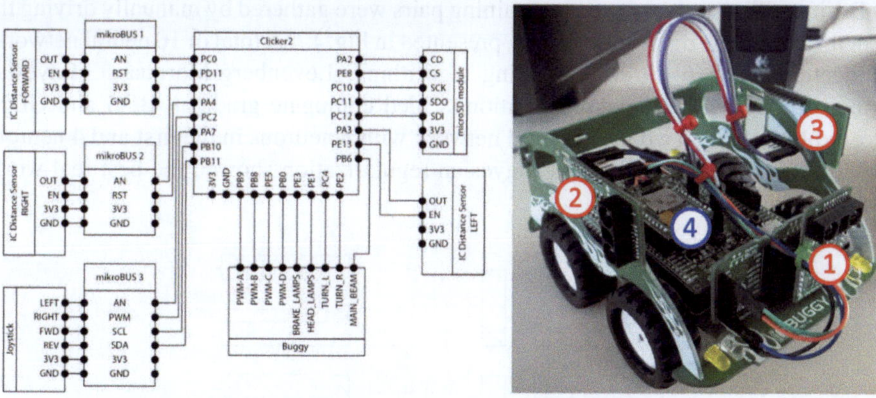

Fig. 15. Left – electronic scheme of the robot; right – mobile robot with IR sensors (1 – front, 2 – right, 3 – left) and microSD module – 4

To accomplish its task, robot needs to go around the obstacle only by analyzing data coming from infrared sensors, with no access to encoders, cameras, dynamometers, etc. We have developed intelligent sensing system that provides information about necessary robot wheels motion to avoid obstacles (Fig. 16). The system is based on infrared sensors' measurements and artificial neural networks that transform raw sensor data into the desired motion. The information about robot motion is transferred to the robot control system that ensures that the robot reaches the goal without collision utilizing simple odometry.

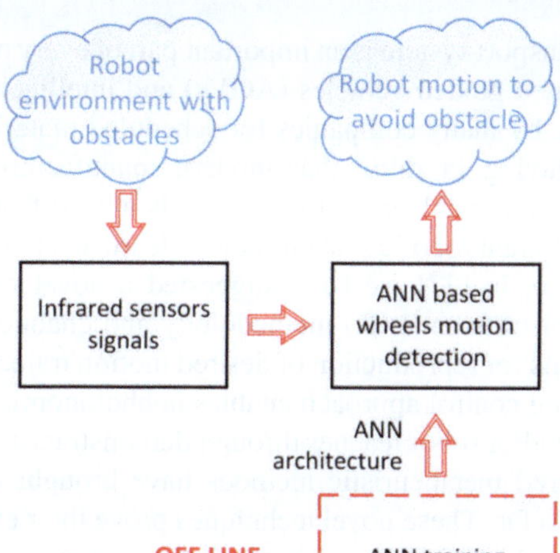

Fig. 16. The structure of intelligent sensing system for detection robot wheels motion

During offline training process, training pairs were gathered by manually driving the robot through some distinct obstacles, presented in Fig. 17. A total of 10 neural networks architectures with 6 different training algorithms (Levenberg-Marquardt, Bayesian regularization, resilient backpropagation, scaled conjugate gradient, BFG and GDX) were explored [26]. Two layer neural network with 8 neurons in the first and 4 neurons in the second layer, trained using Bayesian regularization showed the best real-world performances.

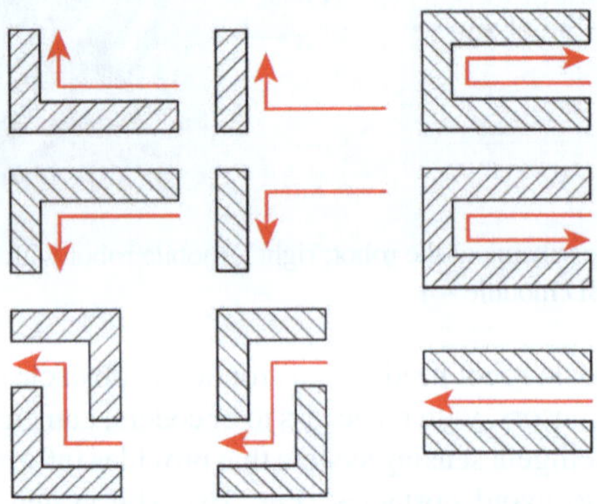

Fig. 17. Obstacles for mobile robot training

6.2 Biologically Inspired Algorithms for Learning and Reproduction of Mobile Robot Motion Trajectories

Intelligent material transport system is an important part of every modern manufacturing environment. Automated guided vehicles (AGVs) and intelligent mobile robots are a common choice made by many companies for scheduled material transport in manufacturing systems. Having in mind that modern manufacturing environments are exposed to various unpredictable conditions, there is a need that manufacturing environments contain intelligent robotic systems able to learn and to improve their behavior in real time. Therefore, in [27] we have suggested a novel method that integrated Learning from Demonstrations (LfD) methodology and chaotic biologically inspired optimization algorithms for reproduction of desired motion trajectories. In other words, nature-inspired learning control approach enables nonholonomic mobile robot to learn from human teacher and/or robot teacher through demonstrations or observations.

Biologically inspired metaheuristic methods have brought much attention in the research community so far. These novel techniques prove their effectiveness in solving many different real world problems, such as process planning optimization [28, 29], integrated process planning and scheduling [30], mobile robot navigation [31], and mobile robot scheduling [32]. Furthermore, many chaotic maps are proposed in order to improve performance of standard metaheuristic algorithms [32, 33]. In our research work, we have proposed and implemented four different chaotic methods, namely Chaotic Bat Algorithm, Chaotic Firefly Algorithm, Chaotic Accelerated Particle Swarm

Optimization, and Chaotic Grey Wolf Optimizer (CGWO) in order to evaluate complex trajectories with different length and unequal number of actuator commands. The optimization objective was to produce such sequence of mobile robot actuator commands that generate minimal error in the final robot pose.

In this paper we opted to present the structure of intelligent sensing system based on Chaotic Grey Wolf Optimizer (Fig. 18). Based on 2D image obtained from camera and using SURF current robot pose is estimated. This pose represents a starting point for generation of optimal trajectory to target position. The trajectory is created by employing CGWO. CGWO parameters and controls that generate desired trajectory are obtained through learning from demonstration.

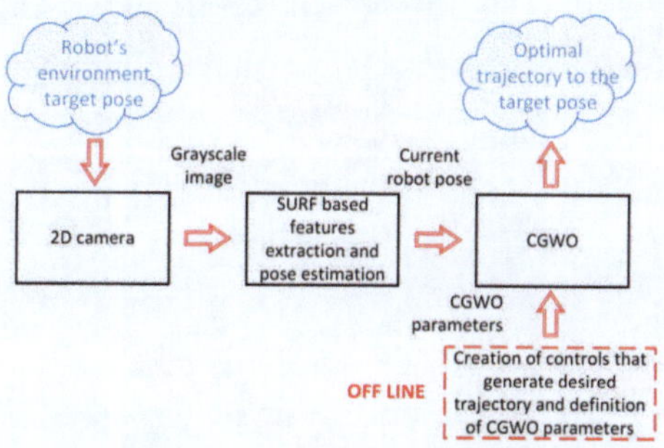

Fig. 18. The structure of intelligent sensing system for generation of optimal robot trajectory

The proposed approach is evaluated through the real world experiments carried out on a real mobile robot in an indoor structured environment. The robotic system used for experimental validation consists of nonholonomic KheperaII mobile robot, low resolution camera, and gripper. The Intel™ i5-2320 3 GHz processor desktop computer with 4 GBs RAM on Windows 7 is employed for robot control and data manipulation. The main objective was to learn the generalized mobile robot actuator commands from demonstrations, so that representative motion trajectory could be reproduced. Using the generated wheel commands, it is noticeable from Fig. 19 that the mobile robot realized wanted trajectory with minimal error in the final robot pose and successfully finishes the transportation task. Evidently, the preferred robot path is ensured in each control iteration which proves the effectiveness and robustness of the proposed approach.

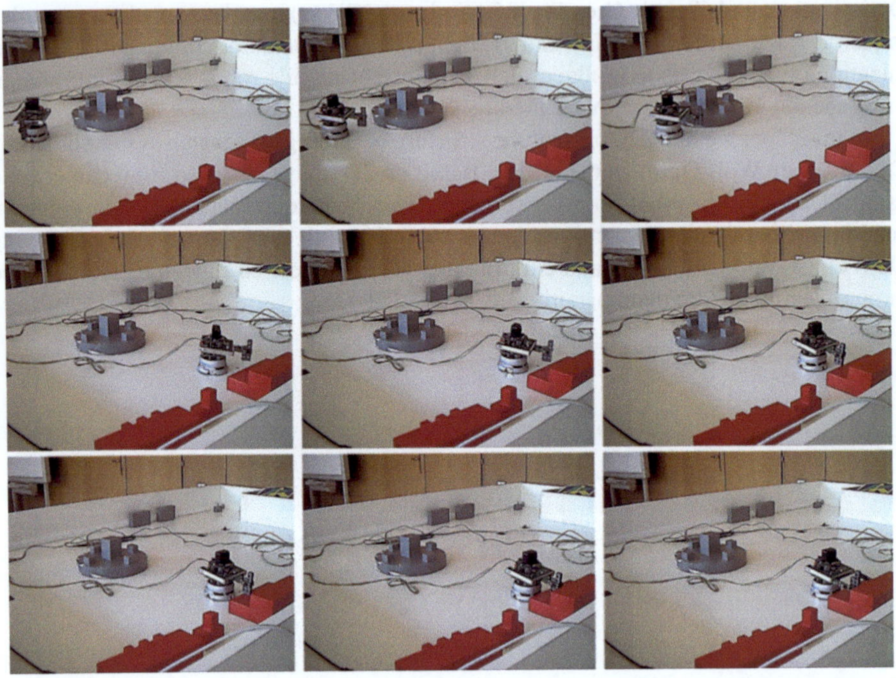

Fig. 19. Real world experiment in a laboratory model of manufacturing environment

7 Conclusion

Recently, in the Laboratory for Manufacturing Automation and in the Laboratory for Robotics and Artificial Intelligence at Faculty of Mechanical Engineering in Belgrade, significant research efforts were put in the development of intelligent sensing systems for various manufacturing application domains. This paper presents some of the results achieved in the fields of machining process monitoring, rubber calendering, robotized assembly and manipulation, mobile robotics and 3D point clouds processing. During development of intelligent sensing systems, we have employed a number of non-stationary signal processing techniques (discrete wavelet transform, Huang-Hilbert transform), machine learning and artificial intelligence methods (support vector machines, artificial neural networks, bio-inspired algorithms, clustering methods, fuzzy inference mechanisms), and implementation was carried out using embedded systems. The developed systems mitigate the border line between cyber and physical world, and they represent cyber physical systems that can be employed in a high level cyber physical manufacturing system.

Although, the applicability of the developed systems is demonstrated in laboratory, innovation potentials of the presented research are yet to be exploited through development activities. Our future research efforts in the field will be not only in the development of new more sophisticated and faster algorithms for intelligent sensing, but also in the area of implementation of intelligent sensing in distributed control of manufacturing resources.

Acknowledgments. We wish to express our gratitude to the Ministry of Education and Science of Serbia for providing financial support under grants TR35004, TR35020 and TR35023.

References

1. ElMaraghy, H., Schuh, G., ElMaraghy, W., Piller, F., Schönsleben, P., Tseng, M., Bernard, A.: Product variety management. CIRP Ann. Manuf. Technol. **62**(2), 629–652 (2013)
2. Thoben, K.D., Wiesner, S.A., Wuest, T.: Industrie 4.0 and smart manufacturing – a review of research issues and application examples. Int. J. Autom. Technol. **11**(1), 4–16 (2017)
3. Berger, C., Hees, A., Braunreuther, S., et al.: characterization of cyber-physical sensor systems. Procedia CIRP **41**, 638–643 (2016)
4. Jakovljevic, Z., Mitrovic, S., Pajic, M.: Cyber physical production systems-an IEC 61499 perspective. In: Proceedings of 5th International Conference on Advanced Manufacturing Engineering and Technologies, LNME, pp. 27–39. Springer, Heidelberg (2017). ISBN 978-3-319-56429-6
5. Mitrović, S., Jakovljević, Ž.: An application of distributed control system based on IEC 61499 and 802.15.4. In: Proceedings of ETIKUM 2017 Conference, Faculty of Technical Sciences in Novi Sad, pp. 37–40 (2017). ISBN 978-86-6022-00-68
6. Nihtianov, S., Tan, Z., George, B.: New trends in smart sensors for industrial applications – part I. IEEE Trans. Industr. Electron. **64**(9), 7281–7283 (2017)
7. Li, X., Li, D., Wan, J., Vasilakos, A.V., Lai, C.-F., Wang, S.: A review of industrial wireless networks in the context of industry 4.0. Wireless Netw. **23**(1), 23–41 (2017)
8. Kagermann, H., Wahlster, W.: Recommendations for Implementing the Strategic Initiative Industrie 4.0 (2013). http://www.acatech.de/fileadmin/userupload/BaumstrukturnachWebsite/Acatech/root/de/MaterialfuerSonderseiten/Industrie4.0/FinalreportIndustrie4.0accessible.pdf
9. Abellan-Nebot, J.V., Romero Subirón, F.: A review of machining monitoring systems based on artificial intelligence process models. Int. J. Adv. Manuf. Technol. **47**(1–4), 237–257 (2010)
10. Teti, R., Jemielniak, K., O'Donnell, G., Dornfeld, D.: Advanced monitoring of machining operations. CIRP Ann. Manuf. Technol. **59**(2), 717–739 (2010)
11. Wegener, K., Bleicher, F., Krajnik, P., Hoffmeister, H.-W., Brecher, C.: Recent developments in grinding machines. CIRP Ann. Manuf. Technol. **66**(2), 779–802 (2017)
12. Colledani, M., Tolio, T., Fischer, A., Iung, B., Lanza, G., Schmitt, R., Váncza, J.: Design and management of manufacturing systems for production quality. CIRP Ann. Manuf. Technol. **63**(2), 773–796 (2014)
13. SPARC: The Partnership for Robotics in Europe. Robotics 2020 Multi-Annual Roadmap for Robotics in Europe (2017). http://www.eurobotics-project.eu
14. SPARC: The Partnership for Robotics in Europe Strategic Research Agenda for Robotics in Europe 2014–2020 (2017). http://www.eurobotics-project.eu
15. Krüger, J., Lien, T.K., Verl, A.: Cooperation of human and machines in assembly lines. CIRP Ann. Manuf. Technol. **58**(2), 628–646 (2009)
16. Jakovljević, Ž.: Comparative analysis of Hilbert Huang and discrete wavelet transform in processing of signals obtained from the cutting process: an intermittent turning example. FME Trans. **41**, 342–348 (2013)

17. Petrovic, P.B., Jakovljevic, Z., Milacic, V.R.: Context sensitive recognition of abrupt changes in cutting process. Expert Syst. Appl. **37**(5), 3721–3729 (2010)
18. Jakovljevic, Z., Pajic, M., Aleksendric, D., Pajic, M.: Wireless sensor network application in monitoring of machining operations. In: Proceedings, 34th International Conference on Production Engineering, Faculty of Mechanical Engineering in Nis, pp. 365–368 (2011). ISBN: 978-86-6055-019-6
19. Jakovljević, Ž., Petrović, P.B.: A New system for textile web feeding at calendering lines in tiremaking industry. In: Proceedings of the Fifth International Conference Heavy Machinery HM 2005, Faculty of Mechanical Engineering in Kraljevo, pp. I B.17–I B.20 (2005). ISBN: 86-82631-28-8
20. Miljković, Z., Mitić, M., Lazarević, M., Babić, B.: Neural network reinforcement learning for visual control of robot manipulators. Expert Syst. Appl. **40**(5), 1721–1736 (2013)
21. Jakovljevic, Z., Puzovic, R., Pajic, M.: Recognition of planar segments in point cloud based on wavelet transform. IEEE Trans. Industr. Inf. **11**(2), 342–352 (2015)
22. Jakovljevic, Z., Petrovic, P.B.: Recognition of contact states in robotized assembly using qualitative wavelet based features and support vector machines. In: Proceedings of the 36th International MATADOR Conference, pp. 305–308. Springer Verlag London Ltd. (2010). ISBN: 978-1-84996-431-9
23. Jakovljevic, Z., Petrovic, P.B., Hodolic, J.: Contact states recognition in robotic part mating based on support vector machines. Int. J. Adv. Manuf. Technol. **59**, 377–395 (2012)
24. Jakovljevic, Z., Petrovic, B.P., Mikovic, V.D., Pajic, M.: Fuzzy inference mechanism for recognition of contact states in intelligent robotic assembly. J. Intell. Manuf. **25**(3), 571–587 (2014)
25. Bøgh, S., Hvilshøj, M., Kristiansen, M.: Identifying and evaluating suitable tasks for autonomous industrial mobile manipulators. Int. J. Adv. Manuf. Technol. **61**, 713–726 (2012)
26. Mitrović, S., Jakovljević, Ž., Dimić, Z., Miljkovic, Z.: Neural networks based control system of a mobile robot for obstacle avoidance in 2D space (In Serbian). In: JUPITER Conference in Belgrade, vol. 40, pp. 4.18–4.28 (2016). ISBN 978-86-7083-893-2
27. Mitić, M., Vuković, N., Petrović, M., Miljković, Z.: Chaotic metaheuristic algorithms for learning and reproduction of robot motion trajectories. Neural Comput. Appl. 1–19 (2016)
28. Miljković, Z., Petrović, M.: Application of modified multi-objective particle swarm optimisation algorithm for flexible process planning problem. Int. J. Comput. Integr. Manuf. **30**(2–3), 271–291 (2017)
29. Petrović, M., Mitić, M., Vuković, N., Miljković, Z.: Chaotic particle swarm optimization algorithm for flexible process planning. Int. J. Adv. Manuf. Technol. **85**(9–12), 2535–2555 (2016)
30. Petrović, M., Petronijević, J., Mitić, M., Vuković, N., Miljković, Z., Babić, B.: The Ant Lion optimization algorithm for integrated process planning and scheduling. Appl. Mech. Mater. **834**, 187–192 (2016)
31. Petrović, M., Miljković, Z., Babić, B.: Integration of process planning, scheduling and mobile robot navigation based on TRIZ and multi-agent methodology. FME Trans. **41**(2), 120–129 (2013)
32. Petrović, M., Vuković, N., Mitić, M., Miljković, Z.: Integration of process planning and scheduling using chaotic particle swarm optimization algorithm. Expert Syst. Appl. **64**, 569–588 (2016)
33. Mitić, M., Vuković, N., Petrović, M., Miljković, Z.: Chaotic fruit fly optimization algorithm. Knowl.-Based Syst. **89**, 446–458 (2015)

Industry 4.0 and Paradigm Change in Economics and Business Management

Dragan Đuričin(✉) and Iva Vuksanović Herceg

Faculty of Economics, University of Belgrade, Belgrade, Serbia
ddjuricin@deloittece.com, ivav@ekof.bg.ac.rs

Abstract. The purpose of this chapter is to discuss the impact of change imperative caused by the fourth industrial revolution and new normalities on economy and industrial organizations inside them as well as on society. There is a real threat that the Great Recession from 2008 and counter recession measures exacerbated the global turmoil and diminished capacity to respond adequately on national and company level to new opportunities. Our intention is to offer a conceptual paper. Model of managed capitalism is a reasonable choice for new normalities. Also, we are treating advanced manufacturing as an unavoidable element of new development model for Industry 4.0. We are advocating for "heterodox approach" as a conceptual platform for new economic policies. Also, we will implement this conceptual novelty in case of Serbia giving the framework with concrete vertical and horizontal industrial policies.

Keywords: Industry 4.0 · Model of capitalism · New normalities
Heterodox approach · Industrial policy

1 Introduction

In the new economy, usually labelled "Industry 4.0", there are two explanatory elements: a set of new normalities and fourth industrial revolution. Elements are not stand-alone, but interrelated. Their interaction influences the structural changes in economy and society. Also, structure of relations and intensity of dependency between their components are pretty unstable. Sometimes they are reinforcing, sometimes they are offsetting each other.

Beside social context, as in any development stage, technology is the main driver of economic growth and social prosperity. Also, it influences the development model, economic policy platform and behavior (and strategy) of basic economic agents. Technology is an enabler, a factor shaping opportunities (inclusive innovations) and threats (disruptive innovations), or both (structural changes). Combining effect of technological change and social change leads to so-called "change imperative", especially in the economy.

The purpose of this paper is to discuss the impact of change imperative on the economy and industrial organizations inside them. This is particularly important because there is a real threat that the Great Recession from 2008 and counter recession measures exacerbated the global turmoil and diminished capacity to respond adequately on

J. Ni et al. (Eds.): AMP 2018, LNME, pp. 37–56, 2018.
https://doi.org/10.1007/978-3-319-89563-5_3

national and company level to new opportunities. Our intention is to offer a conceptual paper. Model of managed capitalism is a reasonable choice for new normalities. Also, we are treating advanced manufacturing as an unavoidable element of new development model for Industry 4.0. We are advocating for "heterodox approach" as conceptual platform for new economic policies. Also, we will implement this conceptual novelty in the case of Serbia giving the framework with concrete vertical and horizontal industrial policies. This logic is a base for structuring the following sections.

2 Industrial Revolution 4.0: The Impact on an Economy

Breakthrough has always been driven by human curiosity. Among all people with the habit of being illuminators, there are some that are in a rare position to imagine the future, and then utilize innovation to become responsible for better reality than today's. The acceleration of technological breakthroughs, including the impact that comes with it, is hard to anticipate and comprehend. But, nor technology or disruption in relation with incumbents is an exogenous factor over which society has any control. Architects of the economic system, policy makers, public authorities, academy members, and institutions of civil society are responsible for guiding such adjustments.

Nowadays, traditional manufacturing is exposed to digital transformation caused by emergence and fast implementation of lead-edge exponentially growing technologies. The pace of change reflects Moore's law on the speed at which information technology driven change happens (Peppard and Ward 2016).

This widespread adoption of IT by the manufacturing industry allowed for disruptive approaches to product development, production and even entire logistics chain (e.g. intelligent robots, drones, 3D printing, etc.).

The new developmental stage in the organization and management of the manufacturing value chain is referred to as the "Industry 4.0" or the "fourth industrial revolution", although disputes still arise around whether it is really a revolution or, actually, an evolution.

The connected terms marking pretty much the same thing are "internet of things" or "internet of everything". Some even use the term "industrial internet".

The industrial revolution is a concept standing for major alterations occurring within a relatively short period that cause fundamental change in economy and society. Since the first mechanical loom, we can distinguish four stages in the ongoing process we call the industrial revolution. Until today, human kind passed through four stages of industrial revolution. At the start of each stage of development the impact of core technology for economic and social development has been enormous.

According to Schwab (2017), exponential growth of opportunities is related with structural changes. Until the advent of the first mechanical weaving loom and water and steam technology in 1784, and its application in emerging industries marking the start of the first industrial revolution, humankind lived in abject poverty. Industry 1.0 was basically powered by coal and moved by iron machines and factories, railways and steamships. The telegraph allowed for communication across oceans and continents. Innovations in agriculture allowed for much higher returns. Industrialization of the economy and expansion of new jobs in emerging industries provoked movement of people from rural to urban areas, and acceleration of modern cities with better infrastructure and social services, including education and health care.

The second stage and the following industrial revolution started in 1870. The second industrial revolution (Industry 2.0) symbol is the conveyor belt. This time, electrical power gave birth to mass production of standardized products. Industry 2.0 was also powered by oil. The new means of transportation were cars, buses, and even planes. Communication became more sophisticated thanks to telephone, TV and radio. Productivity in all sectors rose at unprecedented rate. Industry 2.0 brought modernization and formidable rise in quality of life and thus unprecedented population growth in the world. It also opened the door to the service economy.

The first programmable logic control system in 1969 was just an overture to the third stage (Industry 3.0), also called information revolution, in which symbiosis between electronics and IT led to further automation of the processes and their integration along the value chain. Internet and millions of connected devices gave new meaning to communication. Artificial intelligence is slowly replacing humans in a vast array of activities. The world saw new technologies like flexible robots, 3D printers, nanotechnologies, etc.

Now we are in the midst of the fourth wave of technological advancement, the rise of new digital industrial technology known as Industry 4.0. The fourth industrial revolution is fundamentally different in comparison to the other three. New technologies are fusing the physical, digital and biological worlds.

Industry 4.0 is basically powered by nine technology advances: (1) autonomous robots, (2) big data and analytics, (3) augmented reality, (4) simulation, (5) cloud, (6) horizontal and vertical system integration, (7) internet of things, (8) additive manufacturing and (9) cyber security.

Industry 4.0 is characterized by cyber-physical production systems, symbiosis of the real and the virtual world. Cyber-physical production systems are in fact network of machines organized in a similar way to social networks. Mechanical and electronic components linked by IT communicate via networks.

Character and impact of former industrial revolutions on economy and society has changed dramatically (see Fig. 1). The main consequence of industrial revolution 1.0 was that machines substituted labor. The main impact of industrial revolution 2.0 was automation of production process based on the assembly line and power of electricity. The main impact of industrial revolution 3.0 is augmented automation based on information technology. The main consequence of industrial revolution 4.0 is remote control of whole value chain based on cyber-physical systems.

As the industrial revolutions that came before, the latest industrial revolution is based on core technology. This time the core technology is information and communication technology (ICT). But, Industry 4.0 is more than ICT that lies ahead. What distinguishes the fourth industrial revolution from previous ones is its scale and velocity.

Namely, the latest industrial revolution is so complex that it has the systemic impact on economy and society. It has potential to revolutionize everything, including other technological fields out of ICT and way of life, as well.

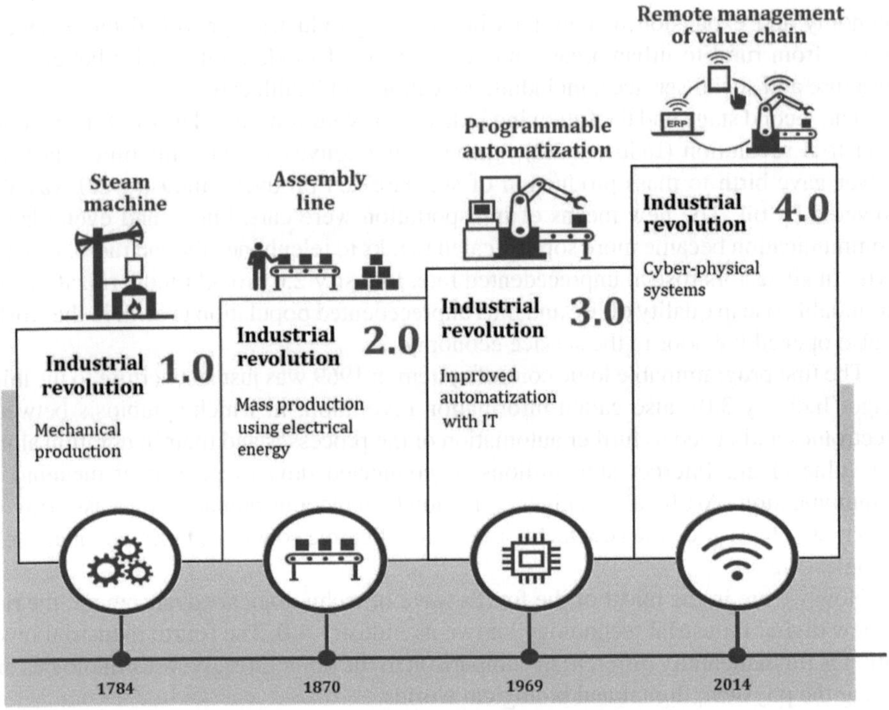

Fig. 1. Industrial revolution impact

Also, there is difference in speed of the breakthrough diffusion. In previous industrial revolutions, with the exception of the third industrial revolution, diffusion of innovations came relatively slowly leaving many potential acquirers (including whole continents) unaffected. The new industrial revolution brought about a profound and systemic change in the economy and society by breaking down barriers between industries as well as between business and private life.

ICT is the core technology in the Industry 4.0. The rate of change and the level of disruption of incumbent technologies driven by ICT are enormous. Large-scale innovation from ICT, fueled by solutions such as internet of things, big data analytics, cognitive computing, augmented reality and other form of consumerization of ICT, continuously drive change in other technological fields.

In Industry 4.0 new possibilities emerge so quickly that it is regularly difficult for industrial organizations to keep pace of progress. Typical innovation is, actually, the amalgam of innovation from cyber (or digital) space, from one side, and physical and biological space, from other side. Also, it changes the character of competition from product to amalgams of products and services. Modern competitive arena is dominated by double amalgams (between different technologies and between product and services) that are usable, available, affordable and accessible to unlimited number of users.

In Industry 4.0, the role of technology-push innovations has been increasing. Technological breakthroughs have potential to accelerate the demand for new products and services, productivity improvement through the value chain, investments in new capacities and pace of growth of the global productivity frontier itself. Cyber-physical amalgams of modern technologies could lead to the supply-side revolution. From the demand side, there are also positive trends because in global networks communication costs tend to be nothing and transportation and logistics costs significantly decline. All of which will open new markets and drive investment and economic growth. Previous changes require that we master and lead in what might be termed as inclusive innovation instead of disruptive innovation[1].

3 New Social Normalities in Industry 4.0

As we already pointed out, in any stage of development of humankind there is interaction between technology and social context. The impact of social context on economic development is more diversified than the impact of technological change. According to the partially modified source (Kohli 2016, pp. xxxvi–xxxviii), until 2050 seven global trends can be expected to affect social normalities. The trends are as follows:

1. Globalization
2. Changes in demographics
3. Urbanization
4. Middle class expansion
5. Fierce competition for finite natural resources
6. Rise of peripheral economies and decline of core economies
7. Emergence of fundamentalism and non-state actors

1. *Globalization.* The Great Recession and, almost equally, the anti-recession measures after its start in 2008, discredited the idea of universal efficiency and applicability of the neoliberal model of capitalism and the market fundamentalism economic policy platform in line with so-called *Washington Consensus* and, along with them, in some sense, the idea of global integration.

Despite the ambitions, the neoliberal model of capitalism did not provide global and relatively smooth growth all over the globe. It exhibits certain deviations from the fundamentals. Deglobalization has emerged as one of the most important deviation of

[1] Christensen (2013) has differentiated sustaining and disruptive technologies when companies are faced with the so-called "innovator's dilemma". Sustaining technologies improve product performance along the dimensions that customers have learned to expect. In contrast, disruptive technologies initially offer lower performance than existing technologies, but their performance improves at higher speed than customers expect. These technologies are ignored by incumbents because they are underperforming in early stages. However, with time, these technologies outperform sustaining technologies causing new entrants to take over business from incumbents. Demand-pull innovation based on rapid customer feedback from early prototypes (or customer relationship management) is critical for rapid and massive spillover of innovations from fertile new technologies.

the market fundamentalism. Ironically, after 2008 the myth of a borderless economy has come crashing down in most developed economies from the West.

In the face of such ambiguity, it is essential to look at the facts. To see how globalization is actually evolving, we can analyze the *DHL Global Connectedness Index*, which combines international trade, capital, people and information flows. In the post-crisis period, two components of the index of the greatest interest for economics, trade and foreign direct investments, were hit hard. Capital flow experienced dramatic drop in the aftermath of the crisis in 2007, but trade experienced a large drop with some delay in 2015. Human flows are continuously but slowly increasing. While the flows of goods, capital and people may have slow down, the flow of information shows continued exponential growth.

Geopolitics is the bold pivot of deglobalization. The proliferation of economic sanctions as a foreign policy tool is a good example of political interference in global integration. Strengthening ties between trade and geopolitics changes the integration paradigm in terms of increase of trade and capital flows between emerging super economic blocks, both West-West and East-East, as well as the growing power of state wealth funds and state-sponsored projects, particularly in infrastructure and prestige sectors like space sector.

Are these recent movements a sign of globalization failing? It is true that a particular type of globalization which is measured in terms of international trade and foreign direct investments is losing ground. In the meantime, the globalization is evolving by itself, from the one where information asymmetry (conventional globalization) was the essence to the other where resource combination is the essence (new globalization). The conventional globalization thrived on capital, trade, people and information flows across national economies. The new globalization focuses the systemic nature of the global economy. Instead of selling to and producing in the global economy, there is possibility of creating from the global economy. The new globalization is more powerful than the old one, particularly because its complementarity with the major ICT breakthroughs.

So-called "double-amalgams" based on disruptive innovations are made in the global economy. The rise of virtual organizations across borders is indication of allocation of some functions to specific countries into "centers of excellence". These organizational units are able to develop what we call a "meta-national advantage". Such advantage comes from amalgams of innovations from physical (and biological) world and cyber (or digital) world as well as innovative business model dedicated to amalgams of product and services.

The orientation toward globalization requires from industrial organization three types of adjustments. First, adaptation by tailoring products/services to local markets. Second, aggregation to achieve economies of scale and scope by extending operations in great many regional market segments. Third, arbitrage to exploit the advantages (comparative, competitive or sustainably competitive). Global market offers aggregation effect by enabling high-perceived value for customers along with high value added for owners. Namely, strategy of global niche players followed simultaneously cost effectiveness and differentiation.

In the post-crisis period, emerging economies from Asia are positioning themselves as globalization's defenders. It is reasonable for another reason. Their own experience shows that integration of local economy in the global market could accelerate productivity growth, facilitate convergence with the best practice and help increase income level and standard of living. Moreover, continuation of the globalization will lead to an even more intertwined world and permit steady growth for both developed and developing world.

The more global the world is, the greater is significance of local matters. We come up with the term "local integration". In many emerging markets, local competitors were winning the game against the incumbent multinationals. We can identify three reasons for that. Firstly, local companies were able to leverage the world at home. Secondly, a speed of adaptation as a consequence of avoidance of "double evaluation" (on the local and the global level). Thirdly, participation in the transformation of the local economy through industrial policy implementation. The last reason is the most significant advantage of "local integration". Namely, digital transformation facilitates previous endeavor.

2. *Demographics.* By 2050, the world will have 9.7 billion people compared to 7.3 billion in 2015. Almost 54% of net increase will be held in Africa. Population ageing and shrinking of workforce will be in all geographies, except Africa and Middle East. Divergence in demographic trends combined with small capacity for job creation will pose a challenge for the least developed part of the world, far beyond current refugee influx in the EU, coincidently from Middle East and Africa.

Previous problem could be exacerbated by the potential of new technologies (in particular, industrial robots) to disrupt labor markets. As automation substitutes for labor, the net displacement of employees by industrial robots might exacerbate the gap between labor income and return to capital. Also, labor markets become increasingly segregated into low-skill, low-pay and high-skill, high-pay segments, which in turn will lead to breaking the social cohesion. Providers of intellectual and financial capital like innovators, investors and shareholders are the larger beneficiaries of technological breakthroughs. Great losers in terms of income are less educated workers with lower competence.

3. *Urbanization.* In 2006 world population reached break even in terms of distribution of population between urban and rural areas. Precisely, more than 50% of the world population lives in cities. These days, on three continents (Europe, North America and South America) two thirds of population is living in urban areas. The pace of urbanization will accelerate between now and 2050 particularly in Asia and Africa. It could be projected that more than 80% of world population will live in 600 mega cities in the time frame 25–30 years from now.

Urbanization is a powerful force for output growth, productivity enhancement, and improvements in standard of living. But this structural change also requires huge investments in urban planning, renewable energy sources, transportation and waste management based on the principles of circular economy.

4. *Middle class expansion.* It is a consequence of sustainable and inclusive economic growth in emerging economies. By 2050, almost 41% of the world's population, or more than 4 billion of people might belong to upper or middle class. As we know

from economic history, emergence of large middle class is a powerful force for economic and social development. Existence of the strong middle class adds pressure on politicians to keep their promises, deliver concrete results, and be held accountable.

5. *Fierce competition for finite natural resources.* Reserves of natural resources are depleting because growth is in relation with growing consumption of them. Under one scenario according to the reference Ghemawat (2017), by 2050 people in as many as 84 countries will be able to enjoy living standards equal or better than those in the Southern Europe today.

The key question in this scenario is whether the world can sustain the demands of the resulting new upper and middle class if they choose to replicate the current lifestyle of Western consumers, or would people throughout the world agree to move to different lifestyles that would demand far less from the earth ("green energy", "circular economy" etc.).

We cannot at this point foresee which scenario is likely to emerge. History teaches that the real life combines elements of different scenarios. But, one is certain, advanced manufacturing will play important role in new transformation.

6. *Rise of peripheral economies and decline of core economies.* Two major shifts in economic power are underway, from the West to the East and from the North to the South. By 2050, three-fourths of global output will be in the emerging economies[2] and just over half of global GDP will be generated in Asia alone. As a consequence, global governance, particularly multinational financial organizations, needs to be transformed to reflect new reality in global economic landscape.
7. *Emergence of fundamentalism and non-state actors.* More than 30% of the global population now uses social media platforms to connect, learn and share information. In the world full of tensions and open conflicts due to slowdown of development, some group of extremists can create and promote unrealistic expectations and offer opportunities for related ideas and ideologies to spread.
   Violent non-state actors are the new phenomenon in geopolitical landscape. They are a serious threat to global security. They may interrupt the process of economic development and growth. To combat serious threat requires actions to constrain influence of geopolitics on economy, to promote more inclusive growth models in order to raise peoples' hope in the rule of law, economic integration and peaceful solutions of deeply rooted conflicts. New industrialization based on advanced manufacturing has a strong positive impact.

Mentioned trends are not stand-alone but interrelated. Sometimes they are reinforcing, sometimes they are offsetting each other. The net effect of these trends on individual national economy will vary from case to case, may change overtime and it strongly depends on the starting position. Great majority of mega trends, with exception of threats from non-state actors, work to the advantage of agile government with "strong economic policy regime" and industrial policies for tradable sectors. In case of "poor

[2] Former term "peripheral economies" is going to be inadequate.

policy regime", when new normalities are not well managed, they could also lead to a negative scenario.

We are convinced of one thing, that in the future human capital with STEM[3] competence rather than physical and financial capital will represent the key driver of growth. It means that people as carriers of human capital are important, but education, including learning-by-doing, research and development, and health care are critical Đuričin (2017).

How does the previous proposition influence the anatomy and the physiology of the economic system in general and industrial organizations specifically? We will offer the answer to this fundamental question in the following chapter.

4 Paradigm Change in Economic Theory and Policy Platform

In each economic system there are components functioning automatically and components that require human interactions. Previous determined two fundamental institutional choices, market forces and state intervention. Market mechanism balancing demand and supply supposed automatic behavior. Industrial policies for tradable sectors respect judgement in human reaction.

Even before the Great Recession, there was a strong dissonance between economics scholars from developed and developing world regarding the dilemma as to which institutional arrangement primarily influenced growth model and economic policy platform in capitalism. Developed or industrialized economies (sometimes called "early developers" or "core economies"), preferred free market, while developing or non-industrialized economies (sometimes called "late developers" or "peripheral economies") opted for government intervention based on the industrial policy doctrine (Vuksanović 2017).

The Great Recession, emerged in the developed world and followed by a slowdown of emerging economies, is a direct consequence of misconceptions of the neoliberal growth model and the related economic policy regime based almost exclusively on market mechanism. The main pathologies of such a system are: (1) financialization, and (2) income inequality.

When the financial sector growth is not connected with the growth in the real economy, it leads to financialization. According to Mukunda (2014, p. 74), financialization is the increase in the influence of financial markets, institutions and elites over both the economy and other institutions of society, including the government.

Disequilibrium between the financial sector and the real economy influences disparity between value creation and value release, emergence of speculative bubbles, bubble burst and, finally, the crisis. The crisis imposes costs on the government in the form of lost tax revenues and fiscal imbalance due to increased spending.

The increase in size of the financial sector has also increased the financial sector's influence on governments, particularly on the ministries of finance, the central bank and international financial organizations like IMF/WB in the area of experimental policies. A typical policy push from the financial industry in the monetary sphere is the negative interest rate policy. Owing to such policy, the cost of capital is going to be insignificant.

[3] STEM - Science, Technology, Engineering and Mathematics.

A dangerous analogy would be a combination of abundant and cheap fast food, which leads to the creation of two pathologies for human beings, obesity and diabetes.

A growth model related with financialization based almost exclusively on services, and predominantly on financial services, is not sustainable, because activities in services are distributive by nature. Rent-seeking is a typical manifestation of a distributive mentality. It involves trying to make value by manipulating regulatory policies. In a system with high financialization, a significant share of transactions is zero-sum, instead of positive-sum (or win-win). There are, at least four weak points of such model.

Firstly, over-proportional development of financial sector increases the economy's exposure to the downside scenario. It is a point made by many influential economists like Kindleberger (1996), Krugman (2009), Minsky (2008) and Rajan (2010), and during the period of rapid securitization before the Great Recession.

Secondly, overdeveloped financial sector easily misallocates resources. A shift from capital expenditures to equity investments generates high private rewards disproportionate to their social productivity. Pay for performance in the financial sector is a sign of the industry's extraordinary clout. Namely, executives in the financial sector made substantially more income in comparison with other sectors, even after adjustment for their eventually higher level of education.

Thirdly, as Orhangazi (2011) brilliantly pointed out in frequently cited reference, investment in financial assets tends to crowd out investment in real assets because the capital market prefers short-term and liquid investments. As a consequence, companies from the real economy face increased pressure to expand returns (dividends and stock buybacks) instead of investing in real assets.

Fourthly, in an economy dominated by the financial sector, distributive mindset has supremacy over a creative one. Rent-seeking is a dangerous form of distributive mindset, which involves a value creation by manipulating regulatory body's policy.

Wealth concentration is a logical consequence of financialization. It is another weak point of the neoliberal model of capitalism. In macroeconomics there is a difference between flows (or income) and stock (or wealth). Income increase is a prerequisite for wealth increase. Despite global growth, relative income inequality has been on the rise. According to *Oxfam* data, almost half of the world's wealth is owned by the global "top 1%" of the population (which includes about 70 million of people), and the bottom half owns as much as the richest 85 individuals.

The changes in the global economy are so radical that they require an ideological discourse in developed economies. Instead of the neoliberal blueprint of capitalism up to 2008, when there was no power to balance it, the new normality is a multipolar world with ideological compromise with the market and the state.

It leads to the existence of some hybrid capitalist systems with the greater role of the state in the economy. This change opens the new question: What are the fundamentals of the new systems and how is it functioning?

In changing the system, technology also matters. The possibilities of ICT as core technology of Industry 4.0 are practically unlimited. Unfortunately, disruptive, too. Namely, new technology destroys position of incumbents and leads to "winner takes all" situation destroying the logic of market forces. "Innovator's dilemma" is transforming in "capitalist's dilemma" (Christensen 2014). The fourth industrial revolution is in infancy stage and it is not easy to forecast what form it will take in the future. The more we can understand its character, the more likely we enjoy benefits. There are two perspectives, demand side and supply side.

As far as demand side is concerned, innovations from the ICT field have the capacity to become an integral part of the product itself and formation of smart connected products (SCP). The phrase "internet of things" (IoT) has risen to reflect the growing power of SCP. On the demand side, a major shift is consumer engagement in design, marketing and delivery of new products/services through costumer's relationship management (CRM). In the near past there was a disconnection between the usability of digital data and physical world in which we can apply it. Namely, decisions remain trapped on two dimensional pages and screens (or 2D), while reality is tree-dimensional (or 3D). The gap between the real and cyber (or digital) worlds limits capability to use billions of information produced from SCP as "actionable information" for business decision. Set of technologies that superimpose digital data and images of physical three-dimensional world known as augmented reality (AR) can close this gap. Smart glasses and screens have potential to increase demand particularly in consumer's goods, construction and retail (Porter and Heppelmann 2017).

Potential for improvement from supply side is even stronger. On the supply side, many incumbents are seeing the introduction of new technologies that create entirely new way of serving existing need as disruption of existing supply-chains. More importantly, demand push innovations will be multiplied if the technology-push innovation in the new technology fields like robotics, 3D printing, artificial intelligence, life science, etc. integrate with them. AR also has a role to play. Disruption is also flowing from competitors using global digital platforms for change in business model and improvements in the way how the products or services are delivered.

There is general recognition that without close integration of virtual innovations from the ICT and industrialization (implementation of physical innovations in advanced manufacturing) through investment and the spillover of emerging amalgams across different industries and formation of the so-called "cyber-physical systems" no national economy in the world has been able to close the development gap between itself and those at the frontier.

Life science and amalgams based on the so-called "cyber-biological systems" also matter, improving the basic human capital (quality and duration of life). Last but not least, every national economy must consider technological base of the energy production, particularly in the fields of renewable energy and energy storage.

Previous theories of growth and related economic policy platforms have developed from macroeconomics perspective. This perspective is good for spotting, for example, correlations between capital expenditures and innovations, and the growth. In that case, the outcome (the growth) is under the impact of these two independent variables. Despite limitations of modeling in macroeconomics[4], to understand what causes growth, you have to crawl inside the microeconomics and form a framework from the ground up to adjust risk appetite of business organizations *vis-à-vis* the market and technological

[4] In such models, a favorite expression is *ceteris paribus*, or with all other things being constant. In reality, other factors hardly ever remain constant.

opportunities. In a search for drivers of growth, microeconomics (or business) perspective, also, matters.

The key presumption of the neoliberal model of capitalism is universal market efficiency. In other words, there is no need for government to intervene in allocation of resources and distributional issues, as well. In addition, the government is not welcome to improve the situation despite well-known failures of market mechanism like asymmetric information, externalities, investment in technology change, risks, etc. Moreover, the government is not welcome even when market prices do not reflect most general tenets of economic development like equal opportunities, technology-driven competitiveness improvement, environmental control, etc. Inefficient market equilibrium presents quite a strong argument why the value maximizing behavior on the micro level does not lead to the general equilibrium and sustainable growth on a macro level.

Empirics, after the implementation of such concept, are signalizing the problem. Namely, evidence based analysis confirmed that universal efficiency of the market is not common, particularly in cases of major macroeconomic distortions like output gap and deflation. In such situations, market forces unleash recession, instead of booming development prospects. Moreover, standard anti-crisis measures based on the same doctrine push the economy to jobless recovery, at best. Consequently, there is a growing consensus among relevant economists that the industrial policy is additional common-sense institutional arrangement (Blanchard et al. 2010; Lin and Monga 2010; Lin 2012; Stiglitz et al. 2013). In that context, the relevance of the so-called "heterodox approach", instead of orthodox (or neoliberal) one is increasing. The heterodox approach integrates macro-economic policies (monetary and fiscal) with industrial policies.

Today, in economic theory there is a consensus that not only in crisis, but also in normal times, the functioning of an economy requires proactive government instead of passive one choosing wait-and-see behavior against what the market forces dictate (Lim 2011). Previous leads to the rejuvenation of industrial policy as a common-sense institutional choice in formulation of economic policies (Stiglitz et al. 2013). The concept is acknowledged by mainstream economists from different sides of ideological spectrum and most influential politicians, as well. Industrial policy enthusiasts have even treated it as a key lever for income convergence and catching up with developed economies (Rodrik 2009).

According to Lim (2011, p. 174), in a later stage of development, growth model based on heterodox approach introduces sustainable competitive advantage which can be seen as a result of synergies between enhanced human capital and new technologies. This approach involves horizontal industrial policies or complementary investments in physical and human capital through science, education, and health care.

Emerging system in which government gives some level of support along with conventional privileges to favored national champions effectuated many times in superior performance and competitive advantage on the global level. Rajan (2010, p. 56) labeled this version of capitalism "managed capitalism". This system is conceptually different from liberal capitalism and state capitalism.

In the heterodox policy approach, industrial policies dedicated to advanced manufacturing can be used to correct market failures as well as previous economic policy failures. But, industrial policies are not just about advanced manufacturing. Support to technological change and support of infant industries (vertical policies) are also critical tenets in agro-food and service industries (for example, in ICT and finance). Competitive exchange rate policy, education policy, health care policy, and infrastructure development are typical examples of sector-neutral (or horizontal) policies. As J. Stiglitz, the most cited thinker from the field, pointed out, "the question is not whether any government should engage in industrial policies, but how to do it right" (2013, p. 9).

The effectiveness of industrial policy in achieving selected objectives depends on three essential conditions: (1) favorable macroeconomic environment, (b) consistent usage of policy instruments, and (c) use of state-owned companies as policy anchor (Stiglitz and Lin 2013, p. 327). Industrial policy instruments depend on whether it focuses on industrial upgrading, propping up declining industries that used to be successful or maintaining social cohesion (Stiglitz and Lin 2013, p. 352). Regardless of the focus, majority of industrial policies were evaluated as the aim for strengthening the institutional framework. This can be achieved through creating institutions and introducing legislation that makes industrial policy instruments more effective. Institutional coordination is necessary on all three levels: among policy agencies, among private entities, and between public and private sectors.

What limits the future of an industrial organization in the age of digitalization? In the past it was its size. Today, it is ability to use the whole world for its development accurately and timely. In the new circumstances any company including local one, must be able to perform their leadership function (for uncertainty), their entrepreneurial function (for innovation) and their stewardship function (for efficiency).

Technological breakthroughs in cyber-physical amalgams in advanced manufacturing would raise perceived value for costumers while reducing costs of productions for suppliers. Technological breakthroughs in cyber-biological amalgams from life science, pharmacy and on-line surgery offer prospects of solutions to many human problems.

Last but not least, new technologies offer solutions for some societal problems including quality of life in urban areas, climate change, energy and food security, leapfrogging of the developing economies to the global best practice, etc. For this promising progress to occur in reality and be sustained and inclusive over the longer term, higher priority must be given to research and development.

5 Serbia's Road of New Industrialization

How to react to a rapidly changing and complex environment full of mutually interrelated risk stressors, particularly if the economy is strongly burdened with effects of past failures? In such challenging situation, policy makers should first examine structural imbalances of the economy and predict its vulnerability caused by risk stressors (see Đuričin and Vuksanović 2012). Also, they should understand where the global economy may be going in the long term. The assumptions of a possible reaction require effective growth model and efficient economic policy platform. Skipping these steps can easily lead to new misconceptions.

The long-term prospects for sustainable and inclusive growth in Serbia depend primarily on implementation of the "strong macro-economic policy regime". To provide a turnaround, this policy originally combines specific drivers of growth (Đuričin 2017).

Firstly, Serbia must improve macroeconomic fundamentals (by imposing competitive FX rate and reasonable level of interest rates). Imposing hard budget constraints requires not only fiscal balance, but also resolution of lasting bleeding points in the economic system. To do that, the government has to continue with structural reforms both for state-owned enterprises from natural monopolies and infrastructure and state-owned commercial enterprises.

Secondly, we need improvement in economic policy platform to unleash new avenues of growth. Today's macroeconomic management in Serbia is much better than in the previous period, but much more can and needs to be done with the growth. Intention to duplicate the level of GDP requires compound average growth rate of 7% for next 15 years. New industrialization based on industrial policies for tradable sector can be an important source for technological change, investment and employment, as demonstrated in the past in many successful emerging economies, particularly in East Asia. Normally, only limited number of sectors need industrial policy.

Thirdly, Serbia desperately needs concerted efforts in health care to improve basic human capital and the development of adequate infrastructure in education to facilitate higher and more inclusive growth, particularly based on advanced human capital. New sources of growth, such as high value-added services (in IT sector, health tourism, and professional services), offer significant potential for growth for countries with well-developed human capital.

Finally, and most importantly, Serbia's economy must close the gap *vis-à-vis* the global technology frontiers. If it intends to unleash new avenues of growth by adopting cutting edge technologies in carefully selected tradable sectors, the economy must implement industrial policies doctrine in order to create new jobs, increase productivity and improve competitiveness. Industrial policy, both vertical and horizontal, should empower technological entrepreneurship as well as research parks and clusters development. Also, previous institutional forms are prerequisite for self-employment in micro and small and medium-sized tech companies.

Following new structural economics, the output is in the center of macroeconomics. The amount of output an economy produces, not large quantity of money and positive expectations about the future, primarily determines the level of its viability (Đuričin and Vuksanović 2014).

We must go back to the basics of macroeconomics, to help keep relevant things in perspective. For that purpose, it is worth to follow the line of reasoning from Moss (2007) and to remember the key pivotal concepts of macroeconomics: output, money and expectations. Key pivotal concepts in microeconomics are: technological change, value management and organizational culture.

Figure 2 demonstrates the relationship between macroeconomic fundamentals and the main strategy choices represented with "microeconomic M". The analogy between the role of output in macroeconomics and the role of technology change on the microeconomic level is important not only in regular times, but also in the period of structural changes when economy is shifted on the road to restructuring, recovery, and catching-up. Harmonization of the strategy of industrial organizations based on technology change with the measures of macroeconomic policies which stimulate output expansion is the prerequisite for prosperity of industrial organizations and economy as a whole. The request for cohesion also exists regarding other key components of the "macroeconomic M" and the "microeconomic M".

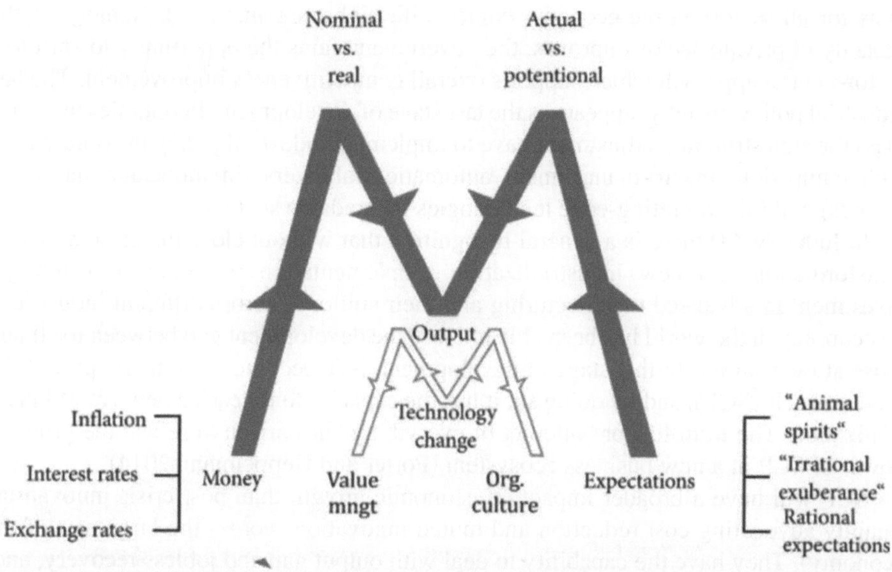

Fig. 2. "Macroeconomic M" and "Microeconomic M": The relationship (Đuričin 2017)

Following the aforementioned request for a minimal level of cohesion in conducting economic policies regarding the impacts on macro and micro levels, the "heterodox approach" in conducting economic policy has imposed a wider set of policy tenets in comparison with orthodox (or neoliberal) approach. According to partially modified source (Blanchard et al. 2010), the main policy tenets are as follows. Firstly, the output gap (low and stable) instead of inflation (low and stable) is a primary policy tenet. Secondly, sustainable employment instead of flexible labor market. Thirdly, balanced GDP structure with the growing role of the real economy instead of financialized economy. Fourthly, price parity between all types of assets (including FX rate) instead of tolerance towards speculative bubbles, primarily in the financial sector. Lastly, dynamic equilibrium between the real economy and the financial sector instead of stability of the financial system.

6 Industrial Policy and Advanced Manufacturing

The new structural economics tends to emphasize policy of "winners picking themselves" by replacing the policy of "picking winners" from conventional structural economics. In such conceptual platform reconsideration, market trial-and-error, namely experimentation and positive reinforcement, is implemented complementary with industrial policy measures.

Industrial policy has three focuses: particular sectors (vertical or sector specific policy), the economy as a whole (horizontal, non-discriminatory, or neutral policy), and future opportunities (creation of new strategically important industries). The vertical policy is most suitable for late developers. Horizontal policy that provides better conditions for all sectors in the economy comes with higher income level. Namely, as the capacity of private sector improves, the government gains the opportunity to shift to a sector-neutral approach which supports overall competitiveness improvement. The last industrial policy, usually, appears as the last stage of development. Economies that wish to go through structural adjustment have to implement industrial policy in coordination with compatible macro-management automatic stabilizers (in monetary and fiscal spheres) and follow cutting-edge technologies for tradable sectors.

In Industry 4.0 there is a general recognition that without close integration of ICT transformation and new industrialization (implementation of innovations through investment in advanced manufacturing and their spillover across different industries), no economy in the world has been able to close the development gap between itself and those at the frontier. In this stage of development, ICT becomes an integral part of the product itself (SCP), and by doing so, it has the capacity to unleash a new era of industrialization. The term IoT or "internet of everything" has arisen to reflect the growing power of SCP in a new business ecosystem (Porter and Heppelmann 2014).

SCP will have a broader impact on economic growth than post-crisis innovations (mainly advocating cost reduction and muted innovations across the large part of the economy). They have the capability to deal with output gap and jobless recovery, and, by doing so, to substantially affect the trajectory of the overall economy toward sustainable development.

As Fig. 3 shows, new industrialization in Serbia has to be based on three pillars (Đuričin 2017). The first pillar refers to vertical industrial policies. Vertical policies are sector-based. They usually refer to tradable sector. The tradable sector consists of industries with some kind of advantage (comparative, competitive and sustainable competitive). Comparative advantage is based on the abundance of factors of production like natural resources, labor force, financial capital and position rent (near to market). Competitive advantage is a difference between the value added and cost of production enabling price premium and/or cost leadership for the same level of price. Sustainable competitive advantage is a long-run advantage, one that accumulates such powerful lead over competitors based on innovations that no one can catch up to.

The second pillar refers to horizontal industrial policy including components such as infrastructure upgrade, state procurement improvement, start-up and scaling up, education and developing skills, investment in science and R&D, etc.

The third pillar of new industrialization represents hard macroeconomic policy regime, including hard budget constraint policy (both macro and micro) and tax collection.

Vertical industrial policy is cultivating tradable industries with sustainable advantage. Each country has some industries that have some sort of advantage based on different sorts of factors from abundance of natural resources, cheap labor, position to

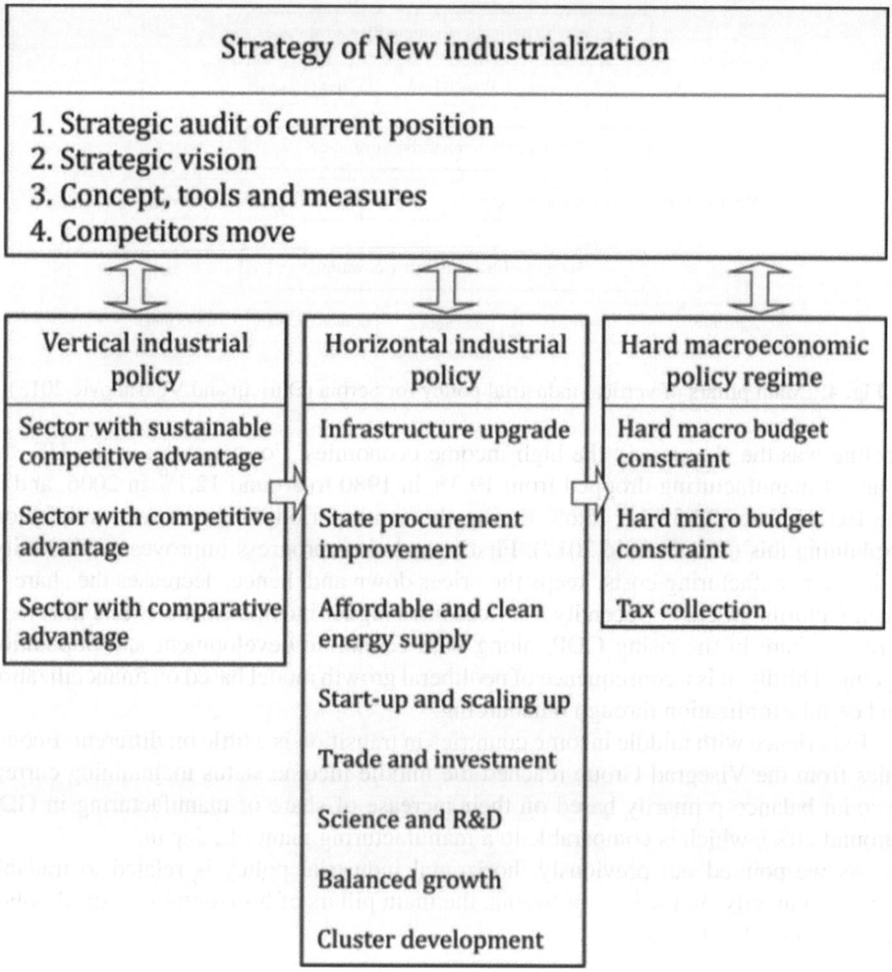

Fig. 3. Heterodox economic policy framework

market to high productivity, high innovativeness, and growth potential. In Serbia, sectors such as ICT, organic food, health tourism, infrastructure and infrastructure related businesses, transport and logistics, energy, manufacturing, automotive, metal industry, wood and furniture, waste management, etc., have potential to achieve advantage and generate significant growth in upcoming years. This is not an exclusive list, and the government should be prepared to work with any industry which can organize behind strong leadership to address shared threats and opportunities for growth. In Fig. 4 we present tradable industries as candidates for vertical industrial policy in Serbia.

As the Fig. 4 shows, the majority of tradable sectors belongs to the field of manufacturing. The role of manufacturing in Serbia's new industrialization is different than in the high income countries. It is true that starting from the 1980s, most of the developed economies in the world experienced a decline in manufacturing as a share in GDP. The

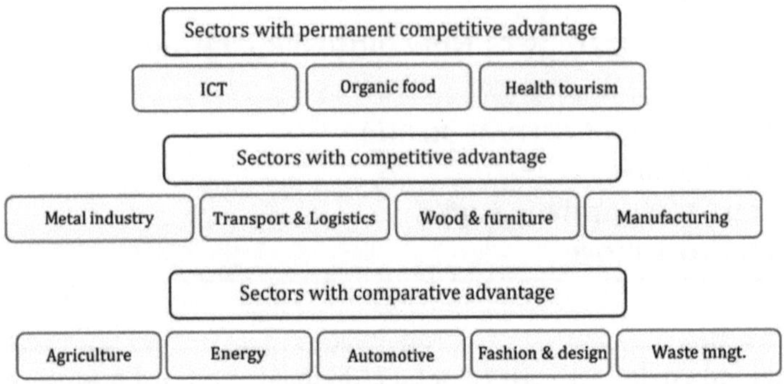

Fig. 4. Main pillars of vertical industrial policy for Serbia (Đuricin and Vuksanovic 2013)

decline was the sharpest in the high income economies. For example, in the US, the share of manufacturing dropped from 19.3% in 1980 to around 12.1% in 2006, and in the EU 15 from 23.5% to 15.6% during the same period. There are several factors explaining this (Vuksanović 2017). Firstly, technical progress improves productivity, reduces manufacturing costs, keeps the prices down and, hence, decreases the share of manufacturing in GDP. Secondly, services have higher income elasticity and thus have a rising share in the rising GDP, along with economic development and population ageing. Thirdly, it is a consequence of neoliberal growth model based on financialization and deindustrialization through outsourcing.

Experience with middle income countries in transition is a little bit different. Economies from the Visegrad Group reached the middle income status maintaining current account balance primarily based on their increase of share of manufacturing in GDP (around 20%), which is comparable to a manufacturing giant like Japan.

As we pointed out previously, horizontal industrial policy is related to tradable sectors indirectly. In the case of Serbia, the main pillars of horizontal industrial policy are represented in Fig. 5.

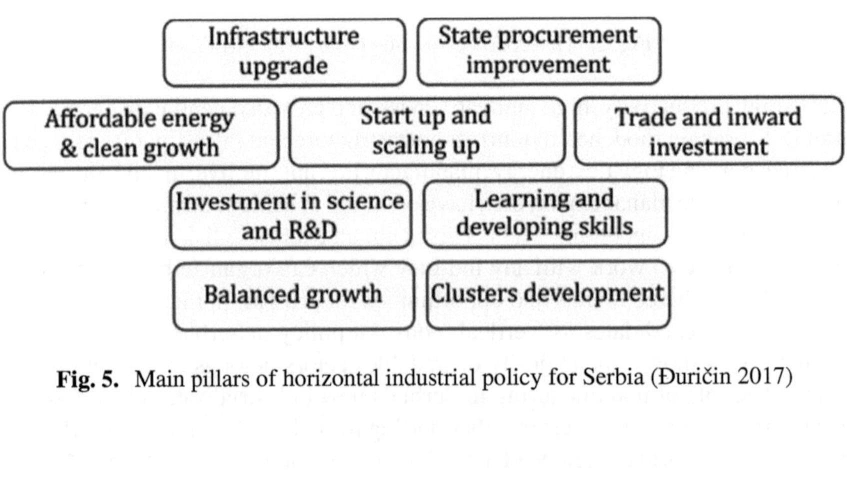

Fig. 5. Main pillars of horizontal industrial policy for Serbia (Đuričin 2017)

7 Conclusion

Long time economists were hostages in the power of theoretical mantra of the "market invisible hand", paying a terrible price for the falsity of economic orthodoxy, which asserts that individuals are capable always making rational decisions, and that markets and regulatory institutions, in the aggregate, are healthy and self-regulating. Unfortunately, irrationality is the real invisible hand that drives human decisions. In the reality of the social and technological context of neoliberal capitalism, humans are primarily irrational and motivated by cognitive biases and technological discovery.

Innovating in digital era is the name of the game of positioning *vis-à-vis* leading trends producing new normalities. By transforming "business as usual" business models, illuminators are in a rare position to imagine future, and then harness innovation.

In today's world, people with STEM competence dominate social animators, including economists. In the long run, the technological change is a good thing for everyone. In the short run, technological advances can be extremely disruptive, and the disruption can persist into the long run if companies and economies as a whole do not have the means to adapt. Define the adequate context for technological innovation to prosper, including industrial policies for tradable sector, is the role of economists.

Each industrial revolution is bringing great opportunities. But it is followed by some challenges. The fourth industrial revolution profoundly impacted not only what we do but also who we are, including our sense of privacy, the time devoted to work and leisure, consumption pattern, etc. The list is endless because it is determined by human imagination. Of course, techno determinism and technological *deus ex machina* could not get results without adequate model of capitalism and economic policy platform. Managed capitalism and heterodox policy approach are our choices.

References

Blanchard, O., Dell'Ariccia, G., Mauro, P.: Rethinking Macroeconomic Policy (IMF Staff Position Note SPN 10/03). IMF, Washington, DC (2010)

Christensen, C.: The Innovator's Dilemma: When New Technologies Cause Great Firms to Fail. Harvard Business Review, Boston (2013)

Christensen, C., Bever, D.V.: The capitalist's dilemma. Harvard Bus. Rev. **92**(6), 60–68 (2014)

DHL Global connectedness index 2016. http://www.dhl.com/en/about_us/logistics_insights/studies_research/global_connectedness_index/global_connectedness_index.html#.VFff5MkpXuM

Đuričin, D.: Escape from Transitionism: What Serbia has Learned from Past Failures and Recommendations for the Future. Serbian Association of Economists and University of Belgrade Faculty of Economics Press, Belgrade (2017)

Đuričin, D., Vuksanović, I.: How macroeconomic policies erode business competitiveness: Serbia's experience. Zagreb Int. Rev. Econ. Bus. **3**(2012), 19–35 (2012)

Đuričin, D., Vuksanović, I.: Reindustrialization strategy of Serbia: how to get it and how to use it. Ekonomikapreduzeća **61**(5–6), 289–308 (2013)

Đuričin, D., Vuksanović, I.: Quest for new development model and economic policy platform for Serbia: the role of industrial policy. Ekonomikapreduzeća **62**(5–6), 229–238 (2014)

Ghemawat, P.: Globalization in the age of Trump. Harvard Bus. Rev. **95**(4), 114–123 (2017)

Kindleberger, Ch.: World Economic Primacy: 1500–1990. Oxford University Press, New York (1996)

Kohli, H.S. (ed.): The World in 2050: Striving for a More Just, Prosperous, and Harmonious Global Community. Oxford University Press, Washington, DC (2016)

Krugman, P.: The Return of Depression Economics and the Crisis of 2008. W. W. Norton and Company, New York (2009)

Lim, W.: Joint discovery and upgrading of comparative advantage: lessons from Korea's development experience. In: Fardoust, S., Kim, Y., Sepulveda, C.P. (eds.) Post Crisis Growth and Development: A Development Agenda for the G-20, pp. 173–237. The World Bank, Washington, DC (2011)

Lin, J.Y.: New Structural Economics: A Framework for Rethinking Development and Policy. International Bank for Reconstruction and Development, Washington, DC (2012)

Lin, J.Y., Monga, C.: Growth identification and facilitation: the role of the state in the dynamics of structural change (Policy Research Working Paper No. 5313). The World Bank, Washington, DC (2010)

Minsky, H.: Stabilizing an Unstable Economy. McGraw-Hill, New Jersey (2008)

Moss, D.: A concise guide to macroeconomics: what managers, execute influence is undermining business. Harvard Bus. Rev. **92**(6), 70–78 (2007)

Mukunda, G.: The price of wall street power. Harvard Bus. Rev. **92**(6), 70–78 (2014)

Orhangazi, O.: Review of financialization at work: key texts and commentary. J. Econ. Issues **XLV**(1), 247–249 (2011)

Oxfam international. Oxfam data on global wealth distribution and projections. https://www.oxfam.org/en/pressroom/pressreleases/2016-01-18/62-people-own-same-half-world-reveals-oxfam-davos-report

Peppard, J., Ward, J.: The Strategic Management of Information Systems: Building a Digital Strategy. Wiley, Chichester (2016)

Porter, M., Heppelmann, J.: Why every organization needs an augmented reality strategy. Harvard Bus. Rev. **95**(6), 46–57 (2017)

Porter, M., Heppelmann, J.: How smart, connected products are transforming competition. Harvard Bus. Rev. **92**(11), 64–88 (2014)

Rajan, R.: Fault Lines: How Hidden Fractures Still Threaten the World Economy. Princeton University Press, Princeton (2010)

Rodrik, D.: Industrial policy: don't ask why, ask how. Middle East Dev. J. **1**(1), 1–29 (2009)

Schwab, K.: The Fourth Industrial Revolution. Crown Business, New York (2017)

Stiglitz, J., Lin, J.Y. (eds.): The Industrial Policy Revolution I: The Role of Government Beyond Ideology. Palgrave Macmillan, New York (2013)

Stiglitz, J., Lin, J.Y., Monga, C.: The Rejuvenation of Industrial Policy (Policy Research Working Paper 6628). World Bank, Washington (2013)

Vuksanović, I.: Industrial policy for Serbia: a matrix approach. Ekonomikapreduzeća **65**(1–2), 175–189 (2017)

Wade, R.: Governing the Market. Princeton University Press, Princeton (1990)

Photogrammetry Applied to Small and Micro Scaled Objects: A Review

Luigi Maria Galantucci(✉), Maria Grazia Guerra, and Fulvio Lavecchia

Dipartimento di Meccanica, Matematica e Management, Politecnico di Bari, Viale Japigia 182, 70126 Bari, Italy
{luigimaria.galantucci,mariagrazia.guerra, fulvio.lavecchia}@poliba.it

Abstract. Digital close range photogrammetry as 3D optical measurement system has achieving a great success due to the several advantages which meet the ever more complex market requests for free-form components. The capability of the sensor to be moved automatically around the scene, varying the distance from the object and the magnification ratio in a single scanning session allows to reconstruct the 3D model of the sample even in presence of undercuts or complex geometries. The latter are typical of most industrial sectors, e.g. automotive and aerospace, biomedical sector, nature science, e.g. entomology, as well as cultural heritage and archeologic field. Photogrammetry based scanners can be designed in a simple, robust and very cheap way, obtaining results similar to more complex and expensive scanners based on other technologies. In the last years, several steps have been carried out to consolidate and metrologically characterize this methodology with the final aim of making it comparable with other ones already accepted. Although, there are some relevant questions which still represent criticalities for this technique.

This paper would provide a review of the main improvements obtained in testing photogrammetry applied to very small object with sub-millimetric features. Firstly, the improvements of using completely automated and controlled scanners were analysed in terms of time saving and precision of the instrument, then, the photogrammetric process has been decomposed in its main phases to allow the emergence of the respective critical issues, and, finally, all the methodological approaches for performance verification have been reported using the existing standards.

Keywords: Measurement · 3D-Image processing · Reverse engineering
Photogrammetry · Micro machining

1 Introduction

Advanced digital technology is already used in manufacturing, but with Industry 4.0 it will radically transform industry and production. Advanced manufacturing is an underpinning technology of Industry 4.0, in the form smart, automated machines, and other technologies and it is ushering in a new age of physical production. Industry 4.0 is also known as Smart manufacturing or Manufacturing 4.0, and it is marked by a shift toward

J. Ni et al. (Eds.): AMP 2018, LNME, pp. 57–77, 2018.
https://doi.org/10.1007/978-3-319-89563-5_4

a physical-to-digital-to-physical connection. It is the leap from digital back to physical-from connected, digital technologies to the creation of a physical object- that constitutes the essence of the Industry 4.0 concept is somewhat unique to manufacturing processes [1]. One of the main issue in this scenario is availability of a complete 3D digital model for the design of products, processes, quality control and production environments. Simulations will be used more extensively in plant operations to leverage real-time data and mirror the physical world in a virtual model. This will allow operators to test and optimize the machine settings for the next product in line in the virtual world before the physical changeover, thereby driving down machine setup times and increasing quality [2].

In this context, for manufacturing of advanced high tech industrial parts, digital and precise measuring methods and equipment are very important and crucial.

Close-range photogrammetry is recognized as a simple, versatile, and effective methodology in surface detection and 3D measurement of components manufactured with micro-machining technology, even if they are characterized by a prevailing dimension respect to the others (e.g., height is much higher than length or vice versa), complex free form geometry, and under-cuts. Through the combination between rotation and translation movements of a camera and the sample to be measured, it can be possible to cover the entire measuring field from several points of view. It is able to provide accurate measurements and 3D photorealistic (thanks to the computation of the texture) surface reconstructions in a simple and inexpensive way, as well as in very short time [3], necessary for the lifecycle management in industrial environment too.

This review has been, then, organized with a brief introduction about photogrammetry applied to small object with sub-millimetric and micrometric feature, explaining all the solution available for meso-micro dimensional range. Then, in Sect. 3, the photogrammetric process has been decomposed in its main parts, which are the most relevant in the small dimensional range. In this context the main advances are presented. Section 4 deals with the critical issue related to the traceability of measurements carried out with a photogrammetric system, and, finally in Sect. 5 "Future perspectives and main potentialities" are presented.

2 Photogrammetric Process Applied to Small Objects

Photogrammetry is usually applied to large scale metrology [4] and more recently to close range [5], while there are still few applications of photogrammetry in micro-range. Small objects, with sub-millimetre or micrometric feature, need to be acquired with high magnification level and a micrometric resolution. There are several solutions proposed to ensure magnification levels higher than 1x or 2x, such as zoom lenses, macro lenses [6, 7] or lenses with extension tubes. Although, there are some important issues to be managed. Zoom lenses are mainly affected by the instability of camera calibration [8], while macro-lenses have the ability to capture a sharp image of a small object with a short working distance, retaining good stability in camera calibration.

In macro-lens photogrammetry, the limited depth of field of these long focal lenses do not allow the retention of the entire image in focus when the lens is very close to the object. This issue can be overcome if many pictures are acquired at different distances, covering the entire object, but with the consequence of high CPU processing times. To solve this limitation, in [9] it is proposed the use of focus stacking extended depth of field in combination with photogrammetry, i.e., using commercial software to create single, completely focused images, from several partially focused images, by combining the focused areas. Then, the completely focused images are used to perform photogrammetric computations (Fig. 1) [9].

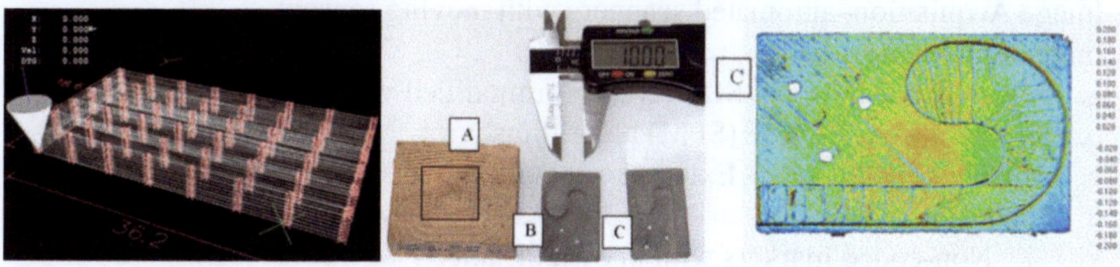

Fig. 1. Example of photogrammetric mapping of benchmark workpieces (18.6 mm × 36.2 mm) and coloured maps describing deviations of specimen C (mm) [9]

In [10] a multi-view passive stereo technique has been applied, as well, for the 3D reconstruction of small sized objects based on a sequence of macro images and an image fusion algorithm has been implemented to extend the depth of field of the images used in the photogrammetric process.

Another method is the use of tilt-shift lenses, which represents a way for controlling and adapting the DoF(depth of field) to the scene of interest, using the so-called camera movements (tilt and shift) [11].

Recently, the usage of extension tubes [12] instead of macro-lenses has been presented. This solution is well-suited for an automated and versatile scanning systems. Depending from the size of the object, it is possible to increase or restrict the field of view (move away or move closer the camera from the object) just choosing a different extension tube length and positioning the digital camera at the correct focusing distance from the object.

In order to get the object in focus, the focus distance of the lens mounted on the extension tube is normally set to infinity, thus it is necessary to operate with a fixed distance from the object, which is unique and depends from the particular coupling of the lens/extension tube length. This criterion allows to get photographs with negligible distortion, and therefore greater precision in reconstructing the 3D digital model, even without a previous internal camera calibration, since the camera lenses are manufactured with the aim to minimize the distortions in the image when focuses to infinity.

Extension tubes have been also adopted in conjunction of macro-lenses [13–15].

3 Main Advances Regarding the Photogrammetric Reconstruction of Small Objects

With the aim to identify the main advances in the photogrammetric reconstruction applied to small objects, the photogrammetric process has been decomposed in its most relevant phases which strongly affect the 3D reconstruction obtained:

1. Camera calibration
2. Image Acquisition–automated scanners with moving sensors
3. Image Processing
   a. Feature Recognition – SIFT, SURF or modified versions
   b. Structure from Motion (SFM) algorithms
   c. Bundle Adjustment -> External orientation
   d. Scale adjustment
      i. Non-coded markers with known distance
      ii. Codified targets with known distances and coordinates
      iii. Calibrated objects-2D or 3D
      iv. Hybrid method
   e. Dense cloud generation

Camera calibration, image acquisition and image processing have been recognized as the most relevant part [16, 17]. In Fig. 2 an example of 3D digital model of a specimen reconstructed using Photogrammetry (Fig. 2b–c) and Conocopic Holography (Fig. 2d).

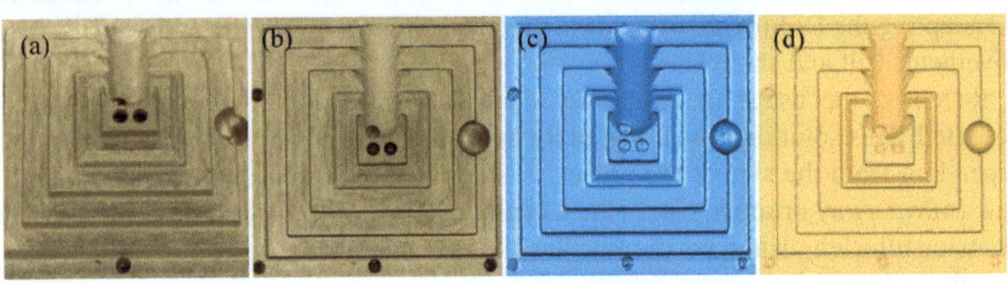

Fig. 2. Pyramidal benchmark: (a) perspective picture, (b) texturized 3D digital model obtained with PSSRT, (c) non-texturized 3D digital model obtained with PSSRT, (d) 3D digital model obtained with OptimetTMConoscan 4000 [17]

3.1 Camera Calibration

Photogrammetry requires at first a procedure of camera calibration, which in literature is known as the set of "internal calibration" and "external orientation" processes. This step is particularly relevant if a reliable measure is needed. Different camera models have been formulated and used in close-range photogrammetry, but generally sensor orientation and calibration is performed with a perspective geometrical model by means of the bundle adjustment (Brown 1971). Internal parameters, according to the Brown's model [18], define the camera model in terms of focal length, principal point coordinates and distortion parameters; while, with the term "external orientation", the spatial orientation of the camera and the camera-object mutual positions during the scan are defined.

Anyway, these two procedures have still some critical issues, heavily worsened at sub-millimetric range. The application of the traditional internal camera calibration model for the estimation of the internal parameters, becomes more difficult. Using zoom lenses implies a bad stability of estimated parameters, while, for ever narrower field of view, the traditional collinearity model could lead to an unstable calibration.

In [19], with the aim to overcome problems related with lenses of very long focal length, two prospective approaches have been adopted, one centred on the traditional collinearity equations and one on an orthogonal projection model. According to this research, under the orthogonal projection model, a fully automatic camera calibration via bundle adjustment with self-calibration is again possible.

Another criticality is related to the size of calibration pattern needed, which has to be of the same order of magnitude of the object to be measured. However, recent researches applied the traditional calibration model to macro-lens equipped with extension tube able to reconstruct small object with micrometric feature [13], also in presence of non-accurate calibration pattern [20].

Recently, with the developing of more advanced software algorithms (see also Sect. 3.3.1 "Software Algorithms"), such as the Structure From Motion (SFM), the estimation of the internal parameters could be conducted without the use of traditional coded targets. It has been demonstrated that the SFM feature-based matching approach, coupled with automated photogrammetric network orientation, can yield camera calibration parameters of greater precision and equal accuracy of automatic self-calibration approach which involves the use of targets [21].

A set of images depicting a scene with a good texture is sufficient for the extraction of natural corresponding image points [22]. These are automatically matched with feature-based approaches and robust estimation techniques. The successive photogrammetric bundle adjustment retrieves the unknown camera parameters and their theoretical accuracies. This does not imply the usage of the traditional calibration grid, while the internal parameters are defined starting from the object's images. A necessary, and in some cases not sufficient, condition for the success of this procedure is the good quality of the images acquired, as well as the object texture with recognizable features. Targetless calibration has been implied in several recent researches with robust results, also if applied to small objects with sub-millimetric features, and in some cases these parameters conduct to a better result.

Most camera modelling approaches are based on additional parameters for modelling deviations between the ideal mathematical model of central perspective and the physical reality of the camera. Several distortion models are known in the literature [23] (e.g. Field-of-View Distortion Model, Division Model or Rational Function Distortion Model) but, for the applications concerning this review, the classic radial and tangential distortion model is considered valid to correct the low distortions produced by the vision system.

Camera calibration involves the estimation of the 10 internal parameters which defines, according to Brown [18], the camera model: focal length, principal point coordinates and distortion parameters and a coefficient called *skew*.

The camera model specifies the transformation from the coordinates of the points in the local camera coordinate system to the image coordinate system (Fig. 3). The local camera coordinate system has its origins in the camera projection center.

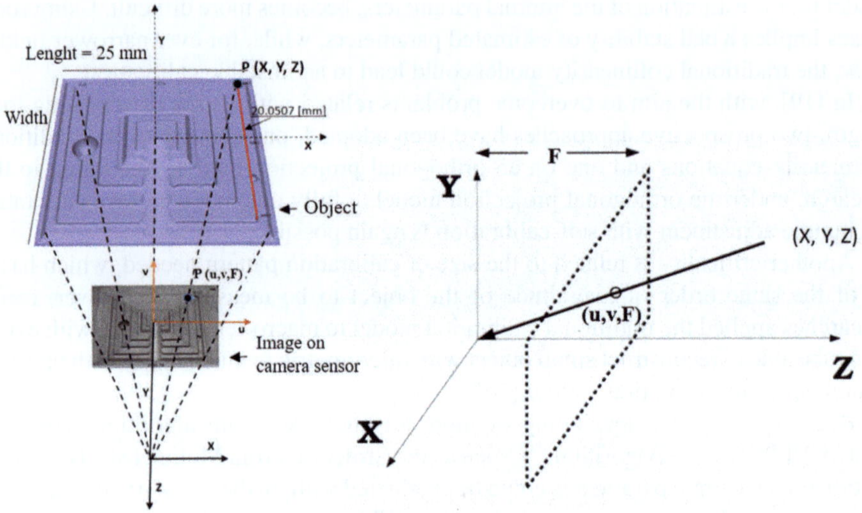

Fig. 3. Projection model [16]

The origin of the image coordinate system is coincident with the principal point.

Points coordinates in the form of X, Y e Z in the local camera reference system can be projected in the image reference system through the following equations:

$$x = X/Z \tag{1}$$

$$y = Y/Z \tag{2}$$

$$r = sqrt(x^2 + y^2) \tag{3}$$

$$x' = x(1 + K_1 r^2 + K_2 r^4 + K_3 r^6) + P_2(r^2 + 2y^2) + 2P_1 xy \tag{4}$$

$$y' = y(1 + K_1 r^2 + K_2 r^4 + K_3 r^6) + P_1(r^2 + 2y^2) + 2P_2 xy \tag{5}$$

$$u = C_x + x' F_x + y' skew \tag{6}$$

$$v = C_y + y' F_y \tag{7}$$

Where:

- (X,Y,Z) - point coordinates in the local camera coordinate system,
- (u,v) - projected point coordinates in the image coordinate system (in pixels)
- (F_x, F_y) - focal lengths

- (C_x, C_y) - principal point coordinates
- K_1, K_2, K_3 - radial distortion coefficients
- P_1, P_2 - tangential distortion coefficients,
- *skew* - skew coefficient between the x and the y axis.

3.2 Image Acquisition

3.2.1 Automated Scanners with Moving Sensors (Motorized Systems)

A great advantage of the photogrammetric technique is the freedom to choose among several strategies thanks to the capability to move the sensor around the object. It allows to reconstruct very complex structure, also in presence of undercuts or free-form surface. The usage of photogrammetry for small objects has been greatly aided by the implementation of motorized systems. Scanners could be designed adopting two main cinematic configurations: Cartesian [9] and polar [19]. The polar configuration is preferred in many cases because it reduces significantly the space needed for the image acquisition and the overall dimensions of the scanner; it is an important factor considering the limited space normally available in deposits and conservation labs.

The implementation of motorized turning table allows to acquire image from multiple views in a controlled way and in a very short time. With this procedure the control of the camera-to-object distance becomes a necessary step to exploit the maximum depth of field allowed, especially in case of small object reconstruction, which requires macro lenses or other optical configurations having the drawback of limited depth of field.

This implementation could be more difficult for very large objects, but it surely represented a great advantage for the acquisition of small objects [3]. In [12] a low-cost Photogrammetric Scanning System with Rotary Table (PSSRT) is illustrated, designed

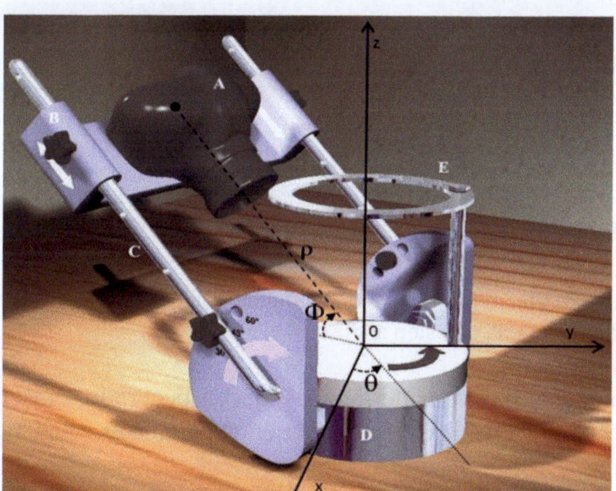

Fig. 4. Design of the prototype realized for the low-cost Photogrammetric Scanning System with Rotary Table: (A) Digital SLR Camera; (B) Platform for focus distance tuning; (C) Rigid tubular frame for tilt angle tuning; (D) Rotary Table; (E) Light [12].

and implemented to return precise 3D digital models of small objects, which presents complex surfaces and sub-millimeter features (Fig. 4). The system comprised a camera mounted on an arm positioned at fixed angles respect to the rotary table. It allows a great process control, as well as, the assurance of a high reproducibility of the measuring procedure.

In [24] has been implemented an improved version of the PSSRT system, completely motorized and automatically controlled, and is currently under experimentation for further developments.

It is important to underline the suitability of the photogrammetric scanning that can be easily implemented as an on-line measuring technique for dimensional verification of manufacturing processes, especially the one applied for small object (it requires just one camera).

3.2.2 Scanning Strategy

The possibility to choose among different acquisition strategies is a great advantage of a photogrammetry-based system. When small objects have to be acquired, several factors must be taken into account. Depending on the shape and geometry of the object, using a polar configuration it is necessary to choose at least two process parameters, tilt angle and step angle values. Tilt angle (Φ) identifies the camera inclination respect to the xy object plane, while, step angle (θ), defined as the rotational angle step of the turning table, determines the number of images and, then, the overlapping level between two subsequent images. These two parameters geometrically define the acquisition strategy (Fig. 5) [25].

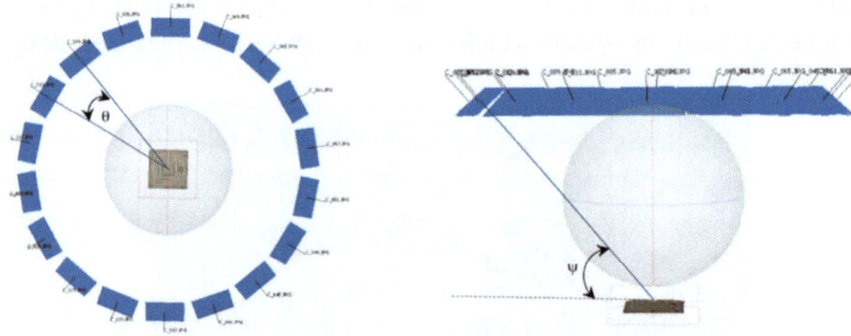

Fig. 5. Reconstructed camera positions (e.g. configuration Tilt Angle 60°, Step Angle 20°). Tilt angle has been indicated as Ψ, and step angle as θ [25].

The best choice involves considering the object aspect ratio, as well as, the depth of field value (Fig. 6) [25]. The latter is a fundamental parameter when high magnification levels are required.

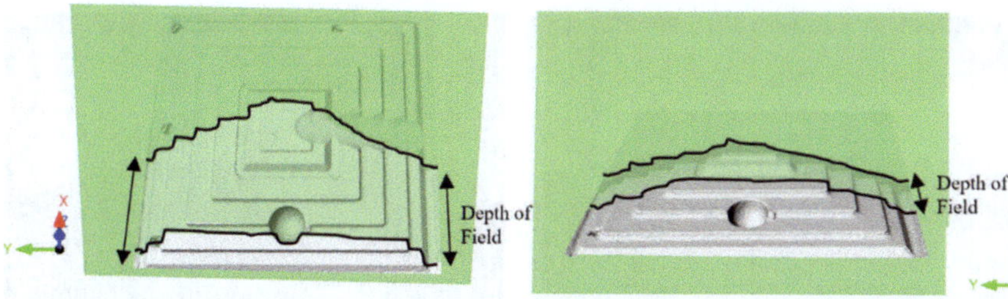

Fig. 6. 3D tilted views. On the left, visualization from camera tilted of 60° from the xy-plane, while on the right, visualization from camera tilted of 30° from the xy-plane [25].

3.2.3 Illumination

Since photogrammetry is an optical-based measuring instrument, if no light grids or patterns are projected into the scene (passive photogrammetry), it is particularly sensitive to the object texture and the environmental illumination, if it is not differently set [26].

Most of the currently available systems tried to overcome this problem, putting the object inside a white box with a diffusely illumination which allows, firstly, to isolate the system from the environmental illumination and its changes over time, but also to ensure a more diffusive light effect. A photogrammetric acquisition is affected by reflections and shadows, because they can change during the multi-view acquisition. The latter could be avoided adopting a lighting system, typically white-light led strips, which moves solidly with the rotating table [12]. This way to operate have been successfully implemented in several experiments regarding the reconstruction of small objects with sub-millimeter feature [12, 13, 17, 27].

In Fig. 7 several artefacts fabricated using additive manufacturing are shown [28].

Another crucial point in passive photogrammetry is the sensitivity of this instrument to the object's texture. Texture less objects, such as e.g. fine uniformly painted and highly reflective surfaces, many injection moulded plastic parts and those obtained with additive manufacturing technologies (Stereolithography or Fused Deposition Modeling with plastic materials), represent a big challenge for this kind of instruments. In [29], authors tried to improve the surfaces texture of the sample to be measured adopting an active photogrammetry through a laser speckle projection system to create a pattern or an artificial texture on the object with good results also when dealing with additive manufactured object [28].

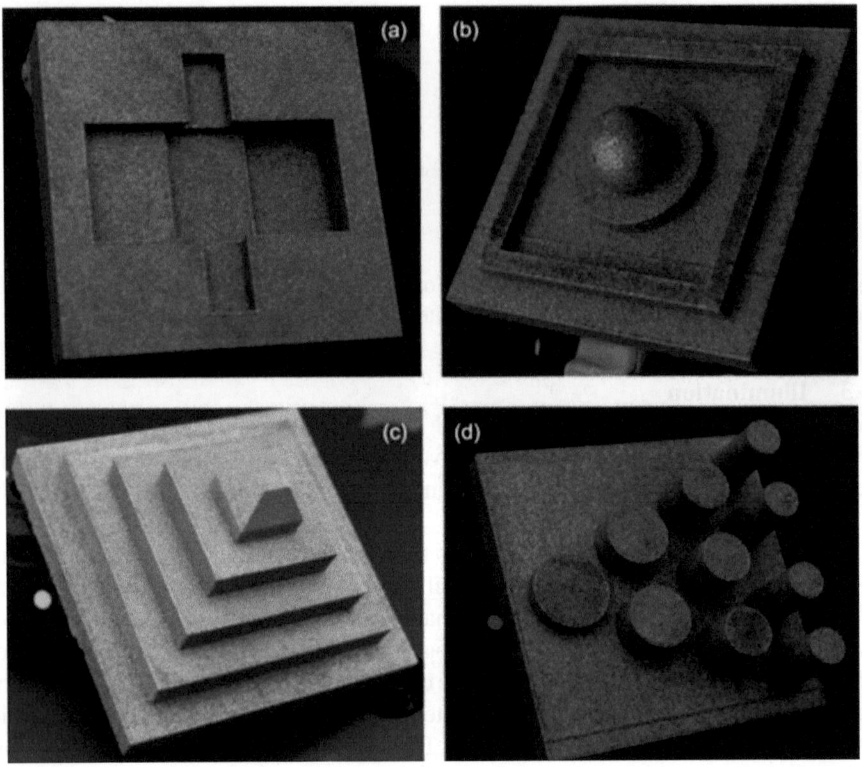

Fig. 7. AM artefacts: (a) recess artefact; (b) sphere artefact; (c) pyramid artefact; (d) pillar artefact [28]

3.3 Image Processing

3.3.1 Software Algorithms

The recent success of the photogrammetric technique coincided with the rise of new algorithms capable to reconstruct objects even if the optimal conditions are still not verified.

The photogrammetric process could be briefly described as the succession of three main phases (Fig. 8).

Among these, the most critical ones are the estimation of the internal parameters, which reduce the lens distortions effect, and the computation of the external parameters, which describe the spatial orientation and the relative camera-object positions during the image acquisition. Several researches have been conducted in the photogrammetric field, but, above all, the development of new algorithms with the aim of simplifying and automating the alignment process has been one of the principal topics of discussion. It is well known, in fact, that the orientation phase is crucial for the success of all the subsequent steps (the dense cloud generation and the meshing process).

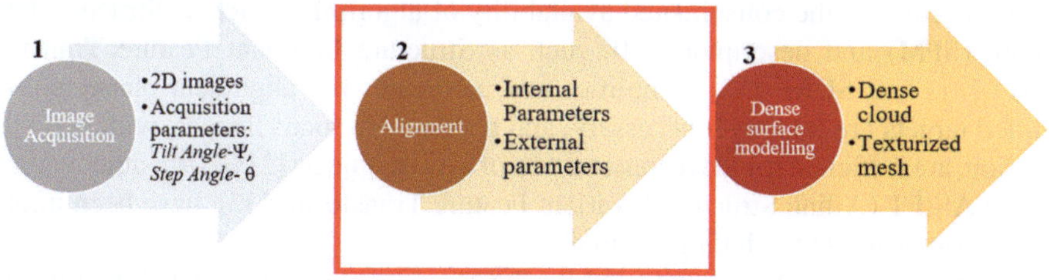

Fig. 8. Description of the main steps of the photogrammetric process [16]

The development of commercial software belonging to the digital imaging and Computer Vision (CV) community, thanks to the high level of automation in data processing, allows time and computational load saving. It could be further decomposed in feature recognition, image matching, and estimation of internal and external parameters. At the start of the process, after the image acquisition, descriptors are used to recognize features on acquired images. The next image matching phase matches common points on images and this is the start of the triangulation process (Fig. 9) [16].

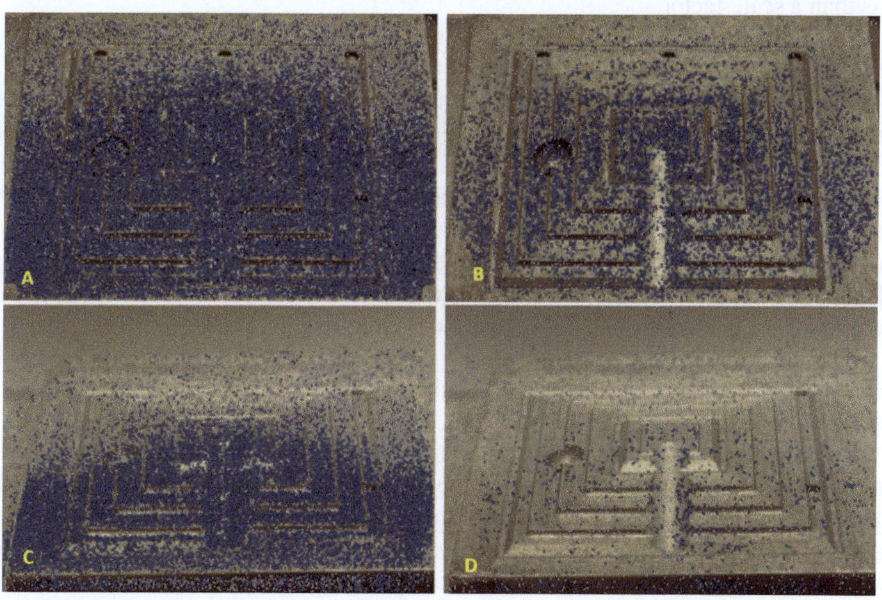

Fig. 9. Tie points: Tilt angle 60° and 72 photos (A); Tilt angle 60° and 36 photos (B); Tilt angle 30° and 72 photos (C); Tilt angle 30° and 36 photos (D) [16].

The quality and the quantity of the recognized features greatly influence the subsequent steps. The data obtained from this phase, as bi-dimensional vectors, become the input for the next one, the computation of internal and external orientation parameters. The estimation of internal parameters and external parameters can be carried out in the same phase, starting from the same input, or in two subsequent ones.

In this contest, the consolidated availability of algorithms, such as Structure From Motion (SFM) and descriptors [30] such as Structure Invariant Feature Transform (SIFT) [31] and SURF [32], simplifies and automates the alignment phase, so that, codified targets are no longer necessary, and good and in focus images, with good illumination, are sufficient for the reconstruction of a lot of objects [21]. Also, other variants, like the ASIFT (Affine Structure Invariant Feature Transform) [33], have been implied as a descriptor in 3D modelling of small.

Other studies have been conducted for the assessment of image matching process.

In [34], a critical review of four dense image-matching algorithms, available as open-source and commercial softwares, SURE (photogrammetric Surface RE construction from imagery), MicMac, multi-resolution and multi-image method, PMVS (Patch-based Multi-View Stereo), has been reported. The algorithms are evaluated according to their ability to produce dense and high-quality 3D point clouds, as well as according to computation time.

3.3.2 Scale Adjustment

After the estimation of the internal parameters, the external orientation is carried out less than a scale factor.

This is one drawback of the photogrammetric process, but several ways to produce a real scaled object have been developed.

Firstly, codified targets with known absolute coordinates and distances, placed around and on the object's surface, could be used to provide the scale factor and contemporary a coordinate reference system. This is the most used procedure in large scale and close-range photogrammetry, but when small objects with sub-millimetric feature have to be acquired, its application becomes more difficult. When the magnification level increases, the object cover more than the 80% of the image, with few possibilities to place codified targets around it. Placing targets on the object surface is also quite difficult because of the presence of small features, as well as free-form shaped parts.

This led to the necessity to find other ways to compute the scale factor. One of this consists of artificially create, after the acquisition, two markers on at least two images, and set their known distance as a scale bar.

Recently, a new hybrid method has been presented for photogrammetric scale adjustment [27] which involves two parameters: the magnification level M, which is the ratio between the size of an object in the image and its true size, and the pixel size which is a specification of the sensor used. Firstly, a pixel distance has to be computed between two points located in the most in focus area, and then this distance has to be converted in millimetres through the knowledge of the magnification ratio and pixel size.

The main drawbacks are represented by the user-dependency and the necessity to know previously the distance between those points. This fact is further worsened in presence of free-form shaped objects.

An alternative way is represented by using reference objects with codified targets as a "calibrator", to compute for each sensor position the external orientation with the scale factor, and then use these parameters for the successive object scanning phase.

The reference object could be a 2D pattern of targets or a 3D geometry with targets with known positions already measured with an optical CMM. To ensure the success of this procedure, the same acquisition strategy must be used for the reference object and for the sample to be reconstructed, which involves the same optical configuration, the same reciprocal position between camera and the object, the same number of images. This condition is ensured by using systems with automatically controlled axis or systems in which camera (or cameras) are fixed and the object is placed always in the same position.

The main advantage of this scale adjustment method is that it is not user-dependent, and it can be used for every kind of objects, independently on their shape. Moreover, it involves the usage of optical coordinate measuring machine to measure targets positions on reference objects and it provides a reliable result.

The first drawback is represented by the realization of these reference objects, because, the higher is the magnification level required, the smaller is the field of view and the smaller is the size of codified targets to use (less than 1 mm). The more simple way is to print them on paper, but commercially available printers are not able to ensure a good printing quality.

In [12] a bi-dimensional pattern of targets was used. According to this research, there is a combined effect of three factors, which strongly affect the computation of the external orientation of the camera during the scan: the camera tilt angle (Φ), the rotational step angle of the turning table (θ) and the geometry of the reference object. 2D reference objects, like a simple pattern of targets printed on a paper, certainly represent the simplest way, but it is not the best solution when the process is performed using a low camera tilt angle ($\Phi = 30°$) and a low number of photographs.

The implementation of 3D reference object overcomes these criticalities and it has been presented in Fig. 10(a) [24]. The reference object used for the adjustment has been designed according to the photogrammetric reconstruction capabilities [24]. The presence of clearly visible edges and a good surface texture, greatly aided the first point recognition phase and made the external orientation more robust, also in presence of low tilt angle value.

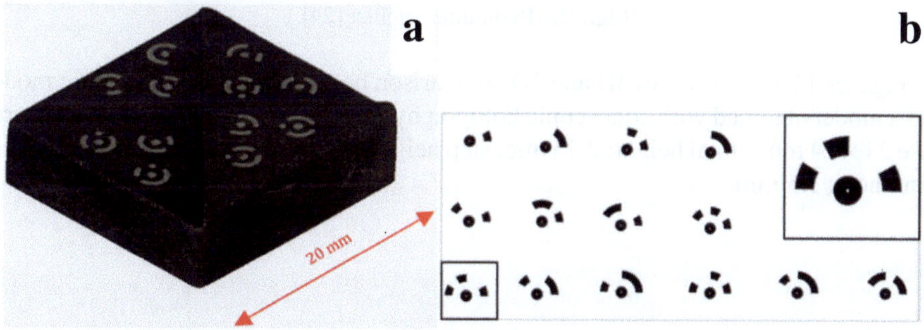

Fig. 10. (a) 3D calibrator texturized model [24] (b) 2D pattern of target

The target were laser ablated on the top surface of the reference object to make them automatically recognizable during the software's markers detection.

4 Measurement Traceability

Photogrammetry is currently used in lots of fields, from cultural heritage, natural science, medicine, biomedical applications with very successful results, also thanks to its characteristic to reconstruct an object with its real colours. However, when it is applicated in the industrial field, for the dimensional verification of industrial components, the need for measurement traceability becomes a fundamental issue.

Accuracy statements

The accuracy statement is very often performed comparing results obtained from the photogrammetric reconstruction with ones obtained with a more consolidated technique. If a 3D or 2D comparison is carried out, as outputs, the average deviation between homologous points is computed together with the standard deviation. Results obtained in recent research works report an average deviation less than 10 micrometers when the external orientation is carried out with a 3D reference object [24]. Figure 11 shows an example of specimen used to analyse the performance of the method.

Fig. 11. Pyramidal artefact [24]

Figures 12 and 13 shows 3D and 2D comparison between a photogrammetric model and a model obtained with conoscopic holography. The main specimen dimensions are: base $24 \times 24\ mm^2$, total height: 2.75 mm, step height: 0.5/0.25 mm, radius of the biggest blind hole: 1.5 mm.

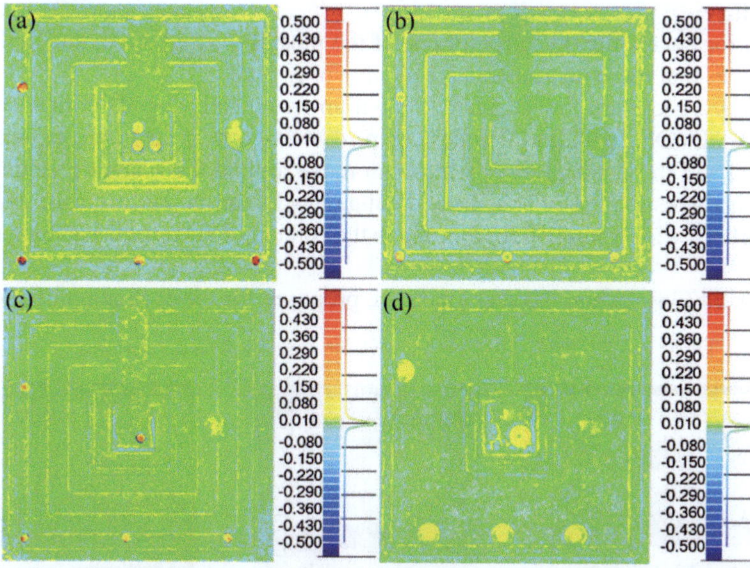

Fig. 12. Example of 3D comparisons between a photogrammetric model and a model obtained with conoscopic holography [17].

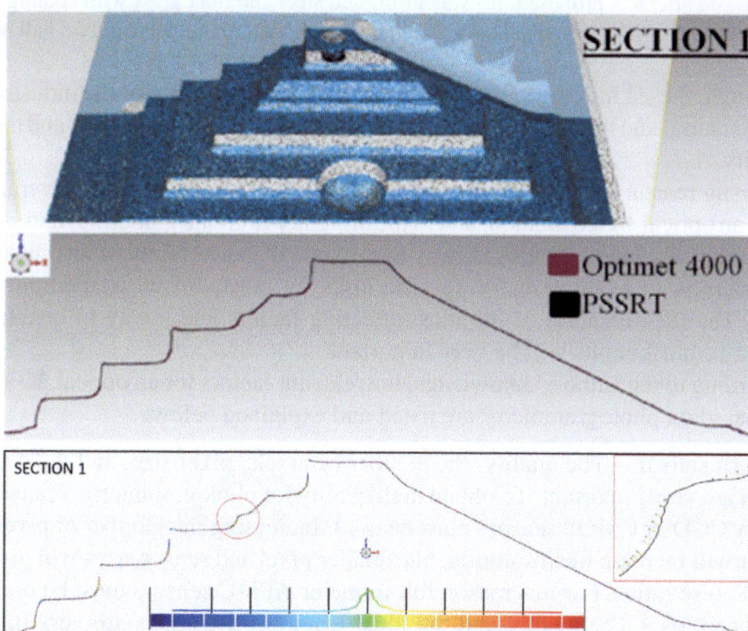

Fig. 13. Example of 2D comparison

Traceability

The need for traceability is currently one of the most discussed topic, due to the great potentialities that these instruments are gaining over time.

Generally, there are few standards drawn up specifically for 3D optical scanners. German VDI/VDE 2634, part 2 and 3 comprise many optical instruments, both those area-based and those with sensor which can move around the scanning volume. This standard involves the use of spheres and gauge blocks for the performance evaluation and it has been widely implied for this aim in the context of 3D optical scanning systems [35–37].

In Fig. 14 [38] several basic artefacts are proposed for the evaluation of 3D optical scanners.

Fig. 14. Basic artefacts proposed. (a) vapour blasted steel and float glass with coating (b) steel spheres and painted balls, (c) Artefacts to evaluate length measurement error: two ball bars [38].

Although, these kind of geometries, are not well representative of the industrial parts actually existing, and the risk is to underestimate the systematic deviations and the whole uncertainty.

The main reason for the lack of standards available is related to the several factors affecting an optical-based method, and to the difficulty to create guidelines and standards able to comprise all these factors. A way to evaluate the uncertainty of an optical measuring system is to evaluate it for specific tasks by means of an adaptation the ISO 15530-3. The identification of the most affecting factors and a way to quantify their uncertainty contributions will be very important.

According to the authors' knowledge, the relevant factors for an optical 3D scanning system based on photogrammetry are listed and explained below:

- **Camera sensor –** The quality, the number of pixels, pixel size, and the size of the sensor are very important. To obtain high quality for photogrammetric scanners good quality CCD or CMOS sensors must be used. Increasing the number of pixels of the sensor will increase the resolution, but smaller pixel and sensor sizes will give lower signal/noise ratios. For this reason full frame or APS-C sensors must be preferred.
- **Camera lens –** The quality and the focal length of the optics are very important. Higher quality lenses produce better images with lower distortions. Higher focal length allows higher camera-object distance and the camera-object distance, increases the magnification ratio and the number of images needed for the photogrammetric reconstruction.

- **Images Overlapping** – Photogrammetry needs an overlapping among adjacent images, at least of 80% in the first direction and 20% in the second one.
- **Camera calibration (estimation of internal parameters)** – It is of great importance to evaluate the stability of this phase when dealing with specific optical configurations, such as zoom lenses, macro lenses or lenses with extension tubes, especially when it is carried out together with the estimation of the external parameters, without codified targets.

  Resolution – The ability to distinguish two objects or features from each other is fundamental in optical 3D metrology. This represents the structure resolution limit of detection of an instrument. It is divided into two component: a lateral part, which is the perpendicular component, and an axial part, which is the parallel component, both with respect to the 3D sensor's optical axis. In photogrammetric field, the lateral resolution (Eq. 7) (GSD: Ground Sampling Distance) is calculated as [24]:

$$GSD\left[\frac{mm}{pixel}\right] = \frac{Flying\ Altitude[mm]}{Focal\ Lenght[mm]} * Pixel\ size[mm/pixel] \quad (7)$$

  A photogrammetric scanning system, if properly equipped, could reach a micrometric resolution [13].
- **Software Algorithms** and their repeatability – software and algorithms developed in 3D image systems played a fundamental role for the application of these techniques. Although, the repeatability of the latter, due to a random component, is not always negligible. In [16], authors analysed the repeatability of algorithms implied in a specific commercial software, regarding a specific procedure for the computation of the external orientation parameters presented in [12]. Although the specificity of the procedures involved, results could be generalised as follows: software repeatability component could be considered negligible in all those cases characterized by a sufficient number of common feature recognized among photographs, which is the essential condition to obtain a robust 3D reconstruction, see Fig. 9.
- **Hardware** – in case of motorized and controlled system, it would be interesting to quantify the influence of any errors in axis positioning on resulting reconstruction.
- **Scale adjustment** – the estimation of internal and external parameters is carried out except for a scale factor. All the procedures, presented in Sect. 3.2, produce errors, which could be divided in a systematic and a random component.
- **Texture-dependency** – Triangulation-based techniques require opaque materials so that testing triangulation system requires reference artefacts that have cooperative optical surface characteristics. In order to quantify a 3D optical scanner's ability to measure different surface finishes (material, reflectance, roughness and colour), National Physical Laboratory's (NPL) developed the 3D material coupon plate, a multi-faceted test artefact [39].
- **Environmental characteristics – (temperature, humidity, lighting conditions)** Their effects could be reduced if measures are carried out in a controlled environment. Anyway, temperature affects not just the object to measure but also the 3D optical scanner. Ways to quantify thermal effect and illumination effect on 3D optical scanners are reported in [40].

With the aim to cover most of criticalities related to 3D optical instruments, many reference objects have been designed. NPL developed a new free-form calibrator geometry [44] specifically suitable for 3D optical measuring instruments (Fig. 15).

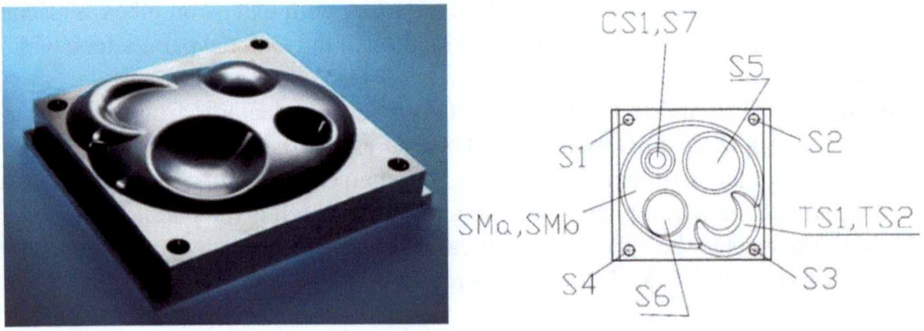

Fig. 15. Photograph and schematic representation of NPL freeform artefact [44]

In [39] four artefacts have been designed and presented to aid understanding of 3D optical scanner main factors. Resolution, orientation, illumination effect, sensitivity to surface color, material and finish.

In [24], a staircase artefact with sub-millimeters heights has been used as reference object for the uncertainty assessment of a photogrammetric scanning system, considering resolution, form errors, reproducibility, with expanded uncertainty values in the order of 10 μm.

For traceability assurance, a way is to apply the substitution method, ISO 15530-3 [41], which transforms traceability from reference objects to real parts.

In this case, traceability is ensured through the usage of a reference object, previously calibrated with an instrument characterized by known uncertainty, typically a mechanical CMM [42, 43], but the additional uncertainty components have to be considered together with the similarity requirements between the calibrated artefact and the workpiece [41].

5 Future Challenges

Future challenges for photogrammetry applied to small objects could be considered, generally, as future demands and requirements for optical dimensional metrology:

- Analysis of the main sources of errors and quantification of their effects on measurement results;
- Accuracy improvement and determination of measurement uncertainty;
- Results have to be comparable with those obtained with contact and optical CMM, with uncertainty less than ten micrometers;
- Reduction or compensation of systematic effects on measurement results;
- Implementation of this technique for in-line measurements in a production environment.

More specifically for a photogrammetric-based system:

- Implementation of strategies in order to avoid lighting variation which can affect the measurement results, increasing the uncertainty;
- Improvements of scanning strategies in order to overcome depth of field limitations.

6 Conclusions

Photogrammetry applied to small objects is a good alternative to contact measuring system in certain conditions. Systems equipped with extension tubes or macro lenses could achieve micrometric resolution up to 2 μm, with magnification levels higher than 2x. Accuracy statements provide micrometric deviations and low uncertainties if the optimal conditions are verified. Although, it is still difficult to achieve a good traceability, because of the large number of factors affecting the system and the difficulty in quantifying their effects on the measuring process.

Several efforts must be done in this direction to enlarge the applicability of optical-based instruments, in this case, photogrammetry, in the industrial field exploiting its several advantages which made this technique of great interest in the last years.

References

1. Sniderman, B., Monika, M., Cotteleer, M.J.: Industry 4.0 and Manufacturing Ecosystems. Deloitte University Press (2016)
2. Pearce, P.: Industry 4.0: How Advancements in Manufacturing Technology Are Transforming the Industry (2017)
3. Nguyen, C.V., Lovell, D.R., Adcock, M., La Salle, J.: Capturing natural-colour 3D models of insects for species discovery and diagnostics. PLoS ONE **9**, e94346 (2014). https://doi.org/10.1371/journal.pone.0094346
4. Peggs, G.N., Maropoulos, P.G., Hughes, E.B., et al.: Recent developments in large-scale dimensional metrology. Proc. Inst. Mech. Eng. Part B J. Eng. Manuf. **223**, 571–595 (2009). https://doi.org/10.1243/09544054JEM1284
5. Luhmann, T.: Close range photogrammetry for industrial applications. ISPRS J. Photogramm. Remote Sens. **65**, 558–569 (2010). https://doi.org/10.1016/j.isprsjprs.2010.06.003
6. Maté González, M.Á., Yravedra, J., González-Aguilera, D., Palomeque-González, J.F., Domínguez-Rodrigo, M.: Micro-photogrammetric characterization of cut marks on bones. J. Archaeol. Sci. **65**, 128–142 (2015). https://doi.org/10.1016/j.jas.2015.08.006
7. Rodríguez-martín, M., Lagüela, S., González-aguilera, D., Rodríguez-gonzálvez, P.: Procedure for quality inspection of welds based on macro-photogrammetric three-dimensional reconstruction. Opt. Laser Technol. **73**, 54–62 (2015)
8. Shortis, M.R., Bellman, C.J., Robson, S., et al.: Stability of zoom and fixed lenses used with digital SLR cameras. Int. Arch. Photogramm. Remote Sens. Spat. Inf. Sci. **36**, 285–290 (2006)
9. Galantucci, L.M., Lavecchia, F., Percoco, G.: Multistack close range photogrammetry for low cost submillimeter metrology. J. Comput. Inf. Sci. Eng. **13**, 044501 (2013). https://doi.org/10.1115/1.4024973
10. Gallo, A., Muzzupappa, M., Bruno, F.: 3D reconstruction of small sized objects from a sequence of multi-focused images. J. Cult. Herit. **15**, 173–182 (2014). https://doi.org/10.1016/j.culher.2013.04.009

11. Nocerino, E., Menna, F., Remondino, F., et al.: Experiments on calibrating tilt-shift lenses for close-range photogrammetry. Int. Arch. Photogramm. Remote Sens. Spat. Inf. Sci. ISPRS Arch. **41**, 99–105 (2016). https://doi.org/10.5194/isprsarchives-XLI-B5-99-2016
12. Galantucci, L.M., Pesce, M., Lavecchia, F.: A stereo photogrammetry scanning methodology, for precise and accurate 3D digitization of small parts with sub-millimeter sized features. CIRP Ann. Manuf. Technol. **64**, 507–510 (2015). https://doi.org/10.1016/j.cirp.2015.04.016
13. Percoco, G., Guerra, M.G., Sanchez Salmeron, A.J., Galantucci, L.M.: Experimental investigation on camera calibration for 3D photogrammetric scanning of micro-features for micrometric resolution. Int. J. Adv. Manuf. Technol. (2017). https://doi.org/10.1007/s00170-016-9949-6
14. Percoco, G., Lavecchia, F., Salmerón, A.J.S.: Preliminary study on the 3D digitization of millimeter scale products by means of photogrammetry. Procedia CIRP **33**, 257–262 (2015). https://doi.org/10.1016/j.procir.2015.06.046
15. Percoco, G., Sánchez Salmerón, A.J.: Photogrammetric measurement of 3D freeform millimetre-sized objects with micro features: an experimental validation of the close-range camera calibration model for narrow angles of view. Meas. Sci. Technol. **26**, 095203 (2015). https://doi.org/10.1088/0957-0233/26/9/095203
16. Lavecchia, F., Guerra, M.G., Galantucci, L.M.: The influence of software algorithms on photogrammetric micro-feature measurement's uncertainty. Int. J. Adv. Manufact. Technol. **93**, 3991–4005 (2017). https://doi.org/10.1007/s00170-017-0786-z
17. Galantucci, L.M., Pesce, M., Lavecchia, F.: A powerful scanning methodology for 3D measurements of small parts with complex surfaces and sub millimeter-sized features, based on close range photogrammetry. Precis. Eng. (2015). https://doi.org/10.1016/j.precisioneng.2015.07.010
18. Brown, D.C.: Close-range camera calibration. Photogramm. Eng. **37**, 855–866 (1971)
19. Stamatopoulos, C., Fraser, C.S.: Calibration of long focal length cameras in close range photogrammetry. Photogramm. Rec. **26**, 339–360 (2011). https://doi.org/10.1111/j.1477-9730.2011.00648.x
20. Percoco, G., Salmerón, A.J.S.: 3D image based modelling for inspection of objects with micro-features, using inaccurate calibration patterns: an experimental contribution. Int. J. Interact. Des. Manuf. **11**, 415–425 (2017). https://doi.org/10.1007/s12008-016-0342-3
21. Fraser, C.S., Stamatopoulos, C.: Automated Target-Free Camera Calibration (2014)
22. Barazzetti, L., Mussio, L., Remondino, F., Scaioni, M.: Targetless camera calibration. Int. Arch. Photogramm. Remote Sens. Spat. Inf. Sci. **XXXVIII:W16**, 335–342 (2011). https://doi.org/10.5194/isprsarchives-XXXVIII-5-W16-335-2011
23. Ricolfe-Viala, C., Sanchez-Salmeron, A.-J.: Lens distortion models evaluation. Appl. Opt. **49**, 5914 (2010). https://doi.org/10.1364/AO.49.005914
24. Lavecchia, F., Guerra, M.G., Galantucci, L.M.: Performance verification of a photogrammetric scanning system for micro-parts using a three-dimensional artefact: adjustment and calibration. Int. J. Adv. Manuf. Technol. (2018). https://doi.org/10.1007/s00170-018-1806-3
25. Lavecchia, F., Guerra, M.G., Galantucci, L.M.: The influence of software algorithms on photogrammetric micro-feature measurement's uncertainty. Int. J. Adv. Manuf. Technol. 1–15 (2017). https://doi.org/10.1007/s00170-017-0786-z
26. Nicolae, C., Nocerino, E., Menna, F., Remondino, F.: Photogrammetry applied to problematic artefacts. ISPRS – Int. Arch. Photogramm. Remote Sens. Spat. Inf. Sci. **XL-5**, 451–456 (2014). https://doi.org/10.5194/isprsarchives-XL-5-451-2014

27. Percoco, G., Modica, F., Fanelli, S.: Image analysis for 3D micro-features: a new hybrid measurement method. Precis. Eng. (2016). https://doi.org/10.1016/j.precisioneng.2016.11.012
28. Sims-Waterhouse, D., Bointon, P., Piano, S., Leach, R.K.: Experimental comparison of photogrammetry for additive manufactured parts with and without laser speckle projection. 103290W (2017). https://doi.org/10.1117/12.2269507
29. Sims-Waterhouse, D., Piano, S., Leach, R.: Verification of micro-scale photogrammetry for smooth three-dimensional object measurement. Meas. Sci. Technol. **28**, 055010 (2017). https://doi.org/10.1088/1361-6501/aa6364
30. Remondino, F.: Detectors and descriptors for photogrammetric applications. Int. Arch. Photogramm. Remote Sens. Spat. Inf. Sci. 2–7 (2006)
31. Lowe, D.G.: Distinctive image features from scale-invariant keypoints. Int. J. Comput. Vis. **60**, 91–110 (2004). https://doi.org/10.1023/B:VISI.0000029664.99615.94
32. Bay, H., Ess, A., Tuytelaars, T., Van Gool, L.: Speeded-up robust features (SURF). Comput. Vis. Image Underst. **110**, 346–359 (2008). https://doi.org/10.1016/j.cviu.2007.09.014
33. Morel, J.-M., Yu, G.: ASIFT: a new framework for fully affine invariant image comparison. SIAM J. Imaging Sci. **2**, 438–469 (2009). https://doi.org/10.1137/080732730
34. Remondino, F., Spera, M.G., Nocerino, E., et al.: State of the art in high density image matching. Photogramm. Rec. **29**, 144–166 (2014). https://doi.org/10.1111/phor.12063
35. Mendricky, R.: Determination of measurement accuracy of optical 3D scanners. 1565–1572 (2016). https://doi.org/10.17973/MMSJ.2016
36. Bernal, C., De Agustina, B., Marín, M.M., Camacho, A.M.: Accuracy analysis of fringe projection systems based on blue light technology. Key Eng. Mater. **615**, 9–14 (2014). https://doi.org/10.4028/www.scientific.net/KEM.615.9
37. Vagovský, J., Buranský, I., Görög, A.: Evaluation of measuring capability of the optical 3D scanner. Energy Procedia **100**, 1198–1206 (2015). https://doi.org/10.1016/j.proeng.2015.01.484
38. Angelo Beraldin, J., Mackinnon, D., Cournoyer, L.: Metrological characterization of 3D imaging systems: progress report on standards developments. Int. Congr. Metrol. **3**, 1–21 (2015). https://doi.org/10.1051/metrology/20150013003
39. Dury, M.R., Woodward, S.D., Brown, S.B., McCarthy, M.B.: Surface finish and 3D optical scanner measurement performance for precision engineering. In: Proceedings of ASPE 2015 Annual Meeting, pp. 419–423 (2015)
40. Dury, M.R., Brown, S., McCarthy, M., Woodward, S.: 3D optical scanner dimensional verification facility at the NPL's "National FreeForm Centre." In: Laser Metrology and Machine Performance XI (LAMDAMAP 2015), pp. 191–200 (2015)
41. ISO 15530-3:2011 Geometrical product specifications (GPS) – Coordinate measuring machines (CMM): Technique for determining the uncertainty of measurement – Part 3: Use of calibrated workpieces or measurement standards
42. Gameros, A., De Chiffre, L., Siller, H.R., et al.: A reverse engineering methodology for nickel alloy turbine blades with internal features. CIRP J. Manuf. Sci. Technol. **9**, 116–124 (2015). https://doi.org/10.1016/j.cirpj.2014.12.001
43. Stolfi, A., De Chiffre, L.: 3D artefact for concurrent scale calibration in computed tomography. CIRP Ann. Manuf. Technol. **65**, 499–502 (2016). https://doi.org/10.1016/j.cirp.2016.04.069
44. McCarthy, M.B., Brown, S.B., Evenden, A., Robinson, A.D.: NPL freeform artefact for verification of non-contact measuring systems. Soc. Photo Opt. **7864**, 78640K–78640K–13 (2011). https://doi.org/10.1117/12.876705

Artificial Neural Networks in Advanced Thermoset Matrix Composite Manufacturing

Pierpaolo Carlone[1], Dragan Aleksendrić[2(✉)], Felice Rubino[1], and Velimir Ćirović[2]

[1] Department of Industrial Engineering, University of Salerno, Fisciano, Italy
[2] Faculty of Mechanical Engineering, University of Belgrade, Belgrade, Serbia
daleksendric@mas.bg.ac.rs

Abstract. Autoclave curing is a common practice to manufacture high temperature thermoset matrix composites. The cycle design and optimization of the temperature-time curve is a key issue for a competitive production. In this paper artificial neural networks (ANN), as a technique of artificial intelligence, were used for prediction of the composite temperature profile during the autoclave curing process. Different neural network models have been investigated regarding their capabilities for prediction of the composite temperature profile. The new neural network model has been developed able to predict the composite temperature profile in the wide range of manufacturing conditions changing.

Keywords: Artificial neural networks · Composite material · Curing process

1 Introduction

Thermosetting matrix polymeric composite materials (PMC) gained a great deal of consideration as a valid alternative to conventional materials.

Among the different processes currently employed in the manufacturing of the thermoset PMC, the autoclave process is the most widespread especially in high cutting edge fields, e.g. aerospace, aeronautic and sporting goods, when superior mechanical properties of the final product are needed [1]. The autoclave process is based on applying of pressure and temperature cycles to compact the composite laminate and promote the resin polymerization. During the curing process, the thermal energy provided to the composite increases part temperature, resulting in an initial decrease of material viscosity, that coupled with the applying of an external pressure allows to obtain an optimal fibers impregnation, and in the activation of the cure reaction. It is generally accepted that the final properties of the manufactured part mainly rely on temperature history experienced by the material (that rules the resin curing progress), residual stresses and void percentage at the end of the cure. What is more excessive temperature peaks or gradients should be avoided in order to prevent the degradation of the resin system and excessive thermal stresses, respectively [2]. The activation of resin reaction strongly affects the progression of the cure front as a consequence of residual stresses and distortions of the processed part. As a general criterion, in-out solidification should be preferred to out-in solidification [3]. Therefore, the planning of the thermal cycle, i.e. definition and optimization of the temperature-time curve plays a key role for the

J. Ni et al. (Eds.): AMP 2018, LNME, pp. 78–88, 2018.
https://doi.org/10.1007/978-3-319-89563-5_5

production of high quality parts. In the recent past, computational approaches, mainly based on numerical simulation [4–8] and optimization [9–13], have been proposed in the inherent literature, as an alternative to expensive trial and error procedures. A computational procedure coupling a finite difference one dimensional model of the process with the Nelder-Mead and Simulated Annealing algorithms has been implemented and tested by Mawardi and Pitchumani [9], for the optimal definition of process parameters. An approach to thermal cycle design, based on the Levenberg-Marquardt algorithm has been proposed by Dofour *et al.* in [10], showing good capabilities in off-line optimization. Ruiz and Trochu implemented a multi-criteria evolutionary algorithm to optimize the thermal cycle in liquid composite molding processes [11]. A hybrid approach, based on the coupling of meta-heuristic optimization techniques with a one dimensional finite element model of the process has been proposed in [12], to evaluate the optimal process parameters under technological constraints. Among others, artificial neural networks (ANNs) proved to be very effective in substitution of direct simulation in several contests, including manufacturing processes [14] and modelling of the composite material performance [1, 15, 16]. As far as the application of ANNs in curing simulation and optimization is concerned, early contributions can be individuated in the works by Rai and Pitchumani [17, 18], who implemented a static neural network of the curing process, recalled by a non-linear programming scheme. More recently their approach has been re-proposed by Jahromi *et al.* [19], who replaced the static ANN with a dynamic ANN, improving the effectiveness of the overall calculations. Further details on the application of soft computing techniques in the manufacturing of composite material can be found in [20, 21].

The present work represents the new step towards development of an integrated system to be applied off-line for process planning and optimization as well as on-line for monitoring and control. In this framework, this paper represents a step further regarding investigation presented in [1]. It deals with the development and implementation of a very fast and efficient procedure for the prediction of some aspects involved with the cure process of a thermosetting matrix composite material. The procedure is also based on the coupling of a finite element thermochemical model of the process with an artificial neural network but with more data used for training and testing of artificial neural networks.

2 Materials and Methods

2.1 Modelling

As it was pointed in [22, 23], rigorously speaking, the curing process modelling requires a multi-physics analysis in order to consider all the involved issues. Indeed, during the process, the initial heating of the material leads to a reduction of the resin viscosity, allowing excess resin to squeeze out from the reinforcing layers. As a consequence, resin volume fraction, material physical properties, and product thickness dynamically change until the onset of gelation. Furthermore, after the solidification, temperature and degree of curing (DOC) gradients and tooling actions cause stresses onset and distortions of the work piece. It is generally accepted that, for the considered case, the influence of the flow and stress-strain models on the thermochemical problem is reasonable negligible, therefore, as a first approximation the curing process can be modelled coupling the heat transfer model and the reaction kinetics model. The thermochemical model is based on the solving of energy balance equation within the laminate expressed by the heat conduction equation between the processing materials [22], where the heat generation term is related to the curing exothermic reactions. The implemented model should include also the dies and other tools in the computational domain to take into account the thermal inertia and the interaction between the tooling and the laminate. However, considering that the present work is the mainly devoted to the development and validation of the optimization procedure, the computing domain have been restricted only to the processing materials, replacing the other elements by means of proper boundary conditions [1, 22], as depicted in Fig. 1.

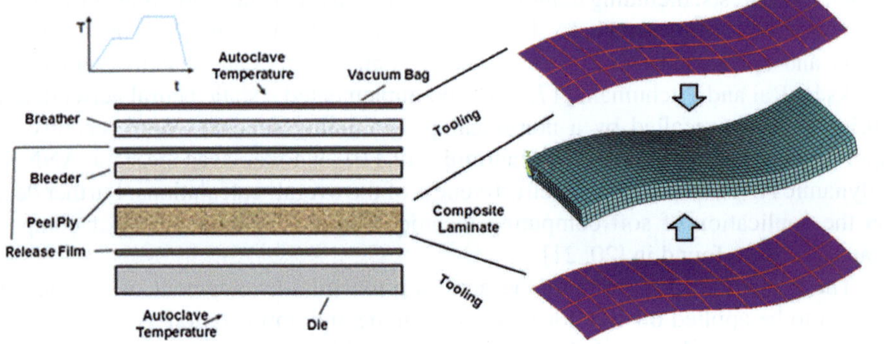

Fig. 1. Process scheme and equivalent finite element model [1, 22]

Other details regarding the thermochemical finite element (FE) model and its validation are extensively reported and debated in [12].

The autoclave curing process of an epoxy based composite material reinforced with graphite fibers was simulated to test the proposed ANN models. The part was modelled similarly to the anterior part of a race car body (see Fig. 2a). The aim was to test the procedure on a reasonably complex geometry and against the potential non-homogeneity (e.g. local variations of volumetric fraction or boundary conditions) that can be observed in a real component. For this reason, distinct values of the fibers volume fraction are considered, 0.50 on the side and 0.58 on the top of the part, resulting in different physical properties depending on the zones of the work piece. The two zones are evidenced by different element colours in the FE model in Fig. 2b. The overall dimensions of the part are approximately $750 \times 360 \times 260$ mm^3, whit a constant thickness equal to 30 mm.

It should be borne in mind that the considered geometry does not correspond to a real product and potential design or structural constraints, not relevant to the present analysis, have not been considered.

Carbon fibers are assumed as unidirectional reinforcement material, while Hercules AS4/3501-6 epoxy resin is considered as the polymeric matrix. Physical properties of

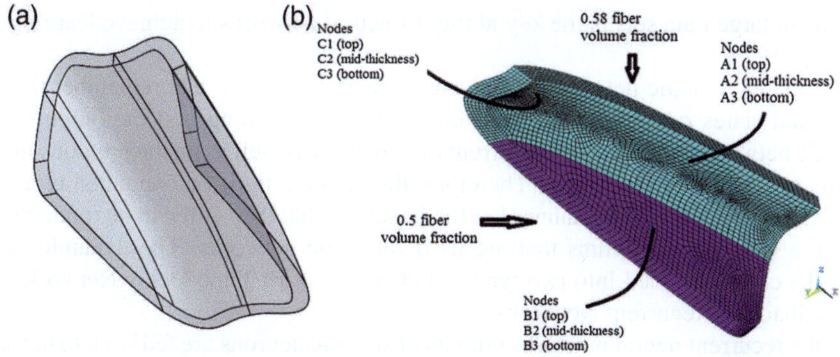

Fig. 2. Work piece solid (a) and finite element (b) models. The control locations monitored and the fibers volume fractions are indicated on the figure [1].

each constituent, resin reaction kinetics models and parameters are defined according to data reported in [4, 6]. Resin vitrification is detected following the relation explained [1] for the considered materials and assuming that the reaction rate vanishes (and the reaction stops) if the computed temperature is lower than the local glass transition temperature.

For symmetry reasons, only one half of the solid model is discretized, applying adiabatic boundary conditions on the symmetry plane (see Fig. 1b). The equivalent FE model of the car body is constituted by 17367 nodes and 14280 thermal elements. The tooling and the vacuum bag are modelled by equivalent heat transfer coefficients to account for all thermal resistances, while external temperature are imposed according to the thermal cycle.

The implemented FE model has been used to simulate 24 different curing cycles, 20 used for the training and 4 for the testing of the networks. Taking into account the aforementioned variations of material properties and boundary conditions, three different locations are selected, considering three control points (placed on the top surface, mid-thickness, and bottom surface, respectively) for each location, as graphically shown in Fig. 5 (right). For the purpose of the illustration of prediction results, more different control points have been observed. In this paper our focus was on points A1, A2, A3, B1, B2, B3, C1, C2, and C3 in the cycles 21 and 24, which showing different composite temperature profiles.

2.2 Artificial Neural Network Model

Inspired to the human nervous system, artificial neural network models are capable to precisely approximate nonlinear behaviour of dynamic processes [1] thanks to their adaptive nature, where 'learning by example' replaces 'programming' in solving the problems. They can be trained to perform a particular function by adjusting the values of weight factors between the neurons, either from the information of outside the network or by the neurons themselves in response to the input. The learning and recognizing patterns in large data sets is the key ability of neural networks to achieve learning and memory [1].

Compared to static networks, the dynamic neural networks can remember the past values and states of the network (dynamic memory) meaning that the output of the dynamic networks depends on the current input values as well as on the previous inputs, outputs or states of the network. Therefore, the gradient must be computed in a more complex way in a dynamic manner but the benefit is that they still can be trained using the same learning algorithms that are used for static networks. The dynamic neural networks can be divided into two types: (1) Feed-forward Time-Delay Networks, and (2) Feedback or Recurrent Networks.

In the recurrent neural networks outputs of specific neurons are fed back to the same neurons or to neurons in the preceding layers.

This dynamic memory is derived from that their outputs at a given instant reflect the current input as well as previous inputs and outputs that are gradually quenched. According to [23], the architecture of the recurrent neural networks can be represented by a graph that contains cycles. Artificial neural networks can be implemented in most cases as part of digital systems (either standard computers or digital circuits for special needs or for signal processing). Accordingly, discrete time systems are a natural framework for testing recurrent neural networks. They are mathematically described by recurrent (return) equations, hence the name of these networks. Discrete-time (or recurrent) equations are discrete-time equations of continuous differential equations. This means that each connection of the neural network with feedback is associated with a delay, which may be equal to zero, in addition to the already associated parameter as in the artificial neural network with signal propagation in advance. Each delay is an integer of an elementary time moment which is considered a time unit. Because the neuron output at a given moment cannot be its own function, the delay at the end of the cycle on the graph must be different from zero. The discrete-time artificial neural network with feedback converts a set of nonlinear discrete-time feedback equations, through the composition of the functions of its neurons, and through time delay associated with its connections. In this paper, the Layer-Recurrent neural network (LRNN) has been employed for dynamic modelling of the autoclave curing process. This layer-recurrent neural network is characterized by a feedback loop, with a single delay, around each layer of the network except for the last (output) layer (see Fig. 3).

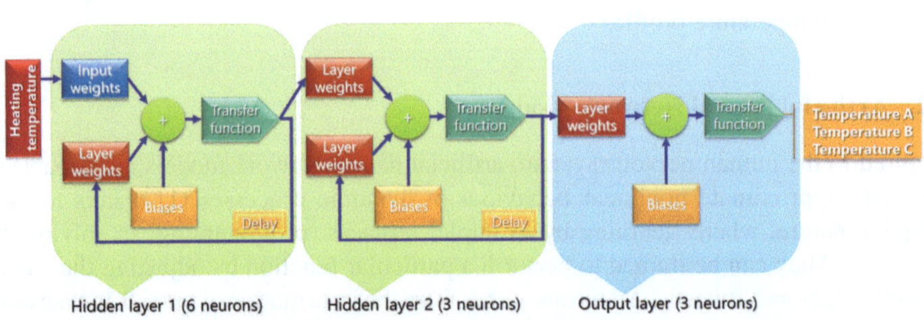

Fig. 3. The structure of dynamic neural model used for modelling of the autoclave curing process.

The appropriate neural network model architecture and training algorithm needs to be found, according to the modelled process.

3 Results and Discussion

In this paper, 24 different layer-recurrent neural networks have been investigated against influence of four training algorithms, namely Levenberg-Marquardt (LM), Bayesian Regularization (BR), Resilient Backpropagation (RP), and Scaled Conjugate Gradient (SCG).

For each network architecture, the "tansig" activation function has been used between the input and the first hidden layer, as well as between hidden layers. The "purelin" activation function has been used between the last hidden and the output layer of the networks. The selection process of the best dynamic neural model can be analysed in two directions: (i) influence of the different neural network architectures on the model prediction performance, and (ii) influence of different training algorithms on the model prediction capabilities regarding the chosen neural network architecture. These comparisons are shown in Figs. 4 and 5. The prediction results are related to the point A2 (see Fig. 2) in the cycle 22. Similar prediction quality was achieved for others control points in this cycle. Figure 4 illustrate the neural network model prediction performance for selected group of different neural network architectures trained by the Bayesian Regularization algorithm. According to Fig. 4 these neural network architectures do not drastically influence the model prediction capabilities, but some difference could be noticed.

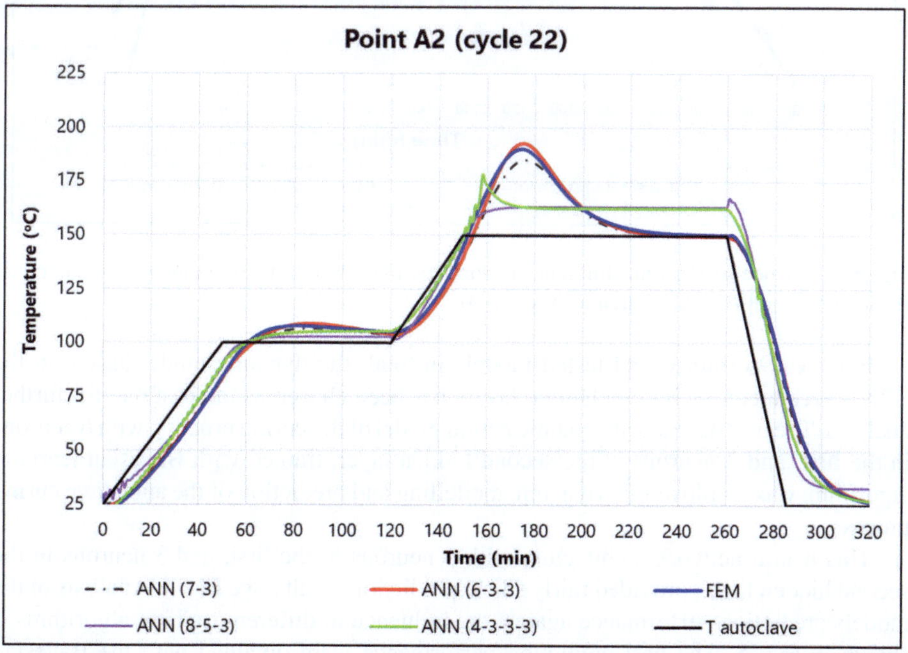

Fig. 4. Influence of the different neural network architectures trained by Bayesian Regularization algorithm on prediction of the composite temperature in the point A2

In this regard, a relatively worse response was provided by the network with two (see ANN 8-5-3 in Fig. 4) and three (see ANN 4-3-3-3 in Fig. 4) hidden layers, which predicted two unexpected and unmotivated temperature peaks at about 155 and 265 min from the beginning of the process, respectively. Obviously, network architectures with three and more hidden layers were not suitable for this kind of modelling. The neural model with one hidden layer (ANN 7-3 in Fig. 4) well predicted composite temperature change during this thermal cycle, but with slightly higher error than the neural model ANN 6-3-3, selected as the best for further dynamic modelling and prediction of the autoclave curing process.

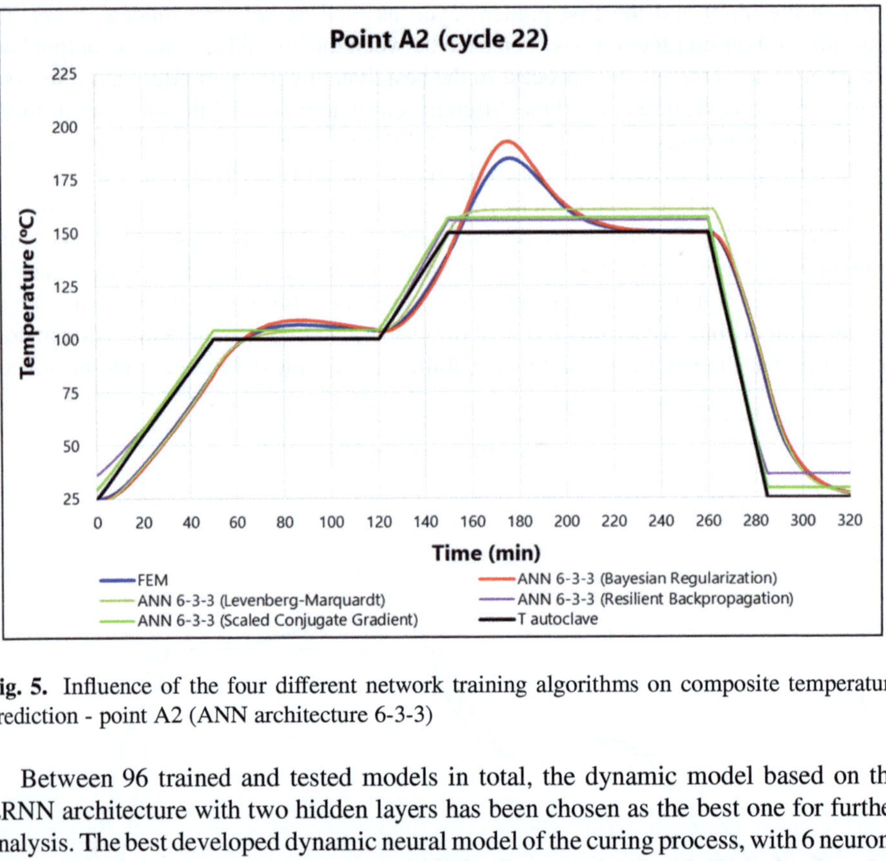

Fig. 5. Influence of the four different network training algorithms on composite temperature prediction - point A2 (ANN architecture 6-3-3)

Between 96 trained and tested models in total, the dynamic model based on the LRNN architecture with two hidden layers has been chosen as the best one for further analysis. The best developed dynamic neural model of the curing process, with 6 neurons in the first, and 3 neurons in the second hidden layer, trained with Bayesian learning algorithm, was employed for dynamic modelling and prediction of the autoclave curing process.

This neural network architecture with 6 neurons in the first, and 3 neurons in the second hidden layer, provided fairly stable prediction results, see Fig. 4. Analysis of the models prediction performance against the influence of different training algorithms is shown in Fig. 5. The best prediction performance were obtained with the Bayesian Regularization (BR) algorithm. The Levenberg-Marquardt (LM) algorithm shows acceptable agreement only during the first 150 min of the cycle, while other two algorithms (the Resilient Backpropagation – RP and Scaled Conjugate Gradient – SCG) preserved the profile of the imposed thermal cycle, shifting it toward the temperature peaks, see Fig. 5. Considering these outcomes and neglecting minor variations, it can be reasonably stated that, as far as the composite temperature prediction is regarded, the model ANN 6-3-3 trained by BR algorithm exhibits fairly stable performance.

The predicted temperature profiles at the specified points are induced by autoclave heating according to the imposed temperature – time cycle, as depicted in Fig. 6 ($T_{heating}$). Figure 6 shows a comparison between temperature profiles provided by the FE model and the profile predicted by the ANN at the control point denoted as A1 in the cycle 21. It was mentioned that the final properties mainly rely on temperature history. Excessive temperature peaks or gradients should be avoided in order to prevent degradation of the resin system and excessive thermal stresses.

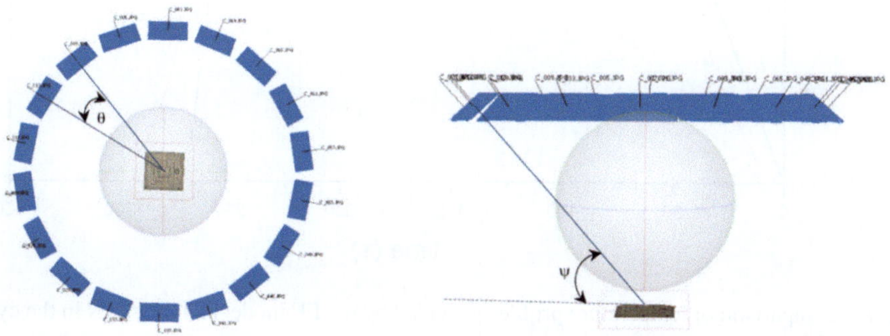

Fig. 6. Comparison of temperature profiles provided by the FE model and the ANN in cycle 21 (point A1).

As can be seen, the dynamic neural model well predicted this temperature peak. Figures 7 and 8 show prediction results related to the composite temperature profiles in the points B2 and B1, in the cycle 24. This further verification is motivated by the need to test the ANN model prediction sensitivity regarding the composite temperature changes across different control points characterized by material properties variations.

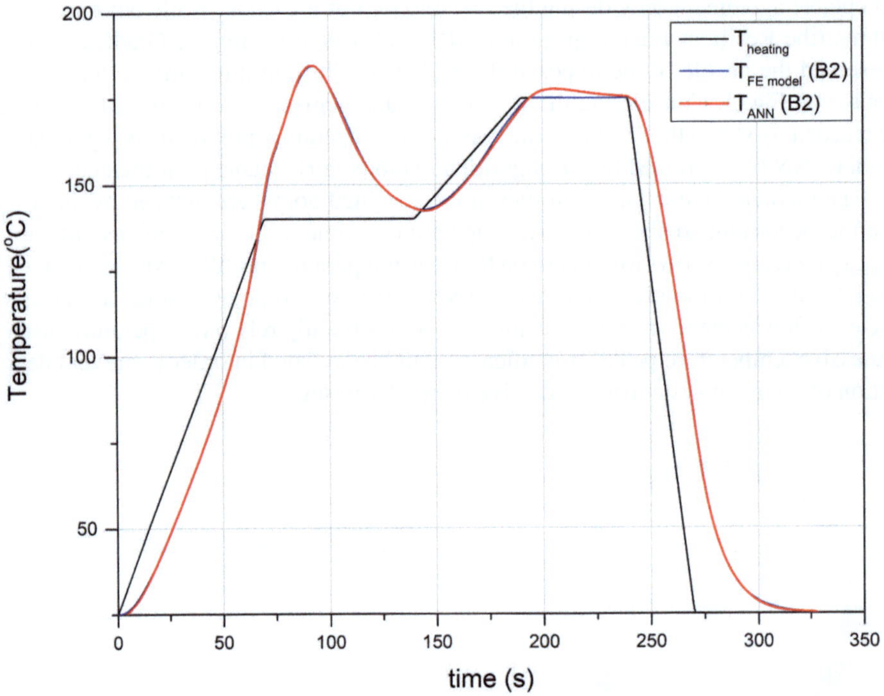

Fig. 7. Comparison of temperature profiles provided by the FE model and the ANN in the cycle 24 (point B2).

According to Figs. 6, 7 and 8, the neural model has been learned about influence of the heating temperature in the specified points of the composite. The neural model has recognized difference in the temperature level at different points in different heating cycles.

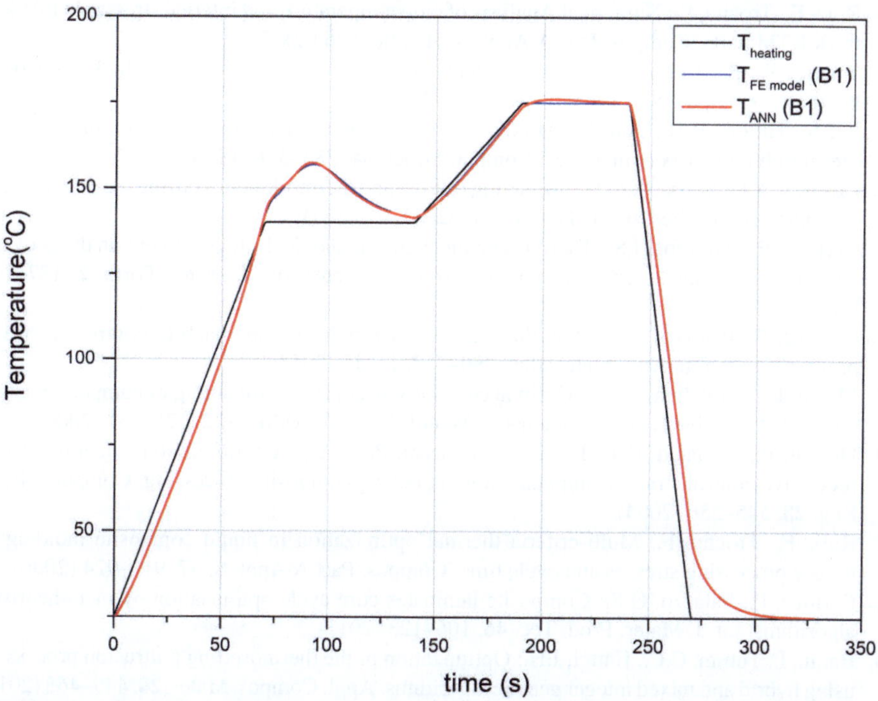

Fig. 8. Comparison of temperature profiles provided by the FE model and the ANN in the cycle 24 (point B1).

4 Conclusions

In this paper an effective procedure for the curing process modelling has been developed. Integration of the finite element thermochemical model and dynamic artificial neural networks is used to develop an improved dynamic neural network model of the autoclave curing process able to predict changes in the composite temperature profiles (at different points and in different thermal cycles). The proposed model is tested versus unknown data related to the autoclave temperature heating over time. It was shown that the developed dynamic neural model of the autoclave curing is able to predict the temperature peaks inside the composite with complex shape in synergy with different fiber volume fractions.

References

1. Carlone, P., Aleksendrić, D., Ćirović, V., Palazzo, G.S.: Modelling of thermoset matrix composite curing process. Key Eng. Mater. **611–612**, 1667–1674 (2014)
2. Oliver, P., Cavarero, M.: Comparison between longitudinal tensile characteristics of thin and thick thermoset composite laminates: influence of curing conditions. Comput. Struct. **76**, 125–137 (2000)

3. Ruiz, E., Trochu, F.: Numerical Analysis of cure temperature and internal stresses in thin and thick RTM parts. Compos. Part A Appl. S. **36**, 806–826 (2005)
4. Young, W.-B.: Compacting pressure and cure cycle for processing of thick composite laminates. Compos. Sci. Technol. **54**, 299–306 (1995)
5. Yi, S., Hilton, H.H., Ahmad, M.F.A.: A finite element approach for cure simulation of thermosetting matrix composites. Comput. Struct. **64**, 383–388 (1996)
6. Costa, V.A.F., Sousa, A.C.M.: Modeling of flow and thermo-kinetics during the cure of thick laminated composites. Int. J. Therm. Sci. **42**, 15–22 (2003)
7. Carlone, P., Palazzo, G.S.: Thermo-chemical and rheological finite element analysis of the cure process of thick carbon-epoxy composite laminates. Int. J. Mater. Form. **2**, 137–140 (2009)
8. Carlone, P., Palazzo, G.S.: A multi-scale non-deterministic approach to composite curing process simulation. Adv. Mater. Res. **750–752**, 11–15 (2013)
9. Mawardi, A., Pitchumani, R.: Optimal cure and current cycles for curing of composites using embedded resistive heating elements. J. Manuf. Sci. E-T ASME **122**, 124–131 (2000)
10. Dufour, P., Michaud, D.J., Tourè, Y., Dhurjati, P.S.: A partial differential equation model predictive control strategy: application to autoclave composite processing. Comput. Chem. Eng. **28**, 545–556 (2004)
11. Ruiz, E., Trochu, F.: Multi-criteria thermal optimization in liquid composite molding to reduce processing stresses and cycle time. Compos. Part A Appl. S. **37**, 913–924 (2006)
12. Carlone, P., Palazzo, G.S.: Composite laminates cure cycle optimisation by meta-heuristic algorithms. Int. J. Mater. Prod. Tec. **46**, 106–123 (2013)
13. Baran, I., Tutum, C.C., Hattel, J.H.: Optimization of the thermosetting pultrusion process by using hybrid and mixed integer genetic algorithms. Appl. Compos. Mater. **20**, 449–463 (2013)
14. Aleksendrić, D., Senatore, A.: Optimization of manufacturing process effects on brake friction material wear. J. Compos. Mater. **46**(22), 2777–2791 (2012)
15. Aleksendrić, D., Duboka, Č.: Prediction of automotive friction material characteristics using artificial neural networks-cold performance. Wear **261**(3–4), 269–282 (2006)
16. Aleksendrić, D.: Neural network prediction of brake friction materials wear. Wear **268**(1–2), 117–125 (2010)
17. Rai, N., Pitchumani, R.: Rapid cure simulation using artificial neural networks. Compos. Part A-Appl. S. **28A**, 847–859 (1997)
18. Rai, N., Pitchumani, R.: Optimal cure cycles for the fabrication of thermosetting matrix composites. Polym. Compos. **18**, 566–581 (2004)
19. Jahromi, P.E., Shojaei, A., Pishvaie, S.M.R.: Prediction and optimization of cure cycle of thick fiber-reinforced composite parts using dynamic artificial neural networks. J. Reinf. Plast. Comp. **31**, 1201–1215 (2012)
20. Aleksendrić, D., Carlone, P.: Soft Computing in Design and Manufacturing of Composite Materials. Elsevier (2015). ISBN 1782421793
21. Rubino, F., Carlone, P., Aleksendrić, D., Ćirović, V., Sorrentino, L., Bellini, C.: Hard and soft computing models of composite curing process looking toward monitoring and control. AIP Conf. Proc. **1769**, 1–6 (2016). https://doi.org/10.1063/1.4963438
22. Carlone, P., Aleksendrić, D., Ćirović, V., Palazzo, G.S.: Meta-modeling of the curing process of thermoset matrix composites by means of a FEM–ANN approach. Compos. Part B Eng. **67**, 441–448 (2014). ISSN 1359-8368
23. Ćirović, V.: Investigation of the possibilities for using artificial intelligence methods in predicting of the performance of the motor vehicles' braking system. Ph.D. thesis (2012). (in Serbian)

Recent Streams of Digital Manufacturing, Its Emerging Trend and Future Directions for SME in Japan

Hideaki Hohnoki(✉)

COHO Consulting, LLC, 2-4-521 Utase, Mihama-ku, Chiba 261-0013, Japan
hideakihohnoki@s6.spaaqs.ne.jp, akihohnoki@gmail.com

Abstract. Small and Medium Enterprises (SME) in Japan has been suffered from the global competition that comes up from manufacturing commoditization as well as from the dissolution of existing affiliation system known as "vertical integration" and "Keiretsu". Of all the industries in Japan 99.7% are SME that has played a great roll in high growth economy and this high ratio provides a useful clue to approach the issue. Statistics shows that Serbia has similar high percentage of SME as well. This study attempts to make positive effects on the said issue by providing new perspectives for SME to obtain driving force to create a survival path with utilizing recent digital technologies. The approach to the issue also covers the point of view to set the product and process strategy for changing market realities including Robotics industries.

Keywords: Digital manufacturing · IT · PLM · SME · Mass customization
3D printer · Robotics

1 Introduction

Japan has many SME (Small and Medium Enterprises) as in 99.7% in industry, and many of them has been suffering from current global competition. This study attempts to introduce a fresh perspective to overcome the current situation and to set a target into new path in the market for SME. This perspective is expected also to apply to Serbia which has the same high percentage of SME ratio as in 99.7% as Japan and Germany.

One of the effective way for the achievement is to utilize a digital technology which has recently made significant progress in revolutionizing manufacturing processes.

This study provides new perspective to attempt to make positive effects on the above issue for the purpose of the items stated below:

a. To changeover from current competition under product commoditization:
   Recent structural change in manufacturing method with the emergence of newly industrialized countries has resulted in downtrend of fabricating industries and then mass-production items such as bulk electronic appliances became just ordinary "commodity", which brought global competition. This study suggests to changeover from the commoditization as in hardware only production situation.

J. Ni et al. (Eds.): AMP 2018, LNME, pp. 89–97, 2018.
https://doi.org/10.1007/978-3-319-89563-5_6

b. To leave away from mass production constraints and shift forward to the needs of the times:
This study suggests SME in Japan to have a new function after the long years of a roll of efficient Sub-Contractors in vertically integrated mass production pyramid as well as a roll of capable affiliated companies among strategic KEIRETSU grouping. Currently still KEIRETSU is active but not work well due to too large organization to move quickly against significant changes in business such as ICT field.

c. To change service activities from routine side toward flexible service depending on the market:
Most IT infrastructure is much less expensive compared with conventional manufacturing industry such as automobile or heavy industries. In view of this, IT investment is highly recommended for SME and digital facility could be powerful weapon for them. Network technology such as utilization of digital data as in big data has high possibility to the reflection of the market demand.

d. Robotics is another field:
In robotics area the interaction field is one of key technologies since more and more life supporting robots are introduced to practical use and it follows that the communication technology with that robot comes to be indispensable. The elemental technologies as in sensing, actuation and interaction used for life supporting robot are the areas of specialties for SME.

2 Background

First of all, SME played a great role in economic growth in Japan after World War 2. In the years of high-growth economy, SME have a function as Sub-Contractors in vertically integrated organization of production. It's a pyramid structure and top of the pyramid is big company, which strongly controlled whole activities and they also educated and trained SME people on the other hand. The pyramid structure has been transformed into more sophisticated style, namely KEIRETSU. KEIRETSU is more strategic and systematic grouping to share common interest. However the structural change of manufacturing industries in Japan has induced a transformation of the relation between dominant big companies and SMEs and then disrupted the ties among them. Currently still KEIRETSU is active but sometimes it is often unable to function for such a huge organization to move quickly against significant changes in business such as ICT field. Therefore, in the near future, current manufacturing structure should be changed to be more flexible style, which means SME should change into more specialized style and the partnership such as divisional cooperation should be more important factor for industry in general. It could be a good opportunity for SME to grow and expand their business. Knowledge and experiences in this innovative change has possibility to create new digital-divide issues. It would not be an overstatement to mention that the proper action to this change can ensure to keep a company and its technology alive, especially in case of SME.

Secondly another structural change in manufacturing field has decreased the overall value of fabricating industries. Fabrication is one of the major manufacturing form and it is located in the middle part of value chain process. It used to be a value-added process but the progress of modularization technology together with the rise of emerging countries has decreased the value of fabrication itself. Afterwards this progress of fabrication technologies gave way to the emergence of newly industrialized countries such as Asian countries and areas, and eventually bulk electronic products primarily became just ordinary "commodity", not special and custom-made product. This commoditization rapidly expanded to various products such as many types of appliances, TV, Personal computer, mobile phone and even automobile, which resulted in battered Japanese SME industries.

A digital technology has recently made significant progress in revolutionizing manufacturing processes. The digitalization continues ever after by creating a new designing system known as PLM (Product Lifecycle Management) which handles each step of the process virtually by using digital data, namely in computer system. PLM usually consists of Designing, 3D Modeling, CAE, 3D Prototyping, Knowledge Engineering, CAT etc. In those process digital data plays a prominent role and is commonly utilized among each steps of the process. This virtual system has been called Digital Manufacturing and widely applied for years mainly in the production of automobile, consumer appliances and other major manufacturing industries. This digital manufacturing process is simply based on the utilization of key characteristics of digital technology which has made a significant contribution to product innovation. In other words, digital manufacturing creates to obtain the right product at the right time, i.e. to get the product which matched the needs of the market. This trend is called Mass Customization which is examined and discussed in the later in this study.

3 Findings

The study reveals the "value linkage" in possible future manufacturing schemes. The author has discussed that recent new trend of digital manufacturing creates interactive movement among manufacturing industry, service business and consumer trend. More specifically that collateral information with product upgrades its own value, and service business values the ideas which has actual reflection from market. Of course, manufacturing industry is based upon market trend [1]. The value linkage is key concept in the study to consider about the possibilities of future products. The concept indicates how to take action against new stream of change showing mass customization movement, that is, how to manufacture as process strategy and what to manufacture as product strategy. In the future from recent IOT trend perspective, it is presented to the discussion about the importance to consider "value" which is liked to the relationship of "product-service-market" coordination and its enhancement, which is called Value Linkage in this study. The recent digital technology is powerful tool in order to ensure effective utilization of the Value Linkage concept, which consists of three types of Value, namely Service Value with Hardware, Diversified Value by Mass Customization and Market Value with Service.

Service Value with Hardware

First "Value" signifies adding service value to simple hardware. This indicates that the incoming trend in market is from hardware only toward the hardware with some related service, i.e. from "simple hardware production" toward "hardware with support and aftercare etc.", in other words, from selling out only toward selling with service, which naturally places high value on market mind. In this case the key factor of the product is not its independent performance but its collaborative capability through other connected hardware. Those collaborative capability produces new service which cannot be performed by original hardware alone.

Diversified Value by Mass Customization

As we touched upon earlier about Japanese economy, "mass production" is a conventional system in industry, this system is suited for manufacturing standardized products in high volume. "High-Mix Low-Volume production" is also widely accepted for creating greater variation of products, but its each volume is usually small. In case of 3D printer, which way does it build products? 3D printer displays great ability to showing variation in building objects. However high volume production by 3D printer is still not be enough under present circumstances and this is one of the point we expect to be improved by digital technology towards the future High-Mix High-Volume manufacturing.

Three ways are possible as key factors to approach the process to manufacture bespoke products en masse (i.e. High-Mix High-Volume manufacturing). The three key factors include "Multi and Continuous machine system" in hardware, "Local Production for Local Consumption" and "Upper Process Dominance" in digital manufacturing process.

First key factor, what we call Multi-Continuous manufacturing system, is connected and/or assembled process by combination of different methods. This is also called Multi-Systemization which we have seen in earlier chapter describing case study of 3D Printer for a bio-fabrication system for artificial tissue engineering. The process is connected with some different methods such as Inkjet, SLA, MPP and ES depending on its cell structure.

Local Manufacturing for Local Consumption, the second key factor, means that the production has been changing from a big factory to local production firm by using network technologies such as IOT. No matter where the production place is, they could deliver the design data through network. They execute production at the necessary point of consumption by using digital manufacturing system including 3D printer, which results in less logistics of finished products but more logistics of materials. Quite simply, for example, in case of space station far from the earth, they deliver materials and build something needed on site in the station. The Local Manufacturing for Local Consumption helps manufacturing bespoke products en masse depending the necessity on site.

The last key factor for High-Mix High-Volume manufacturing is Upper Process Dominance, which means that the progress of PLM method makes upper part of the whole process much more dominant in terms with total optimization. Once digital manufacturing process completely covers the flow, we could control final product at rather beginning part of the total flow and which results in a easier way to manufacture bespoke products en masse [2] (Fig. 1).

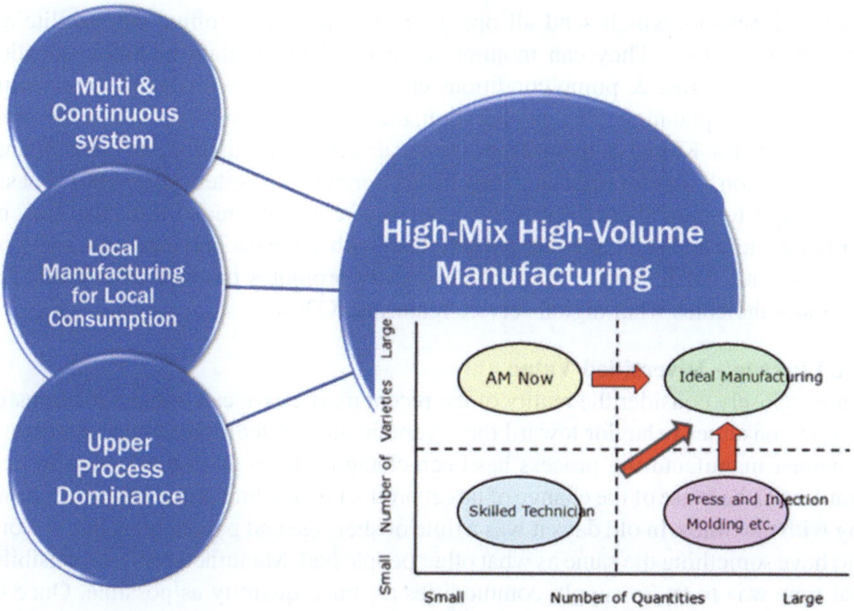

Fig. 1. Process Strategy - High-Mix High-Volume Manufacturing

Market Value with Service

In this case the Value should be collected from the data obtained from the market and the content of the existing service can be modified by that data. In other words, the collected data is considered to be reference of market demand by which the existing service is updated through analysis, estimation of the collected data to be the most recent performance. What we call "big data" is a type of the data through internet service or IOT, which usually has major influence as collected data to the service or activities to be updated. This analysed and/or modified data or information makes original service more flexible to the market trend and, in this manner, existing routine service can be modified to be new service in accordance to the market demand.

4 Discussions

Discussion shows case studies and analysis which confirms relevance of Value Linkage in the same order as described in Findings part, that is, Service Value, Diversified Value and Market Value.

Value Linkage – Service Value

Firstly, the case study about Service Value in manufacturing field is KOMTRAX system by Komatsu, that is Construction Equipment company and they manufacture Dump Truck, Power Shovel, some other construction machine. KOMTRAX is machine tracking system for 300 thousand machines sold already in the world [3]. Each machine has network sensors which send all operation data to remote office via satellite and mobile phone station. They can monitor & accumulate the data including location, operating hour, engine & pump conditions etc. Those data are analyzed for customer for the purpose of planning efficient operation, reducing maintenance cost and downtime etc. as well as for Komatsu themselves for repair service prediction, advance maintenance suggestion, sales forecast etc. They have operated this system since 2001, but still good example to learn about "Hardware + Service" concept which means that they not only manufacture and sell hardware but sell them with some backup service via network and digital data. Their service function apparently promotes product sales itself. This case simply indicates what digital service featuring IOT is.

Value Linkage – Diversified Value

When we closely consider the reality of the recent market, we can witness the transformation in consumer behavior toward the diversification of their preference patterns. In this manner manufacturing process has been changing from mass production to mass customization because of the change of target product depending on each person's desire fitting with the times. In old days it was a time of shortage and people most likely would like to have something the same as what other people had. Manufacturers' responsibility at that time was to try to supply commodities as much quantity as possible. Once the shortage has gone, people began to want different things. Presumably they would like to have different color, shapes, something different from what their neighbors had, in other words, variant products were the target for manufacturers at that time. What happened next is that people started to think that the commodity should be suitable for owner's character, preference, personality……, namely the time of diversity. What we will see when we deepen and develop this personal direction would be something like unique product.

As already noted as above, future manufacturing direction would be mass customization in accordance with evolving customer preferences and manufacturing process should be changed towards High-Mix High-Volume manufacturing. Digital technologies and an equipment such as the 3D printer, as a leading-edge example, have encouraged sophisticated and diversified public demands in terms with manufacturing industries.

At the time when mass production has dominated the world market, an idea was commercialized in a big company and produced in a big factory, and then dump them out into the market, because the market needs the products, in fact, due to the lack of commodities in general. Big companies supply mostly standardized products to clear the supply shortage and people satisfied those ready-made articles. Currently the new method starts to prevail its new process in which an idea is presented publicly on the internet and various people could join on the same platform to collaborate depending on their abilities and roles as in idea-originator, investor, designer, improver, supporter, producer and so on. The original idea could be implemented and realized by various digital tool and equipment such as laser processor and 3D Printer. In this case Digital Technology is the most powerful force for shaping manufacturing methods [4].

The author surveyed specific examples as typical case studies for the above stream of change, namely mass customization. The first typical and simple case example is hearing aid. Hearing aid is a small medical equipment to support hearing-impaired individuals. The equipment should be fit precisely to ear canal. If not, the body of hearing aid is coming loose because one's ear canal itself moves during one is speaking or eating. Each person has a different shape of ear canal, therefore manufacturing process of hearing aid had been fully handmade and time-consuming product for shaping its shell depending on each person's ear canal.

However recent digital manufacturing method made it possible to create full customized outer shell of hearing aid built by 3D printer. Digital technology can make it as precisely as possible.

The second case example is care food, which is prepared mainly for patients in hospital. 3Dprinter is utilized for cooking process and its building object in this case is food. Healthy person can eat everything as far as it is food, but patient needs properly prepared food depending sickness or conditions, not only in terms of nutrition but also easiness of chewing or swallowing. In the kitchen at the hospital they cook food suitable for each person for each day, which is so complexed and sensitive work, therefore it requires proper arrangement with less care. Biozoon in Germany implemented and made it fit for practical applications by using 3D printer technology [5].

The last example is a bio-fabrication system for artificial tissue engineering by Fraunhofer Institute [6]. Building object is cell structure in the human body and they use different method depending on each cell part. For large-bore vein, they employ ink-jet method which makes high-speed building possible and for capillary tube, by Stereolithography (SLA) or Multiphoton-Polymerization (MPP) method because of its requirement of more precise building. The cell structure near the surface of the skin is prepared by porous nano-fiber by electrospun (ES) method. This Multi-Systemization enables totally customized structure such as human cell by a combination technologies with getting strong points of each different methods. The total system applied in this case is still research level and not practical yet.

Value Linkage – Market Value

Digital data is handled and utilized not only in manufacturing field but also in various area of our real society where our activities are. Those activities could be captured through network in the form of digital data, for example, text, image, sound, picture, video at the time when using SNS, purchasing commodities and services, transporting and so on. Those digital data are accumulated, analyzed and estimated and then its result is transformed into the data as a form of added value. The value-added data are given back to real society using application in mobile phone, cloud service, SNS etc. The data recirculation between our real activities and virtual world in our society is effective and rather easy to handle because of nature of digital data as in unlimited replication which enables high productivity and error-less transmission by which we can communicate signals exactly as it is without any change or noise (Fig. 2).

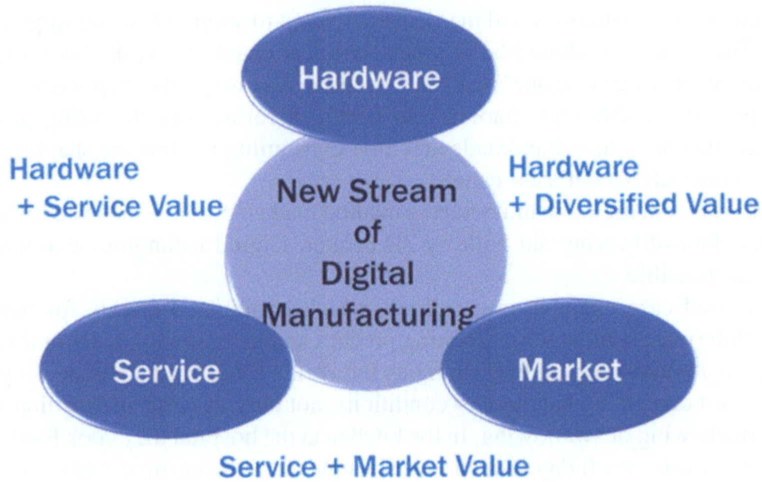

Fig. 2. Value Linkage – Product Strategy

Digital Manufacturing for SME

One of the key characteristics of the digital technology is its inexpensive infrastructure. Most IT infrastructure is much less expensive compared with conventional industry as in steel making, automobile and so on. For this reason, IT investment is relatively easier for SME and digital facility could be powerful weapon for them. SME is usually defined as the company with employee 250 and less and turnover 50 million EUR and less. SME in a comparison with countries indicates that Japan and Germany have some similar factors as in very high SME percentage in industry, R&D framework performed by public institute and long-term employment system. However more different aspects we should pay attention to these 2 countries is that they still try to take advantage in manufacturing industry in the world. That's Industry4.0 and Smart Society in Japan and the policy has come from trying to utilize the abilities and power of SME in each country. The comparison also implies that industry in Japan has rather vertical integration-oriented structure but, on the other hand, Germany is more individualistic and prefer differentiation. Two countries have common factors having come from its own background and different natural tendencies. Serbia, in comparison with the above two countries, shows quite similar figure of very high percentage of SME in industry, which is approximately 99.7%. Since SME in general plays a great role in economic growth, this figure means each of these three countries has strong industrial potential in society. Currently SME has high possibility to grow especially with IT infrastructure.

Japan has survived after damage and loss of WW2 by the growth of manufacturing industries with efficient and capable efforts performed by SME. One of the reasons for the growth is the development of industrial automation. NC machinery, CAD/CAM/CAE play an important role in Digital Manufacturing. Secondly network technology has even greater role, for example, mobile phone and wireless LAN have brought IoT system. To look beyond the surface is the significant function of SME there.

Robotics for SME

Robotics is one of the possible field to apply the Value Linkage concept. Industrial robots have been matters of concern in manufacturing field for a long time and those robots are the machines which mainly observe the simple directions and direct instructions by human. However recent development of digital and sensor technologies has begun to guide them to behave in a similar or in much the same way as human do. Such a felicitous action can be achieved by recent interaction technologies as in using voice and facial recognition technologies appealing to our sense. This implies that mere hardware production is transformed to the hardware with some service to human, for example, advice of today's weather from the information through internet, which means adding service value to hardware in the concept of Value Linkage. Simple hardware comes to be life support robot with that service value. Another possible and desirable feature of life support robot is to individuate it to act differently depending on different person by using some sensing technology such as facial recognition system, which verifies customized value of Value Linkage concept. Network technology also give them some flexible response feature from the market such as various web-sites, which shows market value in Value Linkage concept.

The Robotics field could be better suited for handling by SME due to the facts as mentioned below. Robot generally consists of the technologies including sensing, actuation and intelligence and the first two skill are fit for SME to develop since those are the elemental technology which is generally the area of specialties by SME. The market of life support robot is also suite for SME because of consumer oriented and still have wide area to be developed.

References

1. Hohnoki, H., Tezuka, M., Ishikawa, T.: New trend of Digital Manufacturing and its influences to SME in Japan, pp. 11–12. New Technology Association of Japan (2015)
2. Hohnoki, H., Tezuka, M., Ishikawa, T.: 3D Printing Technology: Its effects and applications to Manufacturing Technology, pp. 72–75. New Technology Association of Japan (2014)
3. Komatsu Ltd. 22 June 2015. Press Release No. 028(2799). https://home.komatsu/en/press/2015/others/1194792_1810.html. http://smartconstruction.komatsu/whats.html. Accessed Dec 2017
4. Anderson, C.: Makers: The New Industrial Revolution. Crown Pub Inc., New York (2014)
5. Biozoon food innovations GMBH, Products/Seneopro. http://biozoon.de/en/products/seneopro/. Accessed Oct 2017
6. Fraunhofer ILT, ArtiVasc 3D, Project Detail. https://www.artivasc.eu/en/projectdetails.html. Accessed Oct 2017

Intelligent Integrated Management and Advanced Metrology for Quality Toward the Factory of the Future

Numan M. Durakbasa[1(✉)], J. Martin Bauer[1,2], Lukas Kräuter[1], Gokcen Bas[1], and Günther Poszvek[1]

[1] Department for Interchangeable Manufacturing and Industrial Metrology, Institute for Production Engineering and Laser Technology, TU Wien, Getreidemarkt 9, BA09, 1060 Vienna, Austria
numan.durakbasa@tuwien.ac.at
[2] UTN Buenos Aires, Buenos Aires, Argentina

Abstract. Intelligence is an essential feature of future development and production systems and intelligent production is a major component of future businesses, also according to the technological developments especially in the field of precision engineering at micro/nano and pico scale production.

The problematic of the high accuracy of the work pieces in micro and nano technologies of the modern production systems according Industry 4.0 concept gained in the last years more and more importance through constantly increasing demands on the intelligent integrated management systems of the production processes. Together with the current trend towards high precision manufacturing and assembly systems, new challenges have emerged through the demand of complex shapes and functional topographies on the micro/nanoscale. On the other hand important charge of the environment, reduced availability of natural resources and the increasing growth of waste require new concepts and strategies to recycle technical consumer goods as there are household instruments, consumer electronics and passenger cars - instead of land filling, burning or steel production a high potential of recycling is necessary. This means that the integrated management system of quality, environment, energy, health & safety and information security in the manufacturing industry plays an essential role for the effective and efficient development of future industrial plants.

Keywords: Integrated management · Production metrology · Industry 4.0
Advanced manufacturing · Artificial intelligence

1 Introduction

Due to the rapid development of technology, new demands are placed on products. This makes products ever more complex and customized. New developments in the industry lead to an increase in the degree of automation and diversity of production, which leads to a new concept, Industry 4.0.

J. Ni et al. (Eds.): AMP 2018, LNME, pp. 98–108, 2018.
https://doi.org/10.1007/978-3-319-89563-5_7

To meet market demands in present and future global industrial world, manufacturing enterprises must be flexible and agile enough to quickly respond to product demand changes. With support of artificial intelligence and modern information technology it is possible to realize modern cost effective customer-driven design and manufacturing taking into account the importance and basic role of quality management and metrology.

For measurement technique can and must be given a key role especially in modern industrial environment. Essential contributions to increase the quality of products and the productive power of plants can be reached furthermore through the aimed application of measurement methods. At present additionally also ecological aspects are to be taken into special consideration.

Innovation of modern manufacturing industry drives demand for advanced metrology applications and quality assurance to support sustainability and improvement of processes [1]. In today's highly competitive world, the businesses that want to survive have adopted total quality management's main objective and customer focused principles in order to have advantage over the competitive companies. The modern industrial facilities have created the trends in advanced applications that have been mapped out [2] and high precision mechanism designs have been studied to establish an interaction between different traditional disciplines [3]. Cooperation of the next-generation manufacturers with technology developers has been an indispensable requirement for developing solutions to the trends of automation, IT and metrology in the production environments. The new aspects of cooperation between the research institutions and the manufacturing industry will provide a development capacity for high quality and innovative products. The development of information highway, technology and advanced engineering data exchange techniques will make the transition to the Industry 4.0 efficient and productive by means of collaborative and interactive environment.

The transition to the Industry 4.0 requires models to be integrated by utilizing advanced information analytics, artificial intelligence and interconnected Industrial Internet of Things (IIoT) as a part of automated and robotic applications, networked intelligent machines and instruments. The intelligent production systems and multi functional integrated factory model [4] as well as extensive use of the IT, AI, simulation, intelligent quality automation, robotics, advanced metrology and advanced engineering data exchange techniques will be able to perform more efficiently, collaboratively and sustainably, that makes possible an agile and optimal industrial production in any kind of industry towards the Industry 4.0.

The new aspects of cooperation between the research institutions and the manufacturing industry will provide a development capacity for high quality and innovative products. The development of information highway, technology and advanced engineering data exchange techniques make global information systems by means of education and training in collaborative and interactive environment [5].

To meet market demands in present and future global industrial world, manufacturing enterprises of any kind and any size must be flexible and agile enough to respond quickly to product demand changes with shorter innovation cycles, resulting in the reduction of life cycle and diversity in the number of variants of products [6].

2 Intelligent integrated Management and Metrology for Design and Production

Intelligence is an essential feature of future development and production systems, and intelligent production is a major component of future business. In times when everyone is much thinking and writing about next generation of factories "Industry 4.0", it is worth to give a working example of the development of production in the field of mechanical engineering. This will be especially possible on the basis of the innovative concept and model for modern enterprises to meet high-level demands for comfortable daily life in the future.

New models for alternative configurations of future industrial organisations in general which are usually applied and especially for small and medium sized enterprises (SMEs) need to be investigated. With support of artificial intelligence and modern information technology as well as advanced metrology it is possible to realise modern cost-effective customer-driven design and manufacturing taking into account the importance and basic role of intelligent integrated management.

The quality assurance process will be carried out simultaneously in all product realisation steps in future factories - from the design process to the assembly procedures. In order to realise intelligent quality assurance and to deal with complex, variable and dynamic quality control problems of the production processes in this environment, the quality assurance system will be enhanced comprehensively through self-optimising processes thrust in the design system and all manufacturing production processes of this environment. Quality management and quality assurance with intelligent, associative, concurrent, interactive, modular, integrative, learning, autonomous, self optimising and self organising functions will be realized especially in such environments of SME collaborations.

The operating models to the manufacturing plants are developing to keep up with the robotic and automation applications for advanced industrial processes [7]. The simulation of the control systems for development of computer aided, automated production are demand driven aspects that states the future of the manufacturing industry [8].

With the support of artificial intelligence and modern information technology as well as advanced metrology it is possible to realise modern cost-effective customer-driven design and manufacturing taking into account the importance and basic role of intelligent integrated management.

There is a need for manufacturing companies to fundamentally change or develop their approach to design to ensure that the final product meets the functional needs of its application. This can be a difficult concept to realize and to do so may mean that there is considerable expense in functional testing to realize the parameters necessary to ensure that the design and its manifestation through manufacture, taking into account the materials selected and its fabrication yields the desired conditions.

The intelligent quality assurance process within the integrated management system will be carried out simultaneously in all product realization steps in the advanced manufacturing industry - from the design process to the assembly procedures. In order to realize automation of quality assurance and to deal with complex, variable and dynamic quality control problems of the production processes in this environment, the intelligent quality assurance system will be enhanced comprehensively through self-optimizing processes thrust in the design system and all manufacturing production processes of this environment. Hence, the design system and other manufacturing processes must be considered as a whole while implementing the self-optimizing process. Quality assurance system with intelligent, associative, concurrent, interactive, collaborative, modular, integrative, learning, autonomous, self-optimising and self organising functions will be realized in the precision production systems. The production of the future will be characterized by a fully integrated and highly flexible production chain with process based integrated intelligent systems under guidance of international standards as represented in the Fig. 1 [ISO 9001: 2015].

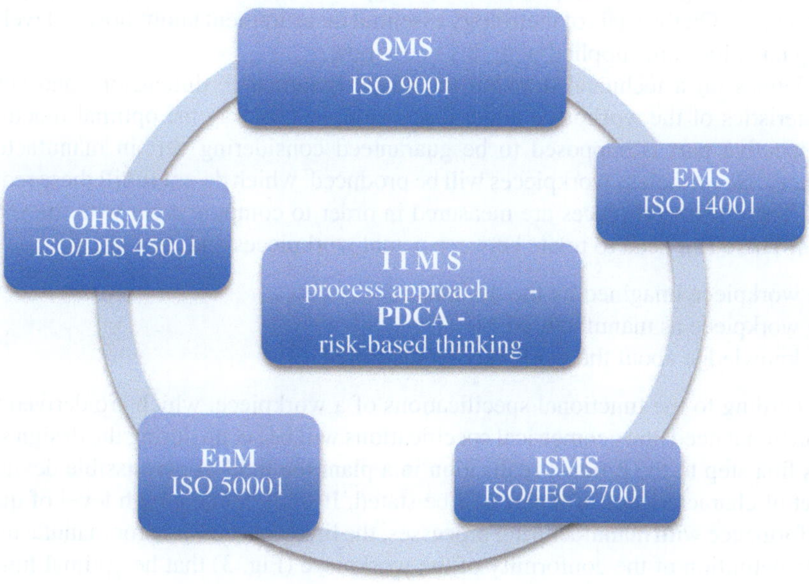

Fig. 1. The intelligent integrated management system of quality, environment, energy, health & safety, security, risk and configuration

The production processes of advanced manufacturing technology can only meet the requirements of quality by means of convenient measuring technologies [9]. The international standards guide this system approach with ISO 9001, ISO 14001, ISO/FDIS 45001, ISO 31000, ISO 27001 and ISO/DIS 50001 through the requirements that address [10–15]:

- Defining the system objectives
- Establishing, implementing and maintaining the management system as a set of processes
- Describing the system
- Continual improvement
- Risk based thinking

- Interconnection, interrelation and sequence of processes
- Maintaining the integrity of the system
- Establishing measurement processes

3 The Model of Precision Manufacturers Based on Intelligent Metrology and Geometrical Product Specifications

The modern methods of quality management [10] environmental management and energy management [11, 15] have been developed and refined in manufacturing organizations and therefore there exists close interaction to intelligent metrology and industrial and technological developments. Metrology is also devoted to the practice of measurement tasks. On the basis of metrology essential measurement know-how is developed and organizations are supplied with this know-how.

It defines on a technical drawing the shape (geometry), dimensions and surface characteristics of the workpiece under discussion. In this way the optimal function of the respective part is supposed to be guaranteed considering certain manufacturing tolerances. Nevertheless workpieces will be produced, which do not fulfill these requirements. Therefore workpieces are measured in order to compare them with the specifications. There is a need to relate between actual workpieces and:

- the workpiece imagined by the designer,
- the workpiece as manufactured,
- the knowledge about the workpiece as measured.

According to the functional specifications of a workpiece, which are derived from the functional need, the geometrical specifications will be set up during the design stage. In this first step to the product realisation in a plant the field of permissible deviations of a set of characteristics of a part will be stated. It shall define a high level of quality in conformance with manufacturing processes, the limits permissible for manufacturing, and the definition of the conformity of the workpiece (Fig. 3) that he optimal function of the respective part is supposed to be guaranteed considering certain manufacturing process. There is a need for manufacturing companies to fundamentally change or develop their approach to design to ensure that the final product meets the functional needs of its application. Verification is the provision of objective evidence that the workpiece fulfills the specification. The definition of the geometrical deviation is used to adjust the manufacturing process. The metrologist begins by reading the specification, taking into account the non-ideal surface model, in order to know the specified characteristics (Fig. 2) [16]. From the real surface of the workpiece, the metrologist defines the individual steps of the verification plan, depending on the measuring equipment. Conformance is then determined by comparing the specified characteristics with the result of measurement.

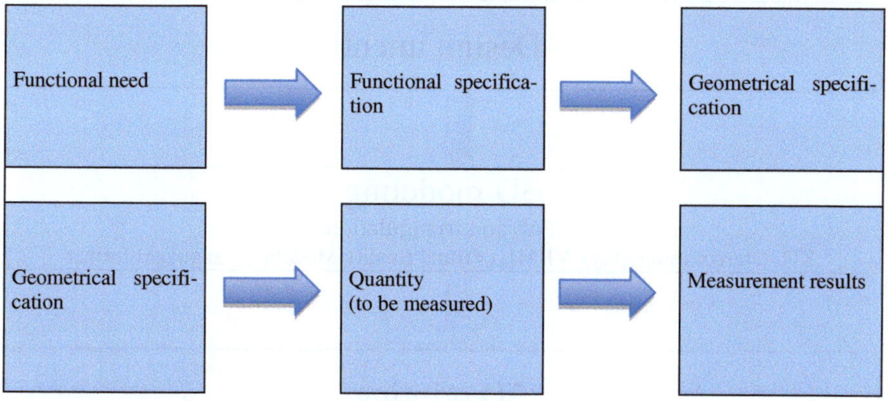

Fig. 2. Conformity approach: workpiece conformity according functional need [16]

4 Linking of Intelligent Integrated Management to Design and Advanced Production

In Intelligent Integrated Management System (IIMS), the design tasks can be carried out intelligently, interactively and concurrently. The exchange quality data and quality assurance data of the design work can be guaranteed by using the quality assurance for the design. Using parallel-processing technologies, the engineers and experts can work on parts of a design task, but the content of the design as a whole is a corporate resource to be managed and secured.

This system makes it possible that the product designers of different function SMEs or factories can work in parallel on all the subtasks of the product.

Collaborative working method in MFIF has the goal to realise not only electronic data exchange function but also an interactive working function on-line, and to work at the different places and at heterogeneous systems out the same product model. Transmission of words, figures and sound by means of multimedia will also be integrated in such interactive CAD systems which could recognise manual drawings, learn the design process of the product, even understand the natural language instruction for the design, and optimise the design process and design quality. This kind of process would be guaranteed by modern data communication technologies and intelligent quality management and quality assurance. A complete 3D model of a workpiece forms the basis for modern production, especially in the 3D printing process, whereby this data set can be created by direct 3D CAD modeling or by physically model and in reverse engineering. The data sets can be generated from measurements on existing workpieces (Fig. 3).A representation based on facets is then generated from a volume or surface model by polygon triangulation and transmitted to the additive manufacturing process in STL (stereolithography) or VRML (Virtual Reality Modeling Language) format [17].

In the Intelligent Integrated Management System (IIMS), the quality management and quality assurance information and programs will be on-line exchanged and modified concurrently, interactively and collaboratively.

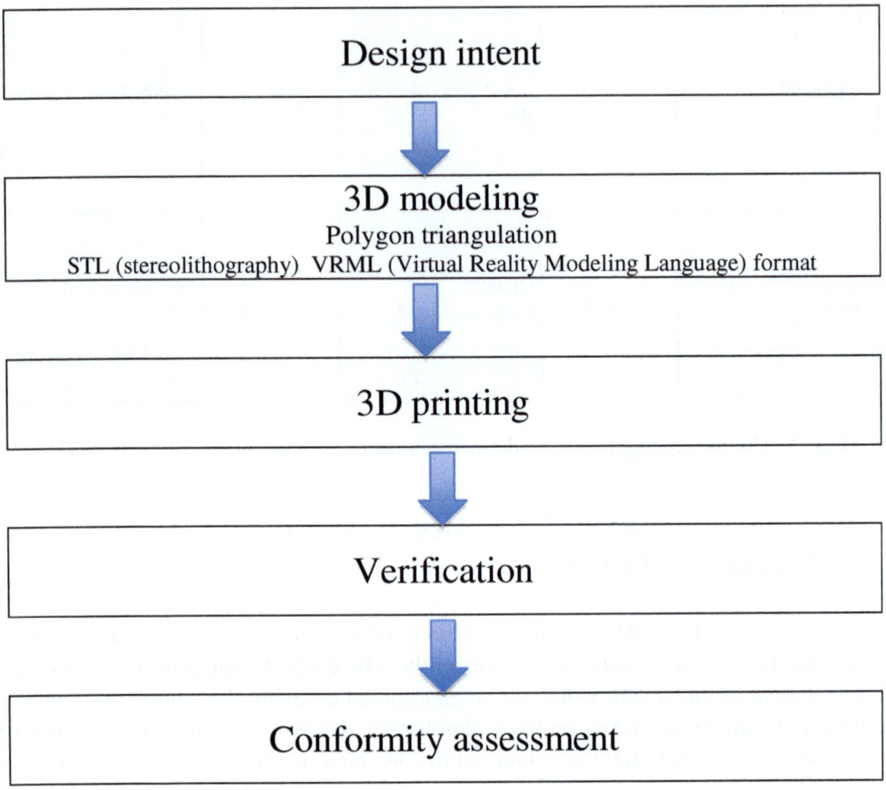

Fig. 3. Product realization process in Additive Manufacturing

It runs autonomously according to the adequate functional quality assurance tasks and organises all manufacturing quality activities and units optimally in the production systems. Learning the processes from the processes, the quality assurance improve the all manufacturing process quality assurance parameters continuously. The implementation of all these properties in an intelligent integrated management and quality assurance architecture is a great challenge, and distributed, decentralised, self-organised and self-optimised concepts will be the main approach for this goal.

5 Intelligent Metrology and Quality Assurance

In the already described advanced manufacturing environment, the integrated quality management system with intelligent coordinate metrology ICM is utilised to test the product or to scan and digitise complex product models with freeform surfaces. This is in order to obtain the digital model of the product and to modify it in intelligent design system and then to create a new modified surface model for 3D printing of the final product by automated and interlinked processes (Figs. 4 and 5).

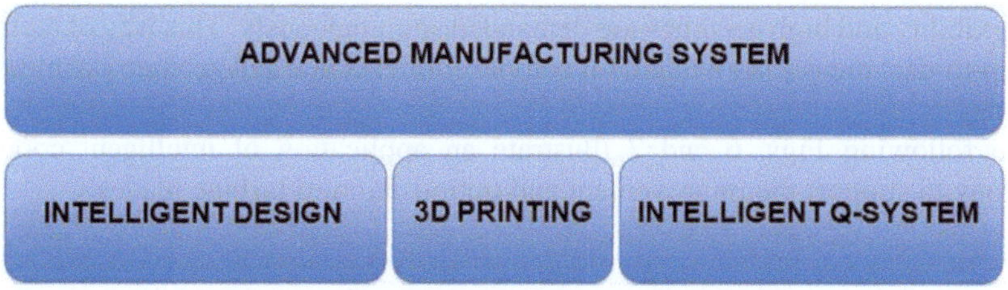

Fig. 4. Advanced manufacturing environment

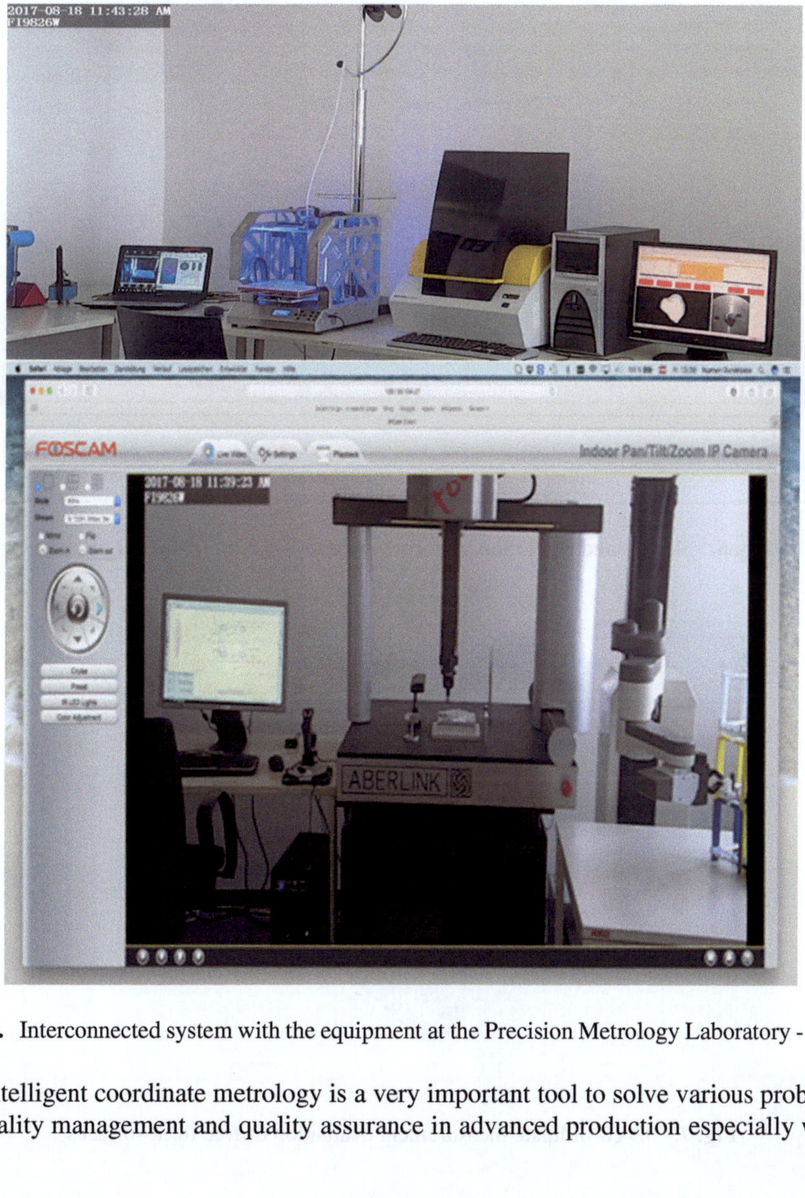

Fig. 5. Interconnected system with the equipment at the Precision Metrology Laboratory - AuM

Intelligent coordinate metrology is a very important tool to solve various problems of quality management and quality assurance in advanced production especially when high flexibility and high accuracy are demanded simultaneously. This way of metrology is the up-to-date measuring method for complex dimensional and geometrical measuring problems.

The following Figs. 6 and 7 illustrate an application of intelligent coordinate metrology to support the improvement and optimisation of turbine blades.

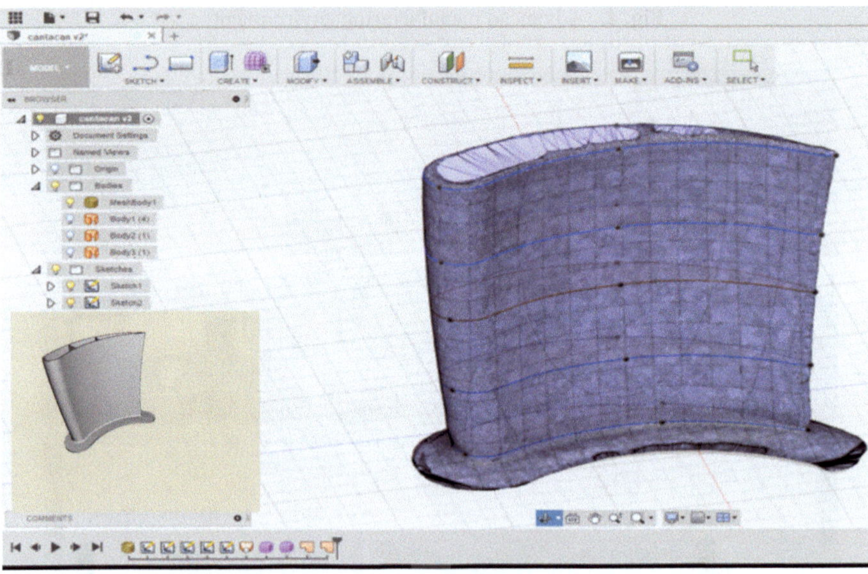

Fig. 6. 3D scanned point cloud file converted to STL format for 3D printing

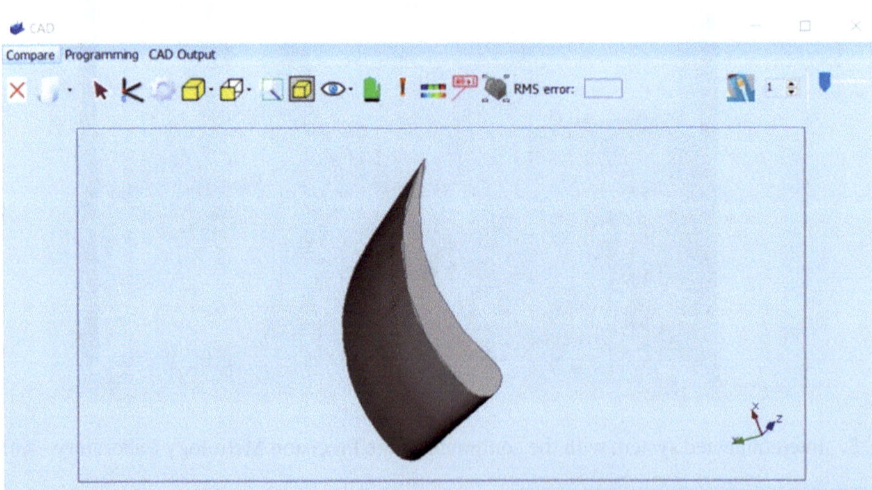

Fig. 7. 3D co-ordinate measurement evaluation of free form surfaces

Figure 6 illustrates the 3D scanning output as a STL file from a 3D laser scanner of the turbine blade whereas Fig. 7 shows the measurement and evaluation of such a free form workpiece using the 3D coordinate measurement technique. The correct function of the turbine blade depends of the numerically defined exact form of the free form surfaces.

In an advanced manufacturing environment, intelligent CMMs are connected by using networks with design and manufacturing.

The goal is to mutually use the data stored in a cloud system, and to realise data parallel-processing. For the concurrent production and the quality management system, it is suitable to use off-line programming technique, through which CNC inspection programs can be worked out without using the CMMs and the products.

By means of this technique, the quality assurance data and inspection CNC programs can be generated simultaneously during the product design. According to the Industry 4.0 concept, a global concurrent quality assurance system must be used with the support of internet and parallel processing information technology. Internet makes it possible to establish a global quality assurance information highway for simultaneous on-line quality assurance data exchange in a network environment and to interact with suppliers and customers and other interested partiesworld wide.

6 Conclusion and Future Work

The challenges of transition to the modern manufacturing industry can be overcome by integrating interconnected intelligent metrological technologies within integrated management models towards realizing agile and optimal manufacturing production. Cooperative activities in the laboratories create systematic knowledge transfer and data acquisition. Each process is carried out step by step in harmony with the required standards.

The quality assurance process will be used in all product production processes in such a system - from the design stage to the final assembly. Quality management and quality assurance in individual activities of different function enterprises - big factories as well as SMEs - in such a collaboration structure play a basic role to ensure the realisation of the concept, e.g. through the intelligent production systems based on quality management and quality assurance in the system to create, to realize and to present the features, such as, concurrent, interactive, modular, integrative, learning, autonomous, self optimising and self organising functions.

References

1. Durakbasa, M.N., Osanna, P.H., Bauer, J.M., Bas, G.: Innovation in production metrology for precision engineering and to support sustainability and improvement of process and product quality in modern manufacturing industry. MSD J. **4**(1), 5–11 (2012)
2. Whitehouse, D.J.: Nanometrology. Contemp. Phys. **49**(5), 351–374 (2008)
3. Smith, S.T., Chetwynd, D.G.: Developments in Nanotechnology. Foundations of Ultraprecision Mechanism Design, vol. 2. Gordon and Breach Science Publishers (1994)

4. Osanna, P.H., Durakbasa, M.N., Crisan, L., Bauer, J.M.: The management and exchange of knowledge and innovation in environments of collaborating small and medium sized enterprises. Commun. IBIMA (CIBIMA), Off. J. Int. Bus. Inf. Manage. Assoc. I **7**, 130–136 (2009)
5. Durakbasa, M.N., Osanna, P.H., Kräuter, L.: The fundamental influence of education, training and certification of persons working in metrology for research and in industry. In: Proceedings of the 7th International Symposium on Tools and Methods of Competitive Engineering, pp. 1409–1410. Delft University of Technology, Eurotech 2 d.o.o., Delft, Netherlands (2008). ISBN: 978-90-5155-044-3
6. Majstorovic, V.: A contribution to the digital quality concept research. Total Qual. Manage. Excellence **33**(3), 27–31 (2005). YU ISSN 1452-0680. International Journal, Special Edition
7. Kopacek, P.: Intelligent manufacturing: present state and future trends. J. Intell. Robot. Syst. **26**, 217–229 (1999)
8. Durakbasa, M.N.: Geometrical product specifications and verification for the analytical description of technical and non-technical structures, Department of Interchangeable Manufacturing and Industrial Metrology, TU-Vienna Austria (2014)
9. Weckenmann, A., Ernst, R.: Anforderungen an die zukünftigeMikro-und Nanomesstechnik-Herausforderungen und Vorgehensweisen (Future Requirements on Micro-and Nanomeasurement Technique Challenges and Approaches), tm Technisches Messen Plattform für Methoden, Systeme und Anwendungen der Messtechnik 67(7-8/2000), 334
10. EN/ISO 9001: Quality Management Systems – Requirements (2015)
11. ISO 14001: Environmental management systems – Requirements with guidance for use (2015)
12. ISO/FDIS 45001: Occupational health and safety management systems - Requirements with guidance for use (2017)
13. ISO/FDIS 31000: Risk management - Guidelines (2017)
14. EN ISO/IEC 27001: Information technology - Security techniques - Information security management systems - Requirements (ISO/IEC 27001:2013 including Cor 1:2014 and Cor 2:2015) (2017)
15. ISO/DIS 50001: 2017; Energy management systems - Requirements with guidance for use
16. Geometrical product specifications (GPS) –General concepts –Part 1: Model for geometrical specification and verification (ISO 17450-1:2011)
17. EN ISO 17296-4: Additive manufacturing – General principles –Part 4: Overview of data processing (2016)

Hybrid Machining: An Industrial Case-Study Comparing Inconel718 Reaming and Drilling with Abrasive Waterjet Technology

B. Wanner(✉), A. Archenti, and C. M. Nicolescu

Department of Production Engineering, Manufacturing and Metrology Systems Division, KTH Royal Institute of Technology, Stockholm, Sweden
bwanner@kth.se

Abstract. Abrasive waterjet technology is one of the fastest growing metal cutting technologies. When used in conjunction with conventional metal cutting methods, abrasive waterjet cutting can be both cost saving and environmentally favorable. This paper shows that when processing hard to cut alloys, abrasive waterjet will be an excellent hybrid alternative. Reaming and drilling have traditionally been used to produce turbine blisks. Reaming is a highly expensive method since it uses very large amounts of cutting tools. Especially when cutting hard materials such as Nickel alloys, tools have to be replaced after only a few minutes of usage. By applying abrasive waterjet cutting to part of the process, the cost for tooling can be almost entirely eliminated. It will also increase the return profit for revert material and greatly reduce environmental emissions. This is because reaming and drilling produce chips while large amounts of cutting fluids are being used. Abrasive waterjet cutting will produce one large metal chunk per cut and uses no cutting fluids. This paper presents a method to combine abrasive waterjet cutting with reaming and drilling.

Keywords: Hybrid processing · Reaming · Abrasive waterjet
Machining vibrations · Sustainability

1 Introduction

1.1 Background

Abrasive waterjet technology has been used for metal cutting for many years [1]. However, it has been used in a limited fashion in hybrid applications. Even in research articles on hybrid processing, it is mostly referred to as an independent method. Here it is presented as a method used in combination with reaming, a conventional cutting method using a very large number of cutting tools. Reaming is not only expensive due to tool wear but also creates long down-times while changing tools. Since the material removed is in the shape of chips, revert material is both bulky and produces low returns of investment. It also uses large amounts of cutting fluids creating a non-favorable environmental situation.

Many industrial factors drive hybrid processing forward. One such factor is cost reductions that can lead to long-term savings in both resources and environment.

J. Ni et al. (Eds.): AMP 2018, LNME, pp. 109–114, 2018.
https://doi.org/10.1007/978-3-319-89563-5_8

Revert material handling is a very large cost to manufacturers. A reduction in revert material handling by increasing the value of revert material extracted during the machining process will increase profits. More importantly, it will have a positive effect on the environment. This industrial case study shows how abrasive waterjet can replace part of the manufacturing process and at the same time contribute to several advantages. It is fast and accurate. There is virtually no tool wear. It produces revert material in the shape of large chunks that can be sold back to the material vendor at a much higher return than for chips. Additionally, abrasive waterjet cutting uses no cutting fluids and thereby has no negative effect on environmental sustainability or on the health of the operators. All these factors contribute to abrasive waterjet cutting as being a viable hybrid method to reaming and other conventional cutting technologies.

2 Approach

In the research presented in this paper, Inconel718 was used for the experiment. A 60 mm thick workpiece was cut at various speeds to document quality and time of cut. Nickel-base alloys are extremely hard to cut using conventional cutting methods, and cause severe cutting tool wear. Therefore, abrasive waterjet cutting was used as a hybrid roughing method for the initial cut, removing some 90% of the material. The remaining 10% was left for the intended reaming finish operation. The striation marks were observed from the abrasive waterjet as the jet tends to spread and deviate from the center cutting line during penetration into the 60 mm thick material. The abrasive waterjet was used to assimilate the reaming roughing operation cutting wedges out of the blisk.

Subsequently, it was used to drill 15 mm diameter holes from the Inconel718 workpiece in order to show drilling capabilities. When cutting the wedges, the waterjet started cutting from the periphery of the material in towards the center and then completing the cut from the center outwards towards the periphery at an angle of 15°. This was done at three different cutting speeds, where the speed is defined as how fast the jet is moving across the material. A total of five holes were drilled using abrasive waterjet. In this case, the jet was cutting through the material from a solid part of the workpiece. In order to compensate for jet drift, the jet was permitted to enter at the center of the drilling hole and then during the penetration cut move it to the intended circumference of the hole. The jet was then moved around the circumference to complete the cutting of the hole.

3 Process Methodology

Abrasive waterjet is capable of producing holes of any shape as well as convergent or divergent cutouts. Extremely small diameter deep holes can be produced which often pose significant problems for conventional drilling methods. When cutting circular holes, however, the tendency of abrasive waterjet is to create tapered walls. By lowering the traverse speed, it will compensate for this tapering effect. If the traverse speed is slowed too much, however, there is a risk for the tapering to start moving outward with increasing depth. Instead of slowing the traverse speed, tilting of the nozzle may compensate for the tapering effect. Controlled correctly, this method can correct the tapering effect without affecting cutting speed [2]. Because of economic and environmental aspects, abrasive waterjet cutting has become a substitute for chemical machining, laser machining, electro discharge machining, and spray etching.

In abrasive waterjet machining, water acts as a carrier that dampens the impact effect of the abrasive on the surface. Abrasive particle grit hardness, size, type, and shape all have an effect both on material removal rate (MRR) and on surface roughness. MRR and surface roughness increase if particle hardness is increased. Harder abrasives act as rigid indentors compared to softer particles, making the hardness ratio between workpiece and abrasive important [4]. By increasing abrasive hardness from 500 to 2500 Vickers, the MRR doubles. However, the increase in MRR is only slight above 1000 Vickers. Similarly, surface roughness increases significantly when the hardness increases from 200 to 1000 Vickers and insignificantly above that. The abrasive shape also has an effect on MRR and surface roughness. More evenly shaped particles produce a lower MRR and a smoother surface. Larger abrasives and higher abrasive mass flow rates cut faster.

Reaming is a finishing operation performed with a multi-edge tool producing high-precision holes or complex geometrically shaped cutouts. When the cutout is in the interior of a component, a hole must first be drilled by a different method prior to applying the reaming operation.

When the cutout is at the periphery of a component as in the case of a turbine blisk, reaming may be used also to initiate the cut. Reaming uses a large number of cutters and produces chips similar to milling. It is a relatively expensive and time consuming cutting method. Especially when processing hard to cut materials such as Inconel718, it becomes a highly expensive method. Standard reaming produces circular holes and pull reaming produces virtually any geometrical shapes [3].

4 Experimental Setup

A 3.5 m straight reaming machine was used equipped with some 400 cutting tools. The photograph in Fig. 1 shows how the tools are fed into the blisk material until the finish dimensions are obtained. The waterjet cutting machine was a 2 × 4 m large table production machine equipped with a 15° tilting head, 0.8 mm orifice, and a sand grit size of 80.

Fig. 1. Reaming setup (left) and waterjet head (right)

5 Results and Discussion

The reaming process cuts a pine tree-shaped opening into the periphery of the blisk. It starts the cut from the outside and works itself into the final geometrical shape, from rough to finish machining. By using abrasive waterjet cutting, the thought is to eliminate the rough machining operation of the reaming process. This will allow for a solid triangular-shaped chunk of material to be salvaged as revert material as opposed to chips. Inconel718 is one of the most expensive alloys used in aerospace applications and the price per pound for solid revert is much greater than for chips. For a blisk with 50–100 machined openings for blades, the savings only in material revert will be very large by using abrasive waterjet cutting instead of reaming. In addition, storage volume for chips is much greater than for solids, thereby saving revert storage space.

Reaming uses large amounts of cutting fluids. These can be completely eliminated during the roughing portion of the machining operation if abrasive waterjet cutting is used. Except for the obvious cost saving, it will contribute to an environmental improvement.

Reaming induces heat and there for stresses into the workpiece with subsequent heat treatment as a consequence. During the abrasive waterjet cutting, virtually no heat is induced into the workpiece. An infrared camera was used to prove the fact (see Fig. 2). The only temperature change during the process is the slight increase in temperature of the water in the tank below the part. This stems from the energy absorbed from the waterjet. The forces and vibrations produced by abrasive waterjet cutting are negligible compared to the reaming operation [5].

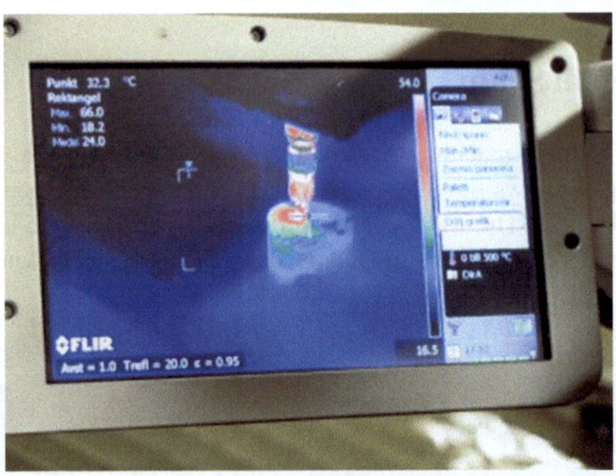

Fig. 2. Temperature measurements during abrasive waterjet cutting

Abrasive waterjet technology uses a crystal and an orifice to converge and shape the jet. For rough machining, an orifice may be used for 50–100 h before replacing. However, the reaming operation requires a large amount of cutting tools, in some cases over 400. When cutting Inconel718, these cutters have to be replaced after only one to four minutes depending on what portion of the cut is being made. Tool replacements are highly expensive and require substantial downtime for the machine. Tool wear also gives rise to emission of unwanted elements into the atmosphere.

Abrasive waterjet cutting is used here as a roughing operation. The quality of cut is therefor not critical. As long as the jet can be controlled to cut relatively straight as to not jeopardize on the final dimensional requirements, the fastest traverse speed possible can be used. Even so, several different traverse speeds were used during the experiments in order to observe quality of the cut surface (See Figs. 3 and 4).

Fig. 3. Waterjet cutting setup and cutting procedure

Fig. 4. Cutting results and surface quality for abrasive waterjet cutting

6 Conclusions

By using abrasive waterjet cutting in combination with conventional cutting methods, increased productivity follows due to reduced downtime from changing conventional tools. Increased material removal rates and improved safety of operations are other benefits. Additionally, the braking of reaming tools or drills at the end of a roughing operation could severely damage the workpiece and therefor have large negative impacts on production costs.

For sustainability, a better control in revert handling, less cutting fluids, and less tool wear will contribute to a better environment. The conventional reaming process produces small chips that require large storage space and complicated cleaning practices when being processed for remelting into new material. Cutting fluids are more environmentally friendly today than ever before. Even so, reaming and drilling requires very large volumes of cutting fluids. These have to be cleaned prior to reuse and it does contribute to a non-desirous environmental situation. The abrasive waterjet cutting method, on the other hand, produces relatively large chunks of revert metal that requires very little processing before remelting. Clear ecological sustainability can therefore be realized by using abrasive waterjet cutting for the roughing operation as opposed to reaming and drilling. As a bonus benefit, chunks of metal have a greater value per pound than chips.

By using an IR camera it was shown that the heat introduced into the workpiece during the abrasive waterjet cutting operation was negligible. There is no additional heat treatment required as a consequence of the roughing operation. This will shorten the process time and decrease the total processing cost.

Tooling cost will be reduced drastically if the hybrid method is used instead of the conventional cutting method by itself. The reaming operation can wear out 300–400 cutting edges within only a few minutes. In the case of abrasive waterjet cutting, the only tooling that wears is the nozzle. There is also a crystal that needs to be replaced at certain interval. However, the tooling cost for abrasive waterjet machining is negligible compared to the reaming machine.

Inconel718 is a very hard to process material. Even so, by using abrasive waterjet cutting, the drilling operation was relatively easy to perform. Virtually no tool wear could be observed. Also in this case, cutting using abrasive waterjet produced a solid chunk of metal.

Acknowledgments. This work was partly carried out at Siemens Industrial Turbomachinery AB, Finspång and Strömbergs Waterjet Company in Gothenburg, Sweden and has been supported by the Centre for Design and Management of Manufacturing Systems – DMMS at KTH.

References

1. Gärdek, H., Ouaha, B.: A State of the Art Report and Comparison with Conventional Methods of Waterjet Machining Technology (2015)
2. Zhang, S., Wu, Y., Chen, D.: Hole-drilling using abrasive water jet in titanium. Int. J. Mach. Mach. Mater. **9**, 47–65 (2011)
3. Wanner, B., Archenti, A., Nicolescu, N.: Hybrid machining: abrasive waterjet technologies used in combination with conventional metal cutting. J. Mach. Eng. **17**(3), 85–96 (2017)
4. Fowler, G., Pashby, I., Shipway, P.: The effect of particle hardness and shape when abrasive waterjet milling titanium alloy Ti6Al4V. Wear **266**, 613–620 (2008)
5. Wanner, B., Eynian, M., Beno, T., Pejryd, L.: Cutter exit effects during milling of thin-walled inconel 718. Adv. Mater. Res. **590**, 297–308 (2012)

Dynamic Definition of Machine Tool Feed Drive Models in Advanced Machine Tools

George Constantin(✉)

University Politehnica of Bucharest, Bucharest, Romania
george.constantin@icmas.eu

Abstract. Machine tools in actual construction, being related to the trend towards to machine tools of high productivity and increased precision and having in view the context of transition to Industry 4.0, have to be studied deeply in the stage of conception and also in that of exploiting. For both situations, the modeling and simulation of the machine tool regarded as a mechatronic system represent methods of analysis, assessment, and optimization for its improvement.

This work proposes the approaching the subject from the point of view of modeling and simulation of one assembly of the machine tool that proves one of the most sensitive kinematic structure responsible for the machine tool precision. The modeling, either Rigid Body Simulation, Digital Block Simulation, Finite Element Modeling or combinations types, implies a concrete and accurate definition of the dynamic parameters stiffness, damping, and friction.

Some mathematical approaches for determining these parameters are presented. Also, some product catalog values and relations coming from the engineering and research experience applied to feed drive components and also to the whole kinematic chain are given. Furthermore, some practical testing and calculation methods are presented.

Keywords: Machine tool · Feed drive · Modeling · Stiffness · Damping
Friction · Dynamic definition · Simulation

1 General Aspects of Machine Tool Dynamics

The constant trend in the evolution of machine tools is the constant increase in speeds and accelerations [5]. Therefore, the study of the dynamic behavior of machine tools and flexibility of structural elements must be considered.

That is why the dynamic models of the axes or the entire machine tool play a very important role in sizing the components and also in designing the control system.

The general way of considering the structures of machine tools that is historically viewed is that of massiveness. Thus, the massive structure leads to the high structural stiffness necessary to reduce the deformations under the influence of the machining forces and the static weight of the machine structure and workpiece. The deformation of the structure, which can be considered as a deformation of the structural loop, leads to errors at the interface between the cutting edge of the tool and the workpiece.

J. Ni et al. (Eds.): AMP 2018, LNME, pp. 115–138, 2018.
https://doi.org/10.1007/978-3-319-89563-5_9

Rigid structures of machine tools tend to transmit vibrations at higher frequencies than deformable (non-rigid) ones. But rigid structures with low internal or external damping will transmit the vibrations caused mainly by the processing process throughout the structure. The transmitted vibrations will cause the structure to vary over time, which can be amplified in the workpiece, if the vibrations have frequencies close to a natural frequency of the machine tool. Therefore, a very rigid machine tool is not associated with small deformations.

The solution of deformation compensation, deformations which could be variable over time that cannot be easily estimated, is to measure deformations and compensate them. This task is almost impossible at high frequencies.

In addition to the rigidity characteristics of a machine tool, its ability to dampen vibrations created or transmitted by high stiffness should be considered. The performance of the machine tool is influenced by the component materials. Internal damping is different for different materials. For example, steel, cast iron and granite have different damping characteristics [22].

Some materials or components could also have nonlinear damping features. It may very well dampen vibration for a short period of time and then increase or maintain high amplitude, etc. Considering all these, all concerned phenomena have to be investigated along with how great the influence of rigidity and damping of the machine tool on the entire behavior of the structure should be.

2 Estimated and Measurable Behavior of Machine Tool

Measurable behavior of a machine tool is a combination of the geometric, thermal, static and dynamic behaviors. The dynamic behavior influences directly the machined part quality being observable through a lot of effects, such as the tracking error, outrunning, self excited vibrations, cross-talk, etc. In general, there are some week pints (bottlenecks) belonging to machine tool structure that influences the dynamic errors in a machine tool. Their identification becomes an essential condition for the desired improvement.

A comprehensive study of the dynamic behavior should be achieved by different methods:

- determination of the frequency answer (natural vibration mode respectively) of the machine tool structure, determined by the mass distribution, stiffness and damping properties in guides and bearings, and also by the degree of structural elasticity (for large frequency ranges);
- experimental modal analysis regarding a set of principal forces, and the acceleration signal distributed on the entire structure;
- simulation models based on rigid bodies and model with finite elements;
- structure testing, including the control loop (control coefficients, acceleration and jerk settings), feed-forward, filters, dead time, measuring system, etc.;
- simulation of the actuating systems, including also the structural machine tool components;

- experiments: evaluation of the input signals, cross-talk measurements, subsequent FFT (Fast Fourier Transform) and cross-talk interpretation;
- stability measurements (process parameters);
- on-line measurements on a running machine tool;
- integration of the process model in a virtual structure.

Given the example of a feed kinematic chain that represents a basic sub model of a machine tool (Fig. 1), we can make some remarks [4]. The constructive variant can be directly from the electric motor rotor to the lead screw through a coupling. One can ascertain that the model consists of components with rotation and translation motions (last element of the drive), with mass properties (mass, mass center, moment of inertia), and also of elastic and damping elements that give the model the characteristic of *dynamic*.

There are many types of modeling used in machine tool. The most developed and used modeling types are Rigid Bodies Simulation – RBS, Digital Block Simulation – DBS, and Finite Element Modeling – FEM. These techniques can be combined [3, 6–8, 29] for obtaining more accurate results in machine tool analysis. All techniques require the dynamic definition of machine tool that means the calculation or determination by tests the dynamic parameters involved in models.

The dynamic parameters, besides the mass ones, strongly influence the dynamic behavior of the assembly of servo axis and are a key element in defining the functional model. Figure 1 presents the structure of a feed drive of a machine tool along with the dynamic parameters required along each axis, stiffness k and damping d.

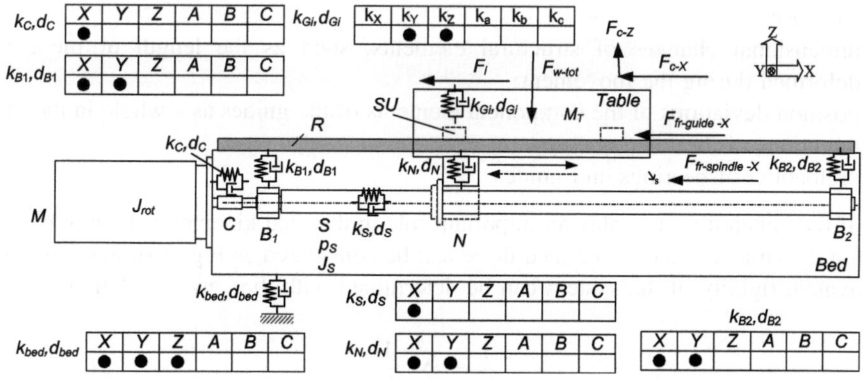

Fig. 1. Feed drive with stiffness and damping parameters of different subsystems and their significant components on axes.

There is a variety of bibliographic sources that propose on the basis of elaborated researches values of these parameters associated with different components in the kinematic chain.

3 Stiffness Modeling

In terms of stiffness, the modeling is characterized by two distinct levels: the model component and the stiffness and damping coefficients. For the first level, a modeling technique is used. In the second one, it is necessary to decide which coefficients and values have to be selected.

The stiffness coefficients of the linear guides, bas screws and nuts are derived from the catalog specifications. Rotational stiffness coefficients are not always available, and can be obtained on request from manufacturers. Some manufacturers use experimental stands to identify rigidity values of components. Other manufacturers focus on virtual calculation models to determine stiffness coefficients.

There is also a way that can be seen as the opposite, namely obtaining reliable information about the stiffness of the linear guides with intermediate elements or bearings by experimental identification that assumes that the tool machine is available for tests and there is also a model RBS or FEM, as demonstrated by numerous works in the field of machine tools [1, 15, 26, 30, 31].

3.1 Stiffness Modeling of Feed Drive

It is accepted that the static stiffness of a machine tool along an axis of the reference frame is variable for different positions occupied by the mobile element (mass of the machine) being influenced by:

- changes in contact stiffness caused by interferences of the mechanical interface (guides);
- backlash;
- dimensional changes of structural elements, such as the length of the screw deformed during the movement;
- position deviations of the component elements of the guides as a whole in the main directions of the reference system;
- preloading of bearings and guides.

If the clutch does not play an important role in the feed kinematic chain dynamics, then a dynamic scheme of the feed drive can be considered as Fig. 3 shows. The axial equivalent rigidity of the system can be determined with the general relation:

$$\frac{1}{k_e} = \sum_{i=1}^{n} \frac{1}{k_i}, \tag{1}$$

where k_e is the equivalent stiffness of the feed drive and k_i – component stiffness (k_{B1}, k_{B2}, k_S, k_N) (Fig. 2).

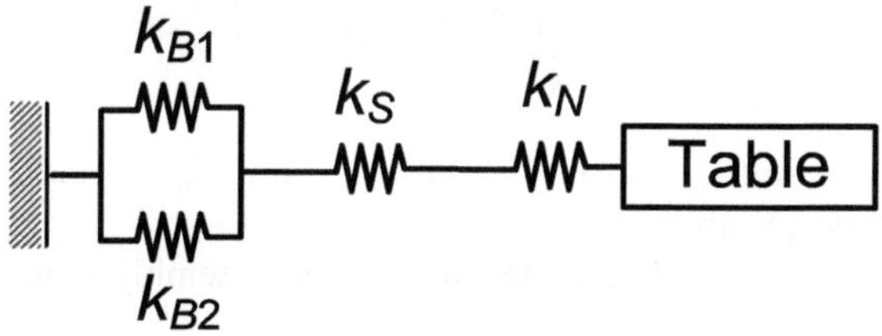

Fig. 2. Dynamic model with springs of the feed drive

Thus, for the case considered it is obtained:

$$\frac{1}{k_e} = \frac{1}{k_{Be}} + \frac{1}{k_S} + \frac{1}{k_N}, \text{ where } k_{Be} = k_{B1} + k_{B2}. \tag{2}$$

3.1.1 Method for Calculating the Ball Screw Stiffness

The stiffness of the ball screw can be determined by considering the screw as a bar loaded by an axial force that generates deformation. The mathematical relation of stiffness is the one known in the field of mechanics of materials [11]:

$$k_s = \frac{\pi\, d^2 E}{4l}, \tag{3}$$

where the bar is considered to have the diameter d (the smallest of the screw) and the length l. The quantity E is the elastic modulus of the screw material, usually steel ($E = 2.1 \cdot 10^{11}$ N/m^2).

The screw nut and mobile element of table type are considered to be halfway, which means that in most cases $l = L/2$, the spindle stiffness becoming

$$k_s = \frac{\pi\, d^2 E}{2L}. \tag{4}$$

Thus, the exact value of the length l for a certain position of the mobile element can be considered. In this case, it is possible to refer to the stiffness of the screw considering the whole length L, obtaining the relation:

$$k_s = \frac{\pi\, d^2 E}{2L} \cdot \frac{L}{l}. \tag{5}$$

3.1.2 Ball Screw Stiffness Given by Catalogues

The catalogues of ball screw nut components usually give information about this assembly type. The assembly stiffness is given by:

$$\frac{1}{k_{bs}} = \frac{1}{k_s} + \frac{1}{k_n}, \tag{6}$$

where k_{bs} is the total stiffness of the ballscrew [N/μm]; k_s – spindle stiffness [N/μm]; and k_n – nut stiffness [N/μm].

The spindle stiffness (see Eq. 4) depends on the assembly type:

$$k_s = 64.7\frac{d_r^2}{L_1} \text{ for fixed - fixed assembly;} \tag{7}$$

$$k_s = 16.8\frac{d_r^2}{L_1} \text{ for fixed - free assembly;} \tag{8}$$

For a single nut, the nut stiffness is:

$$k_n = 0.8 \times k\left(\frac{P}{0.1C}\right)^{1/3}, \tag{9}$$

where k is the stiffness in the table; P – preload; C – dynamic load on table.

3.1.3 Variable Ballscrew Stiffness

A linear axis of a milling machine, namely a feed drive is considered for the study. The slide is driven by a rotary motor and a ball screw-nut mechanism.

For a long ball screw of 1.5–2 m and more, its stiffness characteristic is variable. It is considered an initial length l_0. The total stroke of the slide is h. The machining is done by a milling head and the cutting force component that is involved along the axis is the feed force F_{cut} as disturbance in the system.

In this case the screw stiffness becomes nonlinear due to the dependence of the elastic constant to the length according to (3):

$$k_{sx} = \frac{\pi\, d^2 E}{4 l_x}, \tag{10}$$

where k_x is the ball screw stiffness, E – elasticity modulus of steel, $E = 2 \cdot 10^{11}\ \mathrm{N/m^2}$, A – screw cross section area, having expression $A = \frac{\pi\, d^2}{4}$; d – ball screw diameter in m.

The variable frequency ω_x of the system has the expression [25]:

$$\omega_x = \sqrt{\frac{E\,A}{m} \cdot \frac{1}{l_x}}, \tag{11}$$

where $l_x \in [l_0, l_0 + h]$.

If one considers an initial length l_0 = 0.13 m, for four ball screw sizes (d_1 = 25 mm, d_2 = 40 mm, d_3 = 50 mm, and d_4 = 63 mm) the variation of spindle stiffness with length for a total length of 2 m is presented in Fig. 3.

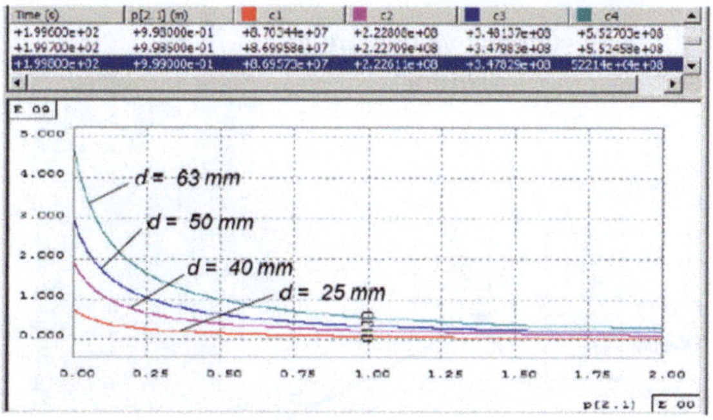

Fig. 3. Variation of ball screw stiffness with length [25]

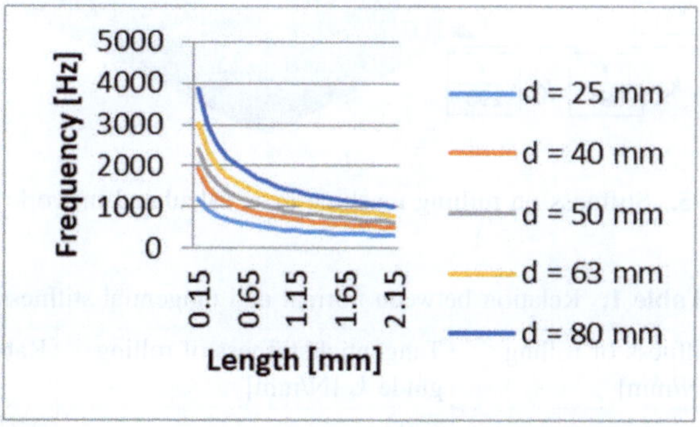

Fig. 4. Variation of ball screw frequency with stiffness and length.

The system frequency also is variable with regard to the active spindle length l_x. For the range of spindle diameters d = [24; 40; 50; 63; 80 mm] the variation is presented in Fig. 4. The mass is considered m = 450 kg.

Wang et al. studied in [27] a CNC milling machine with three controlled axes *X*, *Y*, and *Z*, for which they calculated the stiffness of rolling joint (guide ways and ballscrew nut mechanism) along with the equivalent stiffness of the axes. The used values are given in Fig. 5 for having a global image of the feed drive dynamic definition.

Regarding the rolling guide stiffness calculated in work [27], the data grouped in Table 1 shows that there is a constant ratio between normal and tangential stiffness of k_t/k_n = 0 71. This ratio could be considered for calculating the tangential stiffness in which case the normal rigidity of the rolling guides is taken from catalogs.

3.2 Nut Stiffness

This is more than nut stiffness. It is the stiffness of the assembly spindle-balls-nut. It is obtained based on the Hertz contact theory, which implies three components: spindle through its thread groove, ball, and nut groove. The expression of the deformation of one ball becomes [27]:

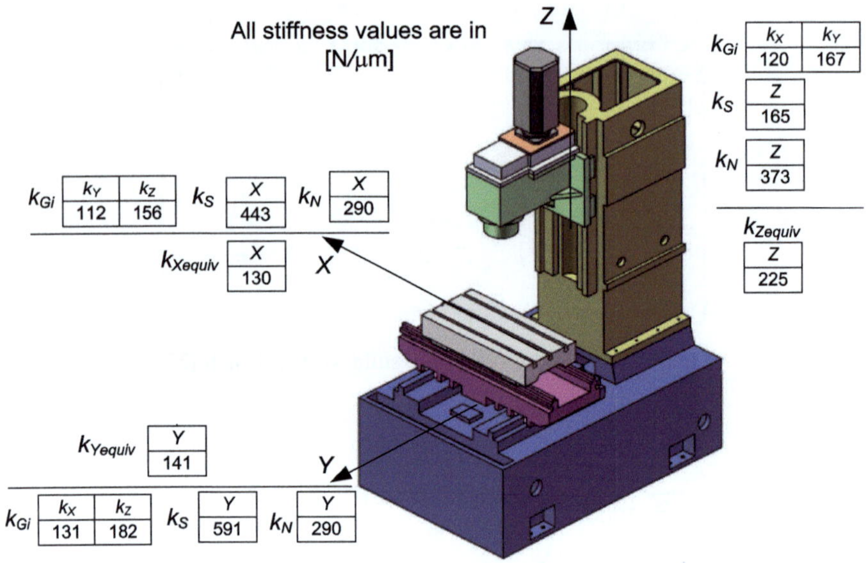

Fig. 5. Stiffness on milling machine axes calculated in work [27]

Table 1. Relation between normal and tangential stiffness

Normal stiffness of rolling guide k_n [N/mm]	Tangential stiffness of rolling guide k_t [N/mm]	Ratio k_n/k_t
167	120	0.718562874
156	112	0.717948718
182	131	0.71978022

$$\delta_i = \frac{K_i}{\pi a_i}\left[\frac{3P}{2}\left(\frac{1-u_1^2}{E_1}+\frac{1-u_2^2}{E_2}\right)\right]^{2/3}\left(\sum \rho_i\right)^{1/3}, i = 1, 2, \tag{12}$$

where P is the normal reaction on each ball; K_i and a_i – Hertz coefficients; β – pressure angle at contact ball-groove; δ_1 – deformation at contact ball-nut; δ_2 – deformation at contact ball-screw; E_1 and E_2 – moduli of elasticity of the two contact components; u_1 and u_2 – Poisson coefficients of the contact objects; $\Sigma\rho$ – synthetic curvature at contact point.

The synthetic curvature for the contact points ball-nut (1) and ball-screw (2) are respectively:

$$\sum \rho_1 = \frac{2}{d_b} + \frac{2}{d_b} - \frac{1}{d_b f_1} - \frac{2\cos\alpha\cos\phi}{d + d_b\cos\alpha}, \tag{13}$$

$$\sum \rho_2 = \frac{2}{d_b} + \frac{2}{d_b} - \frac{1}{d_b f_2} - \frac{2\cos\alpha\cos\phi}{d + d_b\cos\alpha}, \tag{14}$$

where d_b is the ball diameter; f_1 – form factor as ratio nut groove radius to ball radius; f_2 – form factor as ratio screw groove radius to ball radius.

The axial deformation of nut against spindle is:

$$\delta_N = (\delta_1 + \delta_1)/\sin\beta\cos\phi, \tag{15}$$

leading to the nut stiffness k_N

$$k_N = F_a/\delta_N, \tag{16}$$

where F_a is the preload force in screw-nut assembly that at static equilibrium gives the equation:

$$F_a - Pz\sin\beta\cos\phi = 0, \tag{17}$$

with z the number of balls; ϕ – lead screw angle; P – normal reaction of each ball.

The assembly ball screw nut rigidity according to [28] is as follows:

$$\begin{aligned} k_N = {} & \frac{1}{3\pi}[\frac{3}{2}(\frac{1-\mu_1}{E_1} + \frac{1-\mu_2}{E_2})]^{\frac{2}{3}}[\frac{K_n(e)}{m_{na}}(\sum \rho_n)^{\frac{1}{3}} \\ & + \frac{K_S(e)}{m_{sa}}(\sum \rho_s)^{\frac{1}{3}}]Z^{\frac{2}{3}}(\sin\beta)^{\frac{5}{3}}(\cos\lambda)^{\frac{1}{3}}F_P^{\frac{1}{3}}, \end{aligned} \tag{18}$$

where E is the Young's modulus, μ – Poisson's ratio, F_P – the preload of gasket, Z – ball number, λ – helix angle, β – contact angle.

From the product catalog [21], data on the stiffness of nuts with various nominal dimensions of the threaded part can be obtained (Fig. 6).

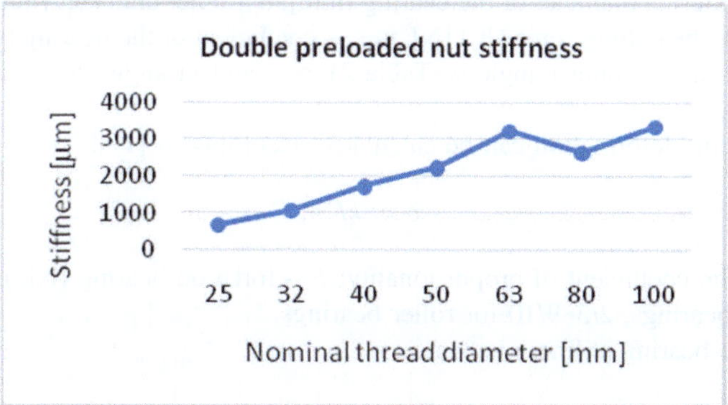

Fig. 6. The influence of nominal diameter on the stiffness of double-nut screw [21].

From the catalog values it is found that for the same nominal diameter d the stiffness is higher for the smaller step (which means a smaller angle of the helix). For example, in the catalog [21], for a diameter of 40 mm, there are two values of 1710 N/μm (p = 10 mm) and 1030 N/μm (p = 20 mm) in ratio r = 1.66. For the same nominal diameter, the ratio is variable because the variants are different in terms of the number of ball circuits and sometimes the diameter of the ball, etc.

3.3 Influence of Bearing Stiffness

The following are some theoretical and practical methods for determining the bearing stiffness with application to angular contact ball bearing. Methods are compared in order to provide an image of methods in order to choose the stiffness values for a dynamic application.

From a theoretical perspective, bearing stiffness is a relation that is based on the contact between the rolling element (ball or roller) and the inner ring.

Obviously, it depends on the material (usually steel on steel or ceramic on steel contact) and also the geometric parameters and the number of rolling elements [2].

Radial stiffness of the bearing is given as the ratio between the force F_r and deflection δ_r by:

$$k_r = \frac{F_r}{\delta_r}. \tag{19}$$

Paper [1] gives the following expression of radial stiffness:

$$k_r = c_{type} \frac{(D+d)^{2/3}}{(D-d)^{1/4}} F_r^{1/3}, \tag{20}$$

where: D – external diameter of the bearing [mm]; d – inner diameter [mm]; F_r – radial component of the cutting force (F_x) [N]; c_{type} – coefficient of the bearing type including the influence of the contact angle α (Table 2); α – contact angle (for series A → 25°, C → 15°).

The bearing deformation can be calculated according to [17]:

$$\delta = kFp, \tag{21}$$

where: k is the coefficient of proportionality; F – force on bearing [N]; p – exponent: 2/3 for ball bearings, 2/3–9/10 for roller bearings.

Thus, the bearing stiffness becomes:

$$c = F/\delta = F^{1-p}. \tag{22}$$

Table 2. Values of the bearing type coefficient

Series type	719…C	719…AC	70…C	70…AC	72…C	72…AC
c_{type}	4.83	5.83	5.43	6.03	5.38	6.03

The ball bearing deformation is a function of external force:

$$\delta_r = \frac{k}{\cos\alpha}\sqrt[3]{\frac{Q^2}{D_W}},\ \text{where } k = 0.436, \tag{23}$$

with deformation given by expression:

$$\delta_r = \left[10^{-4}\cdot\frac{0.002}{(\cos\alpha)^{2/3}}\left(\frac{5}{10}\right)^{2/3}\right]F_r^{2/3}\cdot d_b^{1/3}(i\cdot n)^{2/3}. \tag{24}$$

The Krämer method proposes an expression of rigidity given by

$$k_r = \frac{nF_r}{x_r}, \tag{25}$$

where the deformation for ball bearings is calculated with relation

$$x_r = 1.2\times10^{-7}\times d_b^{-\frac{1}{3}}\times Z^{-\frac{2}{3}}\times F^{\frac{2}{3}}, n = 3/2. \tag{26}$$

Three dimensional angular contact ball bearings were considered for calculation, namely type 7012 (d = 60 mm, D = 95 mm), 7017 (d = 85 mm, D = 130 mm), 7018 (d = 90 mm, D = 140 mm).

Radial stiffness variation diagrams through method [2] are shown in Fig. 7, through Palgrem method [17] in Fig. 8, and Krämer method [14] in Fig. 9.

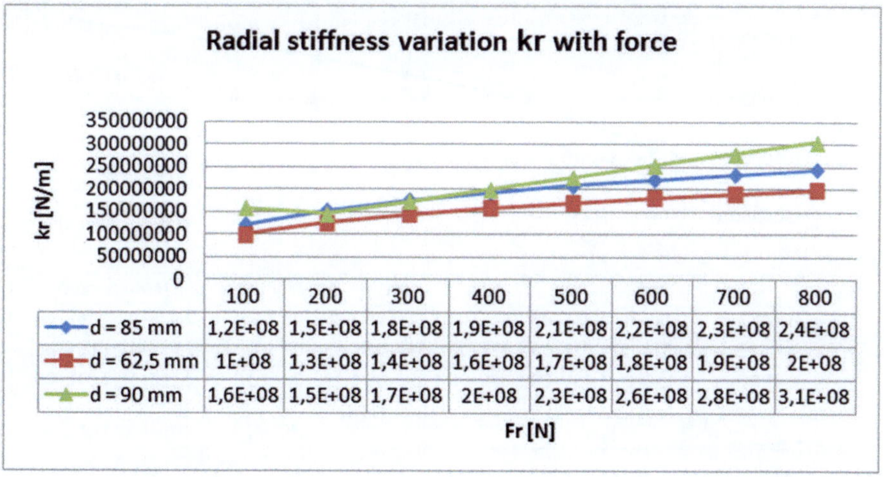

	100	200	300	400	500	600	700	800
d = 85 mm	1,2E+08	1,5E+08	1,8E+08	1,9E+08	2,1E+08	2,2E+08	2,3E+08	2,4E+08
d = 62,5 mm	1E+08	1,3E+08	1,4E+08	1,6E+08	1,7E+08	1,8E+08	1,9E+08	2E+08
d = 90 mm	1,6E+08	1,5E+08	1,7E+08	2E+08	2,3E+08	2,6E+08	2,8E+08	3,1E+08

Fig. 7. Diagram of radial stiffness variation for angular contact ball bearing type 70…C after [17]

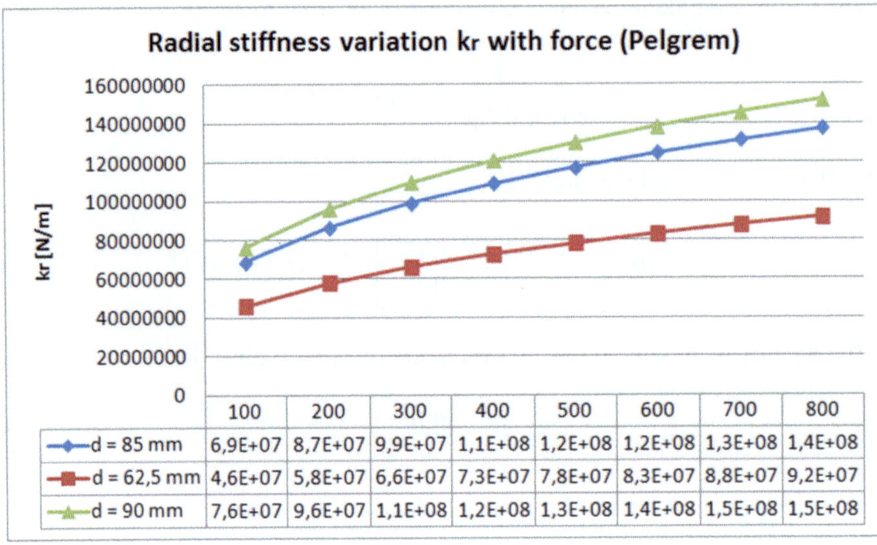

	100	200	300	400	500	600	700	800
d = 85 mm	6,9E+07	8,7E+07	9,9E+07	1,1E+08	1,2E+08	1,2E+08	1,3E+08	1,4E+08
d = 62,5 mm	4,6E+07	5,8E+07	6,6E+07	7,3E+07	7,8E+07	8,3E+07	8,8E+07	9,2E+07
d = 90 mm	7,6E+07	9,6E+07	1,1E+08	1,2E+08	1,3E+08	1,4E+08	1,5E+08	1,5E+08

Fig. 8. Diagram of radial stiffness variation for angular contact ball bearing type 70…C after [17]

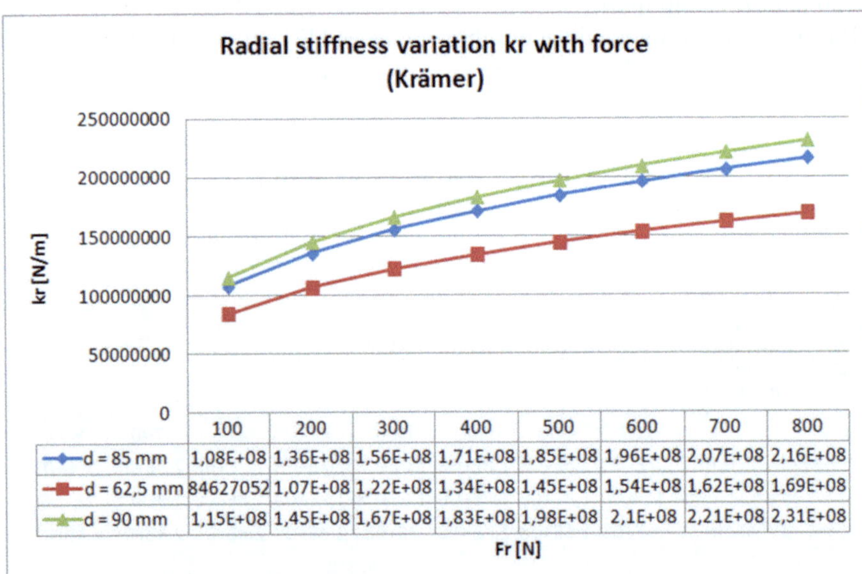

	100	200	300	400	500	600	700	800
d = 85 mm	1,08E+08	1,36E+08	1,56E+08	1,71E+08	1,85E+08	1,96E+08	2,07E+08	2,16E+08
d = 62,5 mm	84627052	1,07E+08	1,22E+08	1,34E+08	1,45E+08	1,54E+08	1,62E+08	1,69E+08
d = 90 mm	1,15E+08	1,45E+08	1,67E+08	1,83E+08	1,98E+08	2,1E+08	2,21E+08	2,31E+08

Fig. 9. Diagram of radial stiffness variation for angular contact ball bearing type 70…C after [14]

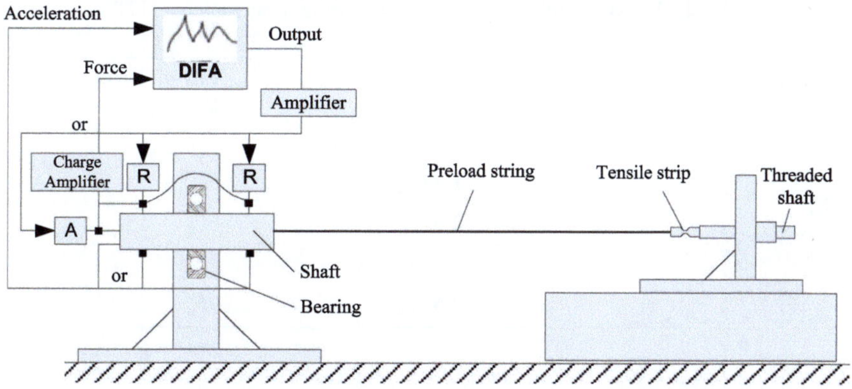

Fig. 10. The measurement diagram of the transfer function [13]

3.3.1 Relationship Between Axial and Radial Stiffness of Bearings Given by Design Experience

In modeling ball bearings, especially in angular contact ball bearings, the following relationships between the radial and axial stiffness are used:

$$k_r = k_a/2 \cdot \cot g^2 \alpha, \tag{27}$$

where α is the contact angle.

3.3.2 Bearing Stiffness Given by Product Catalogs

The bearing catalogs of the manufacturing companies offer relations based on experimental data of radial and axial stiffness that are customized with values specific to each type of bearing, size, load, etc.

The bearing catalog [12] gives the axial rigidity c_a for bearing pairs in an O or X arrangement in tables. It supplies also the relation between radial rigidity c_r and the axial one c_a:

- $c_r = 6 \cdot c_a$ for $\alpha = 15°$,
- $c_r = 3.5 \cdot c_a$ for $\alpha = 20°$,
- $c_r = 2 \cdot c_a$ for $\alpha = 25°$,

where α is the contact angle.

To determine the axial stiffness, the bearing has to be preloaded. The lift-off force up to which the deflection is almost linear is shown in Table 3 along with the axial stiffness c_a of the bearing sets.

Table 3. Axial rigidity for different bearing arrangements [12]

Bearing arrangement	Axial rigidity c_a [N/μm]	Lift-off force K_{aE} [N]
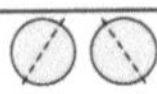	c_a	$3F_v$
	1.64 c_a	$6F_v$
	2 c_a	$6F_v$
	2.24 c_a	$9F_v$
	2.64 c_a	$9F_v$

The axial rigidity is given by the catalog for three cases of preloading: L – low, M – medium; H – high. Table 4 presents axial rigidity ranges for different angular contact bearing series and sizes (*d* – bearing bore diameter [mm]).

Table 4. Rigidity ranges for angular contact bearings (after [12])

d	Axial rigidity c_a [N/μm]		
	L	M	H
10	8.8-35.8	14.1-63.2	19.6-88.8
12	9.4-42.8	15-66.8	20.8-103.1
15	11.4-49.7	18.2-84.4	25.2-117.3
17	17-53.3	19.2-89.9	26.5124.6
20	15-63.7	23.7-106	32.7-146.7
25	17-69.2	26.8-115.2	36.9-158.8
30	21.3-90.4	34.8-148.3	44.5-203.3
35	24.5-103.2 1	38.6-169.4	53-232.2
40	26.6-113.5	41.8-184.3	57.2-251.6
45	30.8-120.8	48.4-196.1	66.2-267.8
50	32.6-138.3	51.2-222.9	70-303.3
55	37.3-154.6	58.6-249.2	80.3-338.8
60	39.5-164.9	62.2-264.1	85-358.3
65	41.5-184.8	65-294.5	88.8-399
70	47-193.7	73.7-308.6	100.6-417.8
75	49.5-203.8	77.5-324.7	105.7-439.4
80	52.3-220.9	81.9-399.3	111.7-471.6
85	55.9-233.1	87.4-368.5	118.9-497.4
90	57.2-257.1	89.4-404.3	121.6-544.6
95	60-272.7	93.7-428.7	127.4-577.3

3.3.3 Experimental Determination of Bearing Stiffness

The experimental determination of the stiffness of bearings used for shaft-bearing assemblies is a tool used for correcting the theoretical models and for assessing the usefulness of the models for estimating the behavior of the shaft-bearing assembly on vibration.

Certainly, experimental assemblies used to determine the rigidity of bearings are based on axial or radial loading with increasing forces in a range and the measurement of displacements in axial or radial directions, respectively.

Knappen proposes in work [13] an experimental stand that is dynamically excited with an exciting hammer or shaker to determine the transfer matrix of the bearing. The preloaded bearing is considered because the non-preloaded bearings are difficult to estimate. Figure 10 presents several solutions of experimental stands that have various advantages and disadvantages. It is considered to be more suitable for experimental tests the stand after Heuvelmans used to determine the linear and rotational stiffness of the bearing (Fig. 10). Preloading is measured by a tensile strip. Measurement of the transfer is done through the Dynamic Signal Analyzer - DIFA, represented in Fig. 10. Two R excitators controlled by DIFA load the shaft in a radial direction. In axial load only the excitator A is used. Both force and acceleration are measured and taken over by DIFA.

It starts from the general dynamic equation and uses identification algorithms that use the Least Square Algorithm to minimize the fit error:

$$\underline{M}\ddot{q} + \underline{B}\dot{q} + \underline{K}q = f, \tag{28}$$

$$\underline{A}^T \underline{A}\underline{\hat{X}}_{ls} = \underline{A}_T \underline{E}, \tag{29}$$

where $\underline{A}$ and $\underline{E}$ contain model characteristics and experimental data. $\hat{X}_{ls}$ contains the unknowns of the dynamic equation.

The differences between the measured and modeled values are made by an iterative solving algorithm:

$$\underline{V}^T \underline{A}\underline{\hat{X}}_{iv} = \underline{V}^T \underline{E}, \tag{30}$$

where the $\underline{V}$ matrix contains variable instrumental outputs.

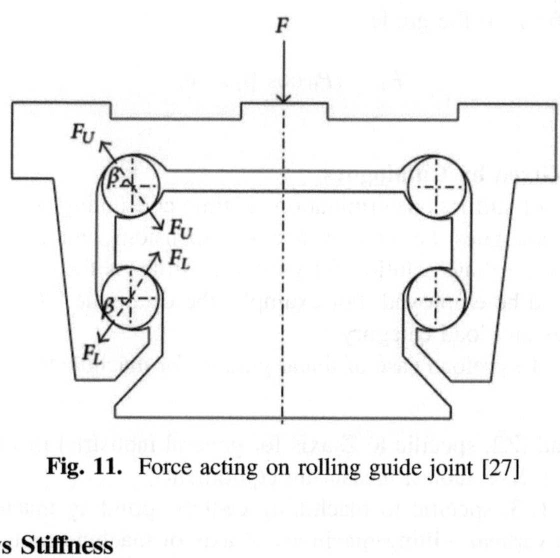

Fig. 11. Force acting on rolling guide joint [27]

3.4 Guideways Stiffness

3.4.1 Linear Rolling Guide Model

The calculation of rigidity of linear guides with intermediate elements is the same as that of ball nut or bearings based on Hertz's theory. The expression of the deformation of one ball given by [27] on the basis of [32] has the same form as (12). $\sum \rho$ is the synthetic curvature at contact point; namely, $\sum \rho = \sum_{i=1}^{2} \sum_{j=1}^{2} (1/R_{ij})$, where R_{ij} ($i = 1, 2; j = 1, 2$) are principal curvatures of the object $\underline{V_i}$.

The rolling guide block-balls-rail is shown in Fig. 11. There are four balls in joint, total number of one row balls being ‡, and contact angle ball-groove is β. The force F generates a normal reactions F_U on the upper row and F_L on the lower row. The equilibrium equation on normal direction is:

$$F + xF_L \sin \beta/2 = zF_U \sin \beta/2. \tag{31}$$

The relation between the preload force and the force distributed on the ball is:

$$F_0 = \sqrt{2}zP_0 \sin \beta. \tag{32}$$

The upper displacement δ_U and the lower one δ_L define the displacement on normal direction δ_n:

$$\delta_n = (\delta_U - \delta_0) \sin \beta = (\delta_0 - \delta_L). \tag{33}$$

where initial deformation δ_0 is given by (12).

The normal stiffness of the guide unit is

$$K_n = F.\delta_n. \tag{34}$$

The lateral displacement without additional force is $\delta_{t^*} = (\delta_U - \delta_0) \cos \beta$ leading to the tangential stiffness of the guide

$$K_t = zP_0 \cos \beta/2 \cdot \delta_t. \tag{35}$$

3.4.2 Stiffness Given by Catalogues

Mathematical way of stiffness determinations is time consuming. It is not independent of the product catalogues because it needs dimensions, numbers and materials. Therefore, for an immediate solution for guideway stiffness for modeling feed drives, the catalogues could be employed. For example, the catalogue [10] gives information for guidways types and load category.

With regard to the preload case of linear guides, for machine tools three classes are of interest:

- medium preload (Z2, specific to Z axis for general industrial machines, EDM, NC lathes, precision X-Y tables, measuring equipment);
- heavy preload (Z3, specific to machining centers, grinding machines, NC lathes, horizontal and vertical milling machines, Z axis of machine tools);
- super heavy preload (Z4, specific to heavy cutting machines).

Some recommendation for stiffness values to be used in modeling applications are given in Figs. 12 and 13 considering different dimensions of the rail width in the range [15, 20, 25, 30, 35, 45, 55, 65 mm] based on catalog [10].

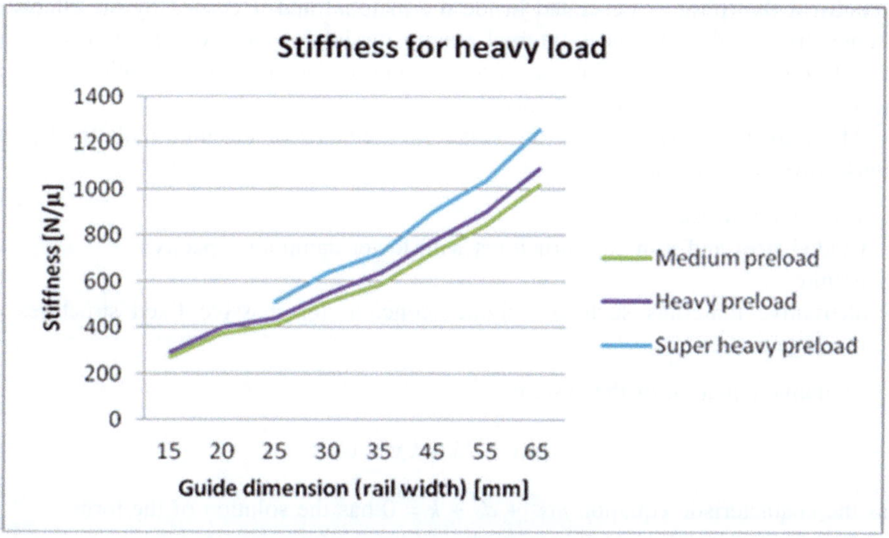

Fig. 12. Linear guide stiffness recommendations for heavy load

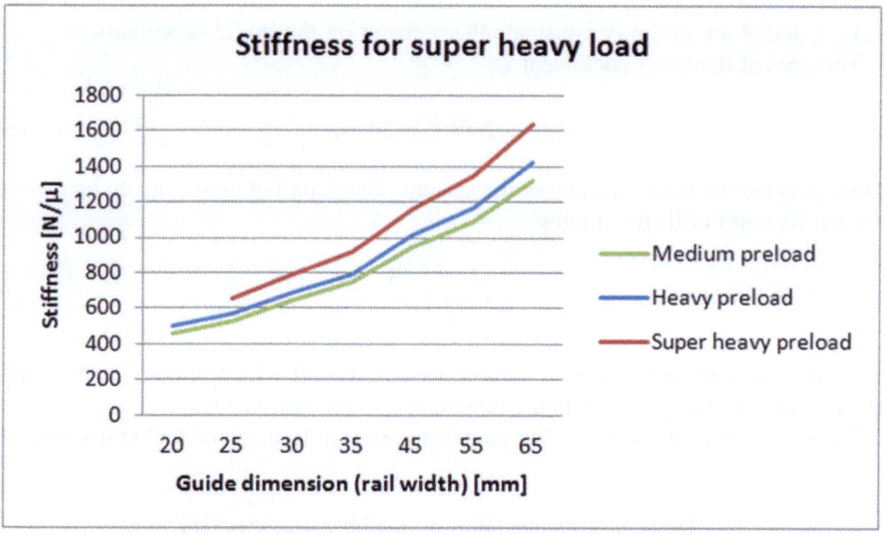

Fig. 13. Linear guide stiffness recommendations for super heavy load

4 Damping Modeling

There are three important sources of damping in mechanical systems, namely:

- structural damping inside the material;
- damping in the joints;
- damping at certain points in the structure that is not well known.

Structural damping is generated inside the material and is caused by the effects of internal friction that produces internal energy dissipation with deformation of components during vibration. Material is a determining factor in this situation and will greatly influence the behavior of machine tools.

The most used materials for the structural elements of machine tools (beds, columns, traverses, etc.) are:

- steel and cast iron;
- welded steel and iron cast structures with lower damping capacity;
- granite;
- alternative materials such as polymer concrete for massive fixed structures of machine tools.

Dynamic equation of the system

$$m\ddot{x} + d\dot{x} + kx = 0. \tag{36}$$

has the characteristic equation $ms^2 + ds + k = 0$ has the solution of the form:

$$x = e^{t\left(\frac{-dt}{2m}\right)}\left[Ae^{t\left[\left(\frac{d}{2m}\right)^2 - \frac{k}{m}\right]^{1/2}} + Be^{-t\left[\left(\frac{d}{2m}\right)^2 - \frac{k}{m}\right]^{1/2}}\right], \tag{37}$$

where A and B are arbitrary constants that depend on the initial conditions.

The critical damping coefficient is:

$$d_c = 2\sqrt{mk} = 2m\omega_0. \tag{38}$$

where ω_0 is the natural frequency of the system. The damping ratio is given by the ratio between real and critical damping:

$$\zeta = \frac{d}{d_c} = \frac{d}{2\sqrt{mk}} = \frac{\Lambda}{2\pi} = \frac{\delta}{\omega} = \frac{\eta}{2}, \tag{39}$$

where: d – damping coefficient; m – mass; k – stiffness; Λ – logarithmic decrement; δ – depreciation constant; ω – angular frequency; η – the loss factor.

Table 5 shows values of the damping ratio relevant to machine tool structures [16].

Table 5. Damping ratios for machine structures [16]

Material	Damping ratios
Steel	0.001–0.002
Cast iron	0.002–0.004
Polymer concrete	0.02–0.0
Welded structure	0.004–0.08
Bed	0.04–0.08

Joint damping of the joints results from the energy dissipation that takes place in the machine parts in contact with the moving parts, such as linear guides, bearings, ball screw nuts, whose operation is based on rolling elements (balls or rollers). These sources are multiple and change the way of approaching damping within a complex machine-tool system.

Table 6 shows values for bearing used for rotation axes and main shafts indicated by the paper [20], which shows a retrospective of the damping parameters for different components of machine tools. Regarding the linear guides, the paper [9] highlights the influence of the experimental results on the identification of the damping parameters (Table 7).

The friction moment in bearing has two distinct components, one involving the viscosity of the lubricant and the speed of rotation, and the second one involving the effect of the loading in the bearing (Table 8):

$$Mr = 10^{-7} f_0 (v \cdot n)^{2/3} d_m^3 + f_1 P_1 d_m. \tag{40}$$

The relation is used for $v \cdot n \geq 2000$, where: Mr is the friction torque [Nmm]; v – kinematic viscosity at working temperature [mm^2/s]; n – speed [min^{-1}]; P_1 – equivalent load on bearing [N]; d_m – average diameter $(d + D)/2$ [mm] with inner diameter d and outer diameter D; f_0 and f_1 – lubrication and time dependent coefficients [–] in reference with [DIN ISO 15312].

Table 6. Damping values for bearings [20]

Component	Damping ratios
Bearings (inner diameter 55 mm)	2 000–9 000 Ns/m
Bearings (inner diameter 90 mm)	5 500–12 000 Ns/m
Bearings (inner diameter 120 mm)	50 000–550 000 Ns/m
Bearings (inner diameter 160 mm)	200 000–1 000 000 Ns/m

Table 7. Damping values for linear guides [9]

Component	Damping value
Ball linear guideway (rail size 35 mm)	4 000–10 000 Ns/m
Roller linear guideway (rail size 35 mm)	3 500–14 500 Ns/m
Ball linear guideway (rail size 45 mm)	4 000–14 000 Ns/m
Roller linear guideway (rail size 45 mm)	6 500–10 500 Ns/m

It is noted that the first term depends on the speed that for feed drives is variable and therefore a specific value for modeling could not be chosen, the process becoming one complex for machine tools. The speed is variable from low (even zero) values, where strong nonlinear phenomena such as stick-slip and hysteresis phenomena may occur, up to values close to the maximum.

Complexity may also increase if loading on bearings or linear guides with rolling elements is considered variable. This can be caused by the continuous change in the position of the center centers, inertial loads produced by the acceleration of the axes, and additional stresses introduced by the forces from the cutting process. Taking them into consideration can cause the friction caused by the instantaneous load not to be determined.

Even if the behavior is speed dependent, it is relatively linear for speeds in the range of 5 000–40 000 mm/min, low speed zone remaining highly complex (Table 9).

Table 8. Friction coefficients for bearing [23]

Component	f_{0r}	f_{1r}
Deep groove ball bearing	1.7−2.3	0.00010−0.00020
Self-aligning ball bearing	2.4−4	0.00008
Angular contact ball bearing	2−7	0.00025−0.00035
Four point bearing	2−3	0.00037
Cylindrical roller bearing with cage	2−4	0.00020−0.00040
Cylindrical roller bearings full complement	5−12	0.00055
Needle-roller bearing	5−10	0.00050
Self-aligning roller bearing	4.5−6.5	0.00017−0.00036
Tapered roller bearing	3−4.5	0.00040
Axial cylinder roller bearing	3−4	0.00150
Axial self-aligning roller bearing	2.5−3.3	0.00023−0.00033

Table 9. Friction coefficients at different force levels for linear guides [16]

Component	Friction force			
Block load (in addition to pretension)	0 kN	10 kN	20 kN	30 kN
v = 5 000 mm/min				
Ball linear guideway (35 mm)	15−20 N	30−35 N	50−55 N	95−100 N
Roller linear guideway (35 mm)	30−35 N	40−45 N	45−50 N	55−60 N
Ball linear guideway (45 mm)	25−30 N	40−45 N	65−70 N	100−105 N
Roller linear guideway (45 mm)	65−70 N	65−70 N	75−80 N	85−90 N
v = 40 000 mm/min				
Ball linear guideway (35 mm)	30−45 N	40−55 N	60−75 N	100−115 N
Roller linear guideway (35 mm)	55−70 N	55−70 N	65−80 N	70−85 N
Ball linear guideway (45 mm)	45−60 N	55−70 N	70−85 N	105−120 N
Roller linear guideway (45 mm)	125−140 N	125−140	130−145 N	135−150 N
Guideway wipers	10−40 N			

As regards other damping sources, they manifest themselves in certain points of the structure of the machine where the energy dissipation occurs, points not so well known and therefore they are included in material damping or even ignored.

5 Determination of Dynamic Parameters of Feed Drives

5.1 Static Stiffness Measurement

The classic method of determining static stiffness is not standardized being just recommended. In most cases, it is minimalist, being made in one point (position of the mobile element) by loading the mobile element with a certain force in the positive and then negative direction of the axes. Practically, it provides only one single-point rigidity value in the XOY plane.

There are large significant differences between the static stiffness measured by the classical method and that obtained by other methods. One method is proposed by Tomáš Stejskal et al. [24] in which the measurements are made immediate after machine stop, in a certain number of points.

In the first method (Fig. 14a), the measurements are made in a number of points (up to 9) evenly displaced along studied axis. For each point, the measurement of displacement is made for a load F which is gradually increased in a range F_1–F_8. Then the measurements are done for a load decreased belonging to the same range.

The second procedure considers measurements from the first position to the last one and back to position 1. The first cycle consists of 1-9-1 displacement without load ($F = 0$ N) and a repeated cycle with a load $F_1 \neq 0$. After the first cycle, the second one is considered with increased load force $F_2 > F_1$.

In the third approach (Fig. 14b), several measurements are made for a load F in each point and also in proximity of the point, moving the slide alternatively in positive and negative direction with a small distance ($l = 5$ mm) (Fig. 14c).

The cycle continues repeating the measurements for an increased force form the range. Then, the slide is moved to the next point resuming the measurement cycle.

The loading force could be considered in the range [20; 40; 60; 80; 100; 120; 140; 160 N].

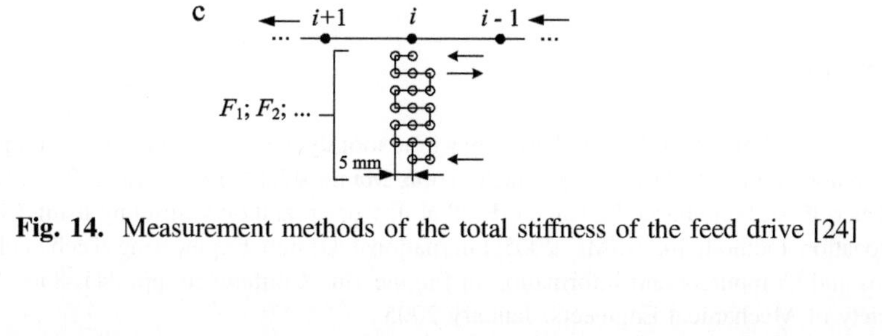

Fig. 14. Measurement methods of the total stiffness of the feed drive [24]

5.2 Dynamic Characteristic Identification

Modeling is a difficult task, especially when trying to consider both linear and rotary motions of the system with dynamic characteristics (stiffness, damping and friction). In DBS, the mathematical processing is complex, the model being very sensitive when introducing the variable position as a time dependent one $x(t)$, $y(t)$, and $z(t)$.

Some applications require the system modeling as a transfer function. The order n plays a very important role due to the fact that reduced orders (n = 1; 2) lead to a model accuracy of 70–80%. A greater precision is obtained for increased orders. For instance, the dynamic stiffness of the feed drive in the work [18] has the form:

$$\mathrm{Sd}(s) = \frac{F(s)}{Xo(s)} = \frac{a_5 s^5 + a_4 s^4 + a_3 s^3 + a_2 s^2 + a_1 s + a_0}{b_3 s^3 + b_2 s^2 + b_1 s + b_0}, \tag{41}$$

in which all coefficients are defined by relations depending on system parameters.

This model including dynamic stiffness is very complex and the definition of all parameters becomes a difficult challenge. This is achieved on the bases of experimental measurements. Based on them, some algorithms of identification available in specific programs are used.

6 Conclusions

The paper has emerged from the need to gather together some concepts related to the modeling of the kinematic chains of the machine tools, which influence decisively the precision of machining.

Methods of determining of dynamic characteristics by mathematical calculations, through measurements, or based on product catalog information, are presented. Values or ranges of the values specific to some components and applications as well as some values specific to some research are presented.

In addition to the dynamic parameters of the feed drive components, some theoretical and experimental methods of setting of stiffness and damping of the feed drive assembly are presented.

Particular aspects related to the dynamic characteristics of feed drives may also be revealed. Thus, the stiffness of the preloaded ball nut is revealed not to vary with loading. It results that, in order to increase the rigidity of the assembly, the rigidity of the bearings can be considered by increasing their number in the arrangement. Also, one can choose a larger diameter of the ball screw.

References

1. Ahmadi, K., Ahmadian, H.: Modelling machine tool dynamics using a distributed parameter tool–holder joint interface. Int. J. Mach. Tools Manuf **47**(12–13), 1916–1928 (2007)
2. Butcher, E.A., Nindujarla, P., Bueler, E.: Stability of up-and down-milling using Chebyshev collocation method. In: ASME 2005 International Design Engineering Technical Conferences and Computers and Information in Engineering Conference, pp. 841–850. American Society of Mechanical Engineers, January 2005

3. Constantin, G., Ghionea, A.: Feed kinematic chain of the bridge in Gantry machine-tools. Archiwum Technologii Maszyn i Automatyzacji **28**(2), 75–84 (2008)
4. Constantin, G., Predincea, N.: Aspectes regarding optimal design of machine tool feed drives. Proc. Manuf. Syst. **10**(4), 171 (2015)
5. Copani, G., Molinari Tosatti, L., Lay, G., Schroeter, M., Bueno, R.: New business models diffusion and trends in European machine tool industry. In: Proceedings of 40th CIRP International Manufacturing Systems Seminar, May 2007
6. Erkorkmaz, K., Yeung, C.H., Altintas, Y.: Virtual CNC system. Part II. High speed contouring application. Int. J. Mach. Tools Manuf **46**(10), 1124–1138 (2006)
7. Fleischer, J., Broos, A.: Parameteroptimierung bei Werkzeugmaschinen–Anwendungsmöglichkeiten und Potentiale. Weimarer Optimierungs-und Stochastiktage, 1 (2004)
8. Fredin, J.: Modelling, simulation and optimisation of a machine tool (Doctoral dissertation, Blekinge Institute of Technology) (2009)
9. Groche, P., Hofmann, T.: Einfluss des dynamischen Übertragungsverhaltens von Stößelführungen auf die Arbeitsgenauigkeit von Umformpressen. EFB Hannover (2005)
10. HIWIN Linear Guideway. http://ruchservomotor.com/Pdf_english/H_L_G_(Low).pdf
11. HIWIN Ballscrews. Technical Information. https://www.hiwin.com/pdf/ballscrews.pdf
12. INA-FAG, Schaeffler Technologies, Super Precison Bearings (2006). https://www.schaeffler.com/remotemedien/media/_shared_media/08_media_library/01_publications/schaeffler_2/catalogue_1/downloads_6/sp1_de_en.pdf
13. Knaapen, R.J.W., Kodde, L., de Kraker, A.: Experimental Determination of Rolling Element Bearing Stiffness. Technische Universiteit Eindhoven, Eindhoven (1997)
14. Krämer, E.: Dynamics of Rotors and Foundation. Springer, Berlin (1993)
15. Lee, W.J., Kim, S.I.: Joint stiffness identification of an ultra-precision machine for machining large-surface micro-features. Int. J. Precis. Eng. Manuf. **10**(5), 115–121 (2009)
16. Maglie, P.: Parallelization of Design and Simulation, vol. 683. ETH Zurich, Zürich (2012)
17. Palgrem, A.: Les Roulments, Description, Theorie, Applications (Bearings, Description, Theory, Applications). SKF, Paris (1967)
18. Pandilov, Z., Milecki, A., Nowak, A., Górski, F., Grajewski, D., Ciglar, D., Klaić, M., Mulc, T.: Virtual modelling and simulation of a CNC machine feed drive system. Trans. FAMENA **39**(4), 37–54 (2015)
19. Riba, R.C., Pérez, R.R., Ahuett, G.H., Jorge, L.S.A., Domínguez, M.D., Molina, G.A.: Metrics for evaluating design of reconfigurable machine tools. In: International Conference on Cooperative Design, Visualization and Engineering, pp. 234–241. Springer, Berlin, Heidelberg, September 2006
20. Schlecht, B.: Maschinenelemente: Getriebe-Verzahnungen-Legerungen. Pearson Studium, ein Imprint von Pearson Education (2010)
21. Servomech, Ball screws and nuts, Catalogue (2010). http://www.servomech.it/Pdf/prodotti/SERVOMECH-ball-screws-and-nuts-catalogue.pdf
22. Slocum, A.: Precision Machine Design. Society of Manufacturing Engineers, Dearborn (1994)
23. Steinhilper, W., Sauer, B., Feldhusen, J.: Konstruktionselemente des Maschinenbaus 1: Grundlagen der Berechnung und Gestaltung von Maschinenelementen. Springer, Heidelberg (2008)
24. Stejskal, T., Svetlík, J., Dovica, M., Demeč, P., Kráľ, J.: Measurement of static stiffness after motion on a three-axis CNC milling table. Appl. Sci. **8**(1), 15 (2017)
25. Stratulat, F., Ionescu, F., Constantin, G.: Dynamic evaluation of a linear axis in milling using modelling and simulation environment. In: Proceedings of the International Conference on Manufacturing Systems, ICMaS, pp. 99–104 (2007)

26. Thurneysen, M., Demaurex, G.: Modalanalyse von komplexen Maschinen, praktische Ueberpruefung. In: Symposium Simulation von Werkzeugmaschinen. IWF/inspire, ETHZ (2008)
27. Wang, D., Lu, Y., Zhang, T., Wang, K., Rinoshika, A.: Effect of stiffness of rolling joints on the dynamic characteristic of ball screw feed systems in a milling machine. Shock Vib. **2015**, 11 (2015)
28. Xian-chun, S., Jian, S., Ming-yuan, C., Yan-feng, L.: Research on axial stiffness of the double-nut ball screw mechanism. In: Proceedings of the 1st International Conference on Mechanical Engineering and Material Science. Atlantis Press, December 2012
29. Zaeh, M., Siedl, D.: A new method for simulation of machining performance by integrating finite element and multi-body simulation for machine tools. CIRP Ann. Manuf. Technol. **56**(1), 383–386 (2007)
30. Zaghbani, I., Songmene, V.: Estimation of machine-tool dynamic parameters during machining operation through operational modal analysis. Int. J. Mach. Tools Manuf **49**(12–13), 947–957 (2009)
31. Zhang, Y.M., Xie, Z.K., Liu, Y.X.: Modeling and calculation of dynamic performances of NC machine tool considering linear rolling guideway. Appl. Mech. Mater. **16**, 510–514 (2009)
32. Zhu, J., Zhang, T., Li, X.: Dynamic characteristic analysis of ball screw feed system based on stiffness characteristic of mechanical joints. J. Mech. Eng. **51**, 72–82 (2015)

Strategic Outsourcing of SMEs in the Context of Industry 4.0: Evidence from Serbia

Bojan Lalic, Tanja Todorovic, Nela Cvetkovic, Nemanja Tasic, and Ugljesa Marjanovic(✉)

University of Novi Sad, Faculty of Technical Sciences, Novi Sad, Serbia
{blalic,ttodorovic,nelacvetkovic,nemanja.tasic, umarjano}@uns.ac.rs

Abstract. The aim of this research is to examine how strategic outsourcing is, along with the use of Industry 4.0 concepts, affecting organizational performances (i.e. lead time) in SMEs. For this purpose multivariate data analysis was performed using the sample from European Manufacturing Survey obtained in Serbia. The empirical results indicate that manufacturing SMEs that have implemented certain Industry 4.0 concepts should outsource assembly process and service offers to decrease lead time. In addition, SMEs in manufacturing sector owning in-house R&D achieve higher performance. However, this is neglected by large number of SMEs.

Keywords: Industry 4.0 · SMEs · Firms performance · EMS

1 Introduction

Manufacturing enterprises are facing the complex environment of Industry 4.0, a new paradigm in industrial production [1]. In the dynamic environment characterized by fast-paced technological advancements, micro, small and medium manufacturing companies (SMEs) in transitional economies are facing complex challenges. Due to specific constraints inherent to the SMEs, the need for revaluation of different activities along the value-chain is of paramount importance. In the light of constant search for sources of competitive advantage, the impact of redefined of manufacturing strategies can increase organizational performance. The optimal outsourcing decision, along with the implementation of Industry 4.0 concepts, is enabling SMEs to leverage their core competencies, focusing on high value-added activities and consequently, reaching increased competitiveness.

This paper analyzes the extent to which micro, small and medium manufacturing firms in Serbia perform in-house or outsource different stages of value creation process. Furthermore, the aim of the paper is to examine whether these firms apply certain Industry 4.0 concepts and whether their usage influences the relationship between the share of in-house contribution to value creation stages and organizational performance.

J. Ni et al. (Eds.): AMP 2018, LNME, pp. 139–145, 2018.
https://doi.org/10.1007/978-3-319-89563-5_10

2 Theoretical Background

2.1 Industry 4.0

Industry 4.0 implies the establishment of intelligent products and production processes based on the combined development of ICT and manufacturing technology [2]. The complex manufacturing environment is characterized by the existence of self-optimizing systems and the digital supply chain in the cyber-physical environment, leading to the cyber physical manufacturing systems that are utilizing internet of things and cloud technologies in order to establish new generation of "smart" factories [3].

2.2 Strategic Outsourcing

Effective strategic positioning in Industry 4.0 manufacturing landscape indicates the need for reassessment of existing manufacturing strategies in order to increase overall competitiveness. In the light of fast-paced technological advancements, SMEs in transitional economies are facing complex challenges. Organizational innovations, as an important constituent of companies' competitive advantage on the market, include implementation of advanced management techniques or a new direction in firm's strategic orientation [4], implying different organizational designs and decisions regarding the insourcing/outsourcing of value-chain activities [5]. The analysis of value-creating activities along the company's value-chain allows the identification of potential sources of competitive advantage [6]. Strategic outsourcing, as a form of vertical disintegration [7], represents a situation where firms rely on intermediate markets as a means of supplementing existing capabilities deployed along a firm's value chain [8]. Improving organizational performance is considered a main reason of a company's decision to outsource, among the other aspects including focusing on the core business of a company and increased flexibility which is often lacking in large enterprises [5, 9].

Flexibility, a capability that can have a positive impact on achieving higher competitiveness level, is one of the aspects to be considered in the process of strategy revaluation [10]. In this regard, strategic outsourcing can be viewed as a basis for leveraging core competences, and especially in the context of SMEs, externalizing value-chain activities can facilitate the achievement of higher level of competitiveness. The expected outcome of strategic outsourcing is an increase in innovative capabilities [8] and consequently, reallocation of resources to additional value-creating activities [7].

2.3 Research Question

Based on literature review, the following research question was proposed in attempt to identify the effect of strategic outsourcing of SMEs and use of smart factory industry 4.0 concepts on performance of manufacturing (Fig. 1):

- RQ: How strategic outsourcing is, along with the use of Industry 4.0 concepts, affecting organizational performances (i.e. lead time) in SMEs?

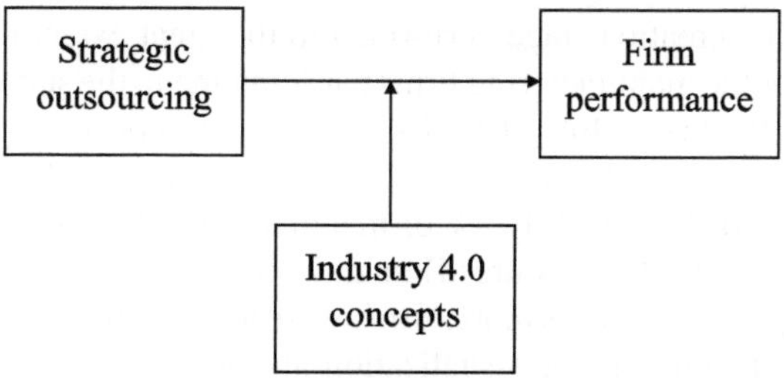

Fig. 1. Research model

3 Data and Methodology

Our analysis used the Serbian dataset from the European Manufacturing Survey (EMS), coordinated by the Fraunhofer Institute for Systems and Innovation Research - ISI [11]. EMS represents the largest European survey on manufacturing activities, development and utilization of technological and organizational innovation and associated performance effects [12].

Our research based on EMS data is from the Serbian subsample from the year 2015. The responding companies were from the manufacturing sector (NACE Rev 2 codes from 10 to 33). This study paper covers 261 observations of micro, small and medium Serbian firms of all manufacturing industries. About 5% of the firms in the sample are micro firms up to 19 employees, another 41% of the firms have between 20 and 49 employees (small firms), and 54% of the firms are medium firms with up to 250 employees. The largest industry in the sample is the manufacturer of food products (NACE 10), followed by manufacture of fabricated metal products, except machinery and equipment (NACE 25) and the manufacture of rubber and plastic products (NACE 22) [13]. Table 1 gives an overview of the sample.

Table 1. EMS database – distribution of firms by size

Firm size	n	%
Up to 19 employees	13	5.0
20 to 49 employees	107	41.0
50 to 249 employees	141	54.0

Multivariate data analysis was used to examine the relationship between share of in-house contribution to the value creation stages, Industry 4.0 concepts and organizational performance.

4 Results and Discussion

Respondents were asked what is the share of their in-house contribution in the following value creation stages: R&D, engineering/design, manufacturing/processing/recycling, assembly, packaging/distribution and service offer. Manufacturing/processing/

recycling is the value creation stage performed to the great extent in-house (85–100% in-house) by most of the manufacturing firms participating in the survey (75% of SMEs). Following, most of the firms (46% of SMEs) have high in-house contribution in service offers. On the other hand, SMEs are mostly outsourcing R&D (40% of SMEs) and engineering/design (30% of firms). Only 4% of manufacturing SMEs reported outsourcing to the great extent (85–100% outsourced) manufacturing/processing/recycling stage in the value creation process. Furthermore, firms were asked if they are applying Industry 4.0 concepts: digital solutions for visualization and software for production planning and scheduling. Figures 2 and 3 demonstrate the results for our research question "How strategic outsourcing, along with the use of Industry 4.0 concepts, are affecting organizational performances (i.e. lead time) in SMEs?". To examine the effect of different shares of in-house contribution to value creation stages and strategic outsourcing on organizational performance (i.e. lead-time), we tested the three models (1) without application of Industry 4.0 concepts, (2) with application of digital solutions and (3) with application of systems for production planning and controlling.

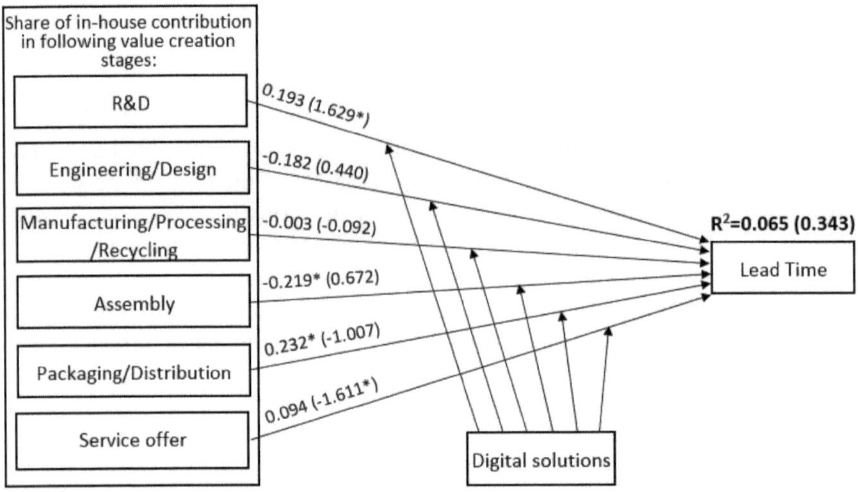

Note. Values w/o digital solutions (w/ digital solutions); *p<0.05.

Fig. 2. Linear regression model (digital solutions).

Figure 2 depicts that in the case of micro, small and medium manufacturing firms not applying Industry 4.0 concepts, higher share of in-house packaging and distribution is positively related to the lead-time decrease ($p < 0.05$). Contrary, results show that firms mostly outsourcing assembly process (85–100% outsourced) report reduction of lead-time ($p < 0.05$). The share of in-house contribution to other value creation stages or outsourcing level is not significantly related to the probability to organizational performance (i.e. lead-time) indicator. Overall, R^2 value (0.065) indicates that there is no statistical significance of the whole model for lead-time when firms are not applying Industry 4.0 concepts. However, with introduction of the variable *digital solutions* to

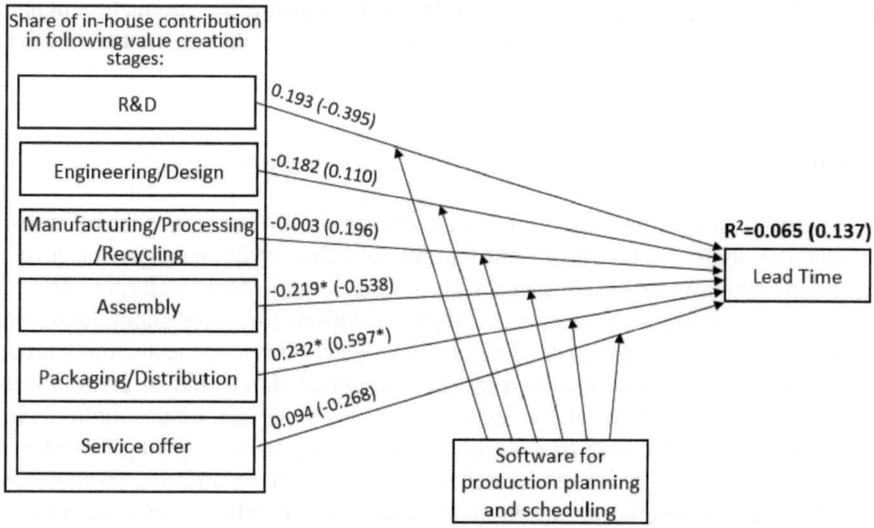

Note. Values w/o software for pp and scheduling (w/ software for pp and scheduling); *p<0.05.

Fig. 3. Linear regression model (software for production planning and scheduling).

the model, results show the significant impact. The R^2 value improves significantly from 0.065 (without application of industry 4.0 concepts) to 0.343 (with the application of digital solutions). Presented model (Fig. 2) explains about 34% of the variances in firm's lead-time. The results for the model with the application of digital solutions within manufacturing firms show that share of in-house contribution or outsourcing level of R&D and service offers are significantly related to the firm lead-time. Higher share of in-house R&D is positively related to the lead-time decrease. This shows that manufacturing firms that have internal research and development obtain reduced lead-time.

Despite this, the lowest share of SMEs in this survey are conducting R&D in-house (only 21% of SMEs), while around 40% of firms is mostly outsourcing (85–100%) and partially outsourcing (25–85%).

This could be explained by the fact that most of the SMEs, although may be well innovative [4], have limited resources and capacities for R&D work [2]. Conversely, share of in-house contribution to service offers is negatively related to the dependent variable, indicating that SMEs outsourcing service offers are more successfully accomplishing decrease of their lead-time. Following, the third research model (with application of systems for production planning and controlling) was tested (Fig. 3).

Examination of the third model (Fig. 3), with variable *software for production planning and scheduling*, demonstrates that model has lower significant effect on dependent ($R^2 = 0.137$) than the model with application of Digital solutions visualization. Nonetheless, the R^2 value still improves comparing to the R^2 value of the model without the application of software for production planning and scheduling (0.065), again implying that Industry 4.0 concepts have statistically significant impact on relationship between the share of in-house contribution to the different value creation stages on lead-time.

This model suggests that in the case if SMEs using software for production planning and scheduling, the high share of in-house packaging and distribution improve the firm's lead-time.

5 Conclusion

This research paper examined the application of Industry 4.0 concept on the relationship between the share of in-house contribution to value creation stages or level of outsourcing and organizational performance (i.e. lead time). Manufacturing SMEs that are not using Industry 4.0 concepts (i.e. digital solutions for visualization or software for production planning and controlling) should mostly outsource the assembly process to improve lead time. Additionally, regardless of application of Industry 4.0 in manufacturing firms, outsourcing the assembly process and service offers improves their organizational performance. In case of manufacturing firms applying Industry 4.0 concepts, high level of in-house research and development significantly promotes lead time. Despite this, results reveal that high percentage of SMEs is outsourcing R&D, confirming previous statements that firms are outsourcing R&D activities due to the lack of human, financial and technical resources [12, 14]. Furthermore, positive effect of in-house packaging and distribution on the reduction of lead time is confirmed.

This research considered only the Serbian subsample. Future studies should widen the sample and include different economies, developed and developing. Nonetheless, this paper includes only certain Industry 4.0 concepts, which implies the need to test this model also taking into account other relevant elements of Industry 4.0 and examine the relationship between strategic outsourcing and performance in different contexts.

References

1. Lalic, B., Majstorovic, V., Marjanovic, U.: The effect of industry 4.0 concepts and e-learning on manufacturing firm performance: evidence from transitional economy. In: Advances in Production Management Systems. The Path to Intelligent, Collaborative and Sustainable Manufacturing, vol. 514, pp. 298–305 (2017)
2. Majstorovic, V., Stojadinovic, S., Zivkovic, S., Djurdjanovic, D., Jakovljevic, Z., Gligorijevic, N.: Cyber-physical manufacturing metrology model (CPM3) for sculptured surfaces - turbine blade application. Procedia CIRP **63**, 658–663 (2017)
3. Majstorovic, V., Stojadinovic, S., Jakovljevic, Z., Zivkovic, S.: Cyber-physical manufacturing metrology model (CPM 3) – big data analytics issue. In: Procedia CIRP (2018)
4. Lalic, B., Anisic, Z., Medic, N., Tasic, N., Marjanovic, U.: The impact of organizational innovation concepts on new products and related services. In: International Conference on Production Research - ICPR 2017, pp. 110–115 (2017)
5. Windrum, P., Reinstaller, A., Bull, C.: The outsourcing productivity paradox: total outsourcing, organisational innovation, and long run productivity growth. J. Evol. Econ. **19**(2), 197–229 (2009)
6. Noke, H., Hughes, M.: Climbing the value chain. Int. J. Oper. Prod. Manag. **30**(2), 132–154 (2010)

7. Sousa, R., Voss, C.A.: Operational implications of manufacturing outsourcing for subcontractor plants: an empirical investigation. Int. J. Oper. Prod. Manag. **27**(9), 974–997 (2007)
8. Quinn, J.B.: Strategic outsourcing: leveraging knowledge capabilities. Sloan Manage. Rev. **40**(4), 9–21 (1999)
9. Di Gregorio, D., Musteen, M., Thomas, D.E.: Offshore outsourcing as a source of international competitiveness for SMEs. J. Int. Bus. Stud. **40**(6), 969–988 (2009)
10. Lalic, B., Medic, N., Delic, M., Tasic, N., Marjanovic, U.: Open innovation in developing regions: an empirical analysis across manufacturing companies. Int. J. Ind. Eng. Manag. **8**(3), 111–120 (2017)
11. Fraunhofer Institute, European Manufacturing Survey (2017). http://www.isi.fraunhofer.de/isi-en/i/projekte/fems.php. Accessed 02 Feb 2017
12. Lalić, B., Tasić, N., Marjanović, U., Delić, M., Cvetković, N.: Inter-organizational collaboration for innovation in manufacturing firms. In: Annals of DAAAM and Proceedings of the International DAAAM Symposium, vol. 27, no. 1, pp. 721–729 (2016)
13. Marjanović, U., Lalić, B., Delić, M., Tasić, N.: Industry 4.0: evidence from transitional economy. Int. J. Glob. Bus. **10**(1), 26–36 (2017)
14. Konsti-Laakso, S., Pihkala, T., Kraus, S.: Facilitating SME innovation capability through business networking. Creat. Innov. Manag. **21**(1), 93–105 (2012)

Fundamental Requirements for CAPP Software Design Focusing on Industry 4.0 Specific Features

Peter Monka[1], Katarina Monkova[1(✉)], Milan Edl[2], Helena Zidkova[2], and Vladimir Duchek[2]

[1] Faculty of Manufacturing Technologies with the seat in Presov, Technical University of Kosice, Kosice, Slovakia
{peter.pavol.monka,katarina.monkova}@tuke.sk

[2] Faculty of Mechanical Engineering, University of West Bohemia, Pilsen, Czech Republic
edl@kpv.zcu.cz, {zidkova,duchekv}@kto.zcu.cz

Abstract. The article deals with the basic requirements imposed on software applications for Computer Aided Process Planning (CAPP) with respect to specific features of Industry 4.0. There are described two basic steps in the article that preceded the design of a new information system created by authors in the way to be suitable for enterprises that operate in the period of industrial revolution "Industry 4.0". The first step was in analysis of European and Slovak industrial market, where the composition of enterprises from several points of view were investigated. The second step was focused on specification of basic requirements that have to be kept by information system applied into plant within Industry 4.0. On the basis of realized analyses and on the multivariant theory, it was designed new information system that has been successfully applied into real company conditions.

Keywords: Software requirements · Market analysis · Information system
Process plans · Multivariant approach

1 Introduction

The objective of product design is to raise or assure the technical value of product not only from the view of the design, but also due to the elaboration of drawings, process plan preparation and data processing. The term "product design" describes the comprehensive application of methods and work aids during design process of a product. Every step, operation and manufacturing equipment are written in information system of the plant. The most important part of information system (IS) is a process plan for part production. The objective of manufacturing design is to systemize preparation manufacturing or assembly processes and to build up the planning data for rational realization of the different partial assignments within the framework of the concept. The aim of work study and wage structure is to optimize the working conditions at the work place and to increase output in accordance with scientific methods and processes [1, 2].

J. Ni et al. (Eds.): AMP 2018, LNME, pp. 146–155, 2018.
https://doi.org/10.1007/978-3-319-89563-5_11

In time decision making about Computer Aided Process Planning (CAPP), manufacturing companies can select information system that is between two border zones: at one side is a complex system and at opposite side is a specific software, which represents independent solution for each enterprise action. First one is difficult to get for small enterprises due to an expensive price. Second of them generally dispose only by flat possibility of interconnection to related information systems. Inside area of these diametric positions lays requirements of production companies majority [3].

The goal of authors has been to design new manufacturing information system that would include advantages of both extreme: integration of IS in specialized systems environs (CAD/CAM …) and the possibility to share data by every ring in supplier chain. One of the first steps at software application design was detection of typical demands at European and Slovak industrial markets from the view of representative enterpriser composition. After this specification, it will be possible to process key requests for manufacturing information system satisfied the requirements of Industry 4.0. Both these steps are presented in the article.

2 European and Slovak Industrial Markets Analysis

In terms of reaching the Strategy 2020 targets, most countries in the EU continue making modest improvements closing up on the 2008 pre-crisis indicator levels. Thus, they are falling behind on the planned 2020 targets, which are planned to be achieved in the short time span of the coming next years.

The statistical studies (Figs. 1, 2 and 3) pointed to the fact that European micro companies constitute a substantial part of the European market. They form the basis – 92% (17.82 million business units) of the overall number of companies and employ 39% of the employees at EU market. Small and medium size companies together comprise 7.5% of the overall number and employ 30.3% of the employees.

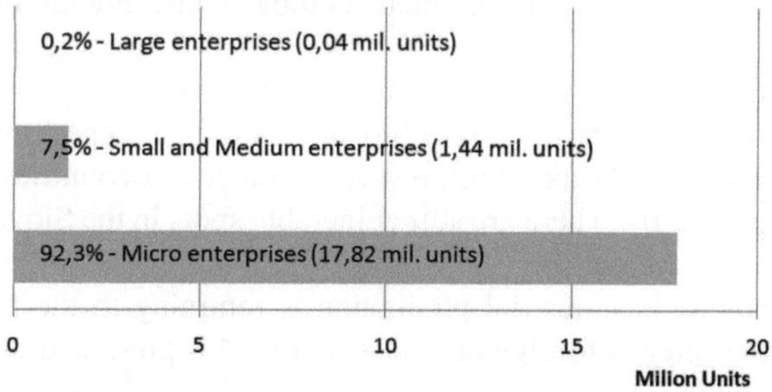

Fig. 1. Numbers of enterprise units at European market in conformity with largeness of enterprises [4]

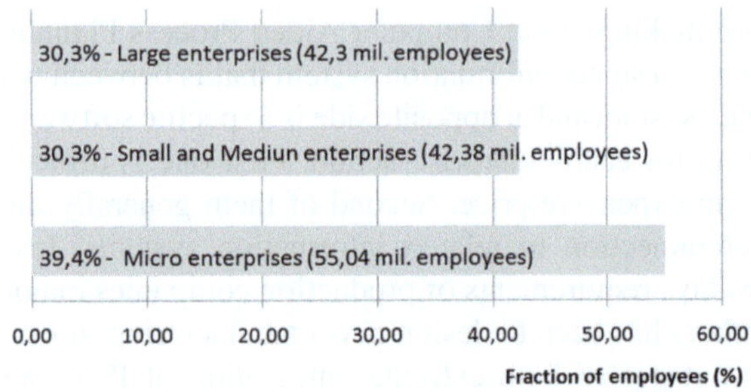

Fig. 2. Fraction of employees in European enterprises [4]

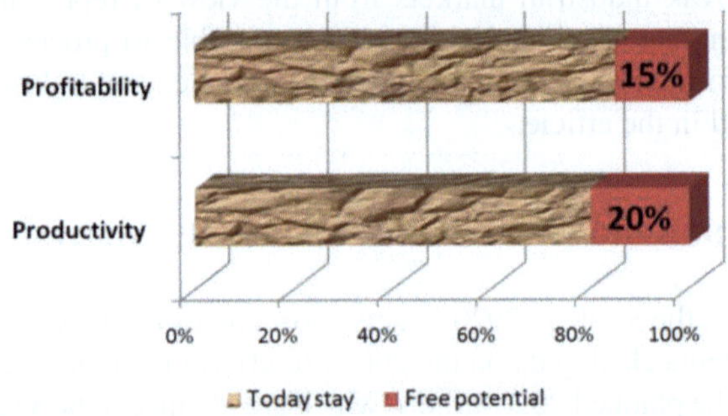

Fig. 3. Potential of European microenterprises [4]

The rest - 0.2% production unites and 30.2% employees - is covered by large companies [4, 5]. From this same study it can be seen that micro companies have disposal of free potential of 20% of the productivity and 15% profitability. These are very important characteristics which describe distinct ability of production dynamic growth and the possibility of effective evaluation of micro company instruments basically "over a night".

The positive development in Slovakia follows the overall improvement of the EU labour market. However, in the context of Slovak labour market performance within the EU area, the position of Slovakia with respect to other EU countries might not have improved that significantly. There are still vulnerable spots in the Slovak labour market that need continued attention [5].

At present, interest in industrial production is returning to the Slovak market in connection with the entry of foreign investors. In Fig. 4 is presented SWOT analysis of production in Slovakia.

Strengths	Weaknesses
• strategic location of Slovakia, • availability of international airports • (BA, KE, PP, ZA, Vienna, Krakow, Ostrava), • stable macroeconomic environment, • strong history in engineering industry, • favourable labour costs and high productivity.	• lack of funding for the educational system, • lack of financial involvement of the private sector, • weaker cooperation and use of research & development potential of local universities.
Opportunities	**Threats**
• relocation of traffic flows on environmental transport modes (inland, maritime waterways), • relocation of global automotive and electronics industries towards Central Europe - new supply opportunities, • strengthening cooperation between manufacturers, suppliers, science and research, • opportunities for the creation of new research institution / organizations, • the benefits of research activities in the EU programs.	• unbalanced external dependence of the industry, • concentration of the industry in the western part of Slovakia, • shortage of skilled labour, • congestion of road infrastructure.

Fig. 4. SWOT analysis of production in Slovakia

From the analysis also follows that Slovakia is specialised in low value-added manufacturing goods - the share of domestic value-added in exports is low, because of mainly assembly or components manufacturing. Slovakia's score on the Services Trade Restrictiveness Index (STRI) is better than the OECD average in 14 out of 17 service sectors, pointing to the relative liberal regulatory regime applicable to foreign entry in services. One of the pillars of Slovakia's economy, bringing investment, employment and innovations is the automotive industry as it is shown in Fig. 5, where the position of car producers in Slovakia is visible [6, 7].

Fig. 5. Automotive industry in Slovakia

Slovakia is already home to three car-makers, Volkswagen Slovakia, PSA Peugeot Citroën Slovakia and Kia Motors Slovakia. Over the past 20 years, car production increased from 2,952 vehicles in 1993 to more than 970,000 in 2014. The latter data, equalling 178 cars per 1,000 citizens, has made Slovakia the biggest per capita producer in the world. In the Fig. 6 it is presented preview that expresses the number of vehicles yearly produced per thousand inhabitants in EU countries [8].

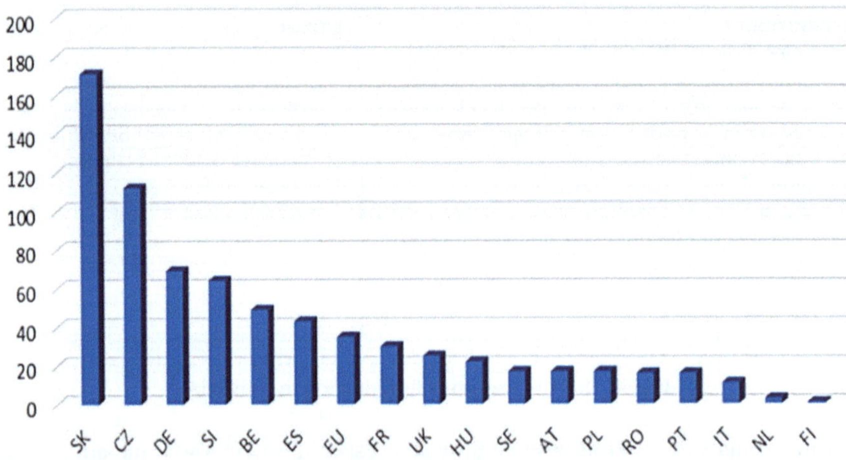

Fig. 6. Automotive industry – the number of vehicles yearly produced per thousand inhabitants (EU countries) [8]

As it was said, the philosophy of Computer Aided Process Planning (CAPP) represents activities leading to a creation of production documentation and the details of equipment for the production process.

The cost structure analyses for small and medium series production indicate a significant ratio of CAPP in the production costs composition. From the point of view of these analyses it is very important to pay a considerable attention to the CAPP area which can, as a result, influence the output costs of a product and its quality in a great measure [9–11].

On the basis of the objectives and structural concept specified above, the requirements for information system design related to the production in the field of mechanical engineering was defined by authors.

3 Requirements for Process Plans Design Within Industry 4.0

Based on the analysis described in previous chapter it can be said that they are the micro companies that constitute the significant part of the enterprise subjects in European and Slovak markets. The specifications of this type of enterprise units imply diametrically different demands on information systems from the normal setting of IS appropriate for large and medium size companies. This part of companies requests mainly very simple implementation, possibility of modular concept for covering all necessary areas [12], reliable and secure data formats and structures, possibility of flexible data sharing among partners in supplier chain, possibility of a trouble-free extension to the needed modules, securing the possibility of a relatively fast transfer to IS higher levels when necessary and last but not least reasonable price.

Next demands on information system can be summarized into the following points:

- The system has to be able to work with the possibility of a user view on the production process from several angles.
- Enterprise subject should be limited when launching new products to the production process as little as possible.
- It should be applicable for a wide range of business.
- It should be modular.
- The simplified structure production system design divided into parts according to can be to advantageously applied for designing of information system composition.

Layout of this simplified structure is shown in Fig. 7.

Production system design					
1. Product design		**2. Manufacturing design**		**3. Studies of labour and production costs**	
1.1	Design procedures	2.1	Technological structure planning	3.1	Work analyses
1.2	Design methods	2.2	Process planning	3.2	Work measurement
1.3	Value analysis	2.3	NC programming	3.3	Wage schemes

Fig. 7. Simplified structure of production system design

The system designed in this way can be built for the wide product definitions and from the view of three technological approaches, as are: Individual, Type and Group Approaches. These approaches have to be accessible in information system, because every producer chooses a process plan of the same product according to the specific conditions of the plant.

One of the next necessities for selection of suitable CAPP system is its flexibility and the ability to quickly adapt to user requirements. For example, when purchasing a comprehensive program package from a general contractor, the user is usually not able to make adjustments to the structure of the system. If necessary, such adjustments require the supplier to contact again, which is usually time and cost-intensive. Appropriate software to support the creation of technological documentation can be obtained from these points of view:

(a) complete delivery of the software application by the general contractor,
(b) purchase of a software application from the general contractor and subsequent implementation by the user,
(c) purchasing all software application components from different vendors,
(d) development of the software application and purchase of other components from different suppliers, but self-integration,
(e) development of software application by the company in itself power.

With the production process as a homogenous unit, which includes technological and labour processes organised via various possible parallel phases, deals so called theory of multi-variant process planning. Within this theory, the final product could be processed in optimized way from the view of conditions while fulfilling all the demands required by a consumer. The main objectives of this theory are [13]:

- Creation of the unified definition environment for all the factors immediately influencing the result of production process.
- Flexible interface which enables bidirectional exchange of the required information with all surrounding systems.

Via the unified definition of environment, the philosophical and conceptual unity is secured within the whole issue falling into the formation area of multi-variable process planning, a distinct classification product constituent and the laws of production sequence for operation projection which allow the use of several possibilities designed by information system based on this theory.

Keeping the principles of this theory, it is also possible to create combination possibilities of various techniques used at individual process planning, which are based on the strategy aimed at achieving the specific goal of the production unit (Fig. 8).

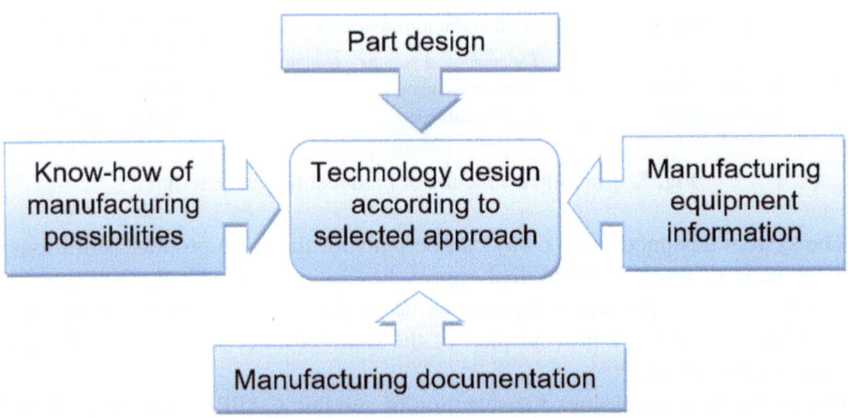

Fig. 8. Simplified scheme of multi-variant process planning

The analysis has shown that flexible interface of the designed system has to enable an effective work in the production environment so to have all individual relevant systems, that create heterogeneous information system (CAD/CAM application, wage records, accounting, material management,…), the inter-connection secured via the suitable interfaces. This is due to preventing the errors caused by data redundancy, or by human factor, but also due to reducing response time to minimum.

Really tested Multi-Variant Process Planning system, designed by authors, was originated so to be possible to interconnect not only wide variety of CAD/CAM systems (models, CL data & NC programs, etc.), but also various methods of technological approaches, which correspond to requests of European plants (Fig. 9).

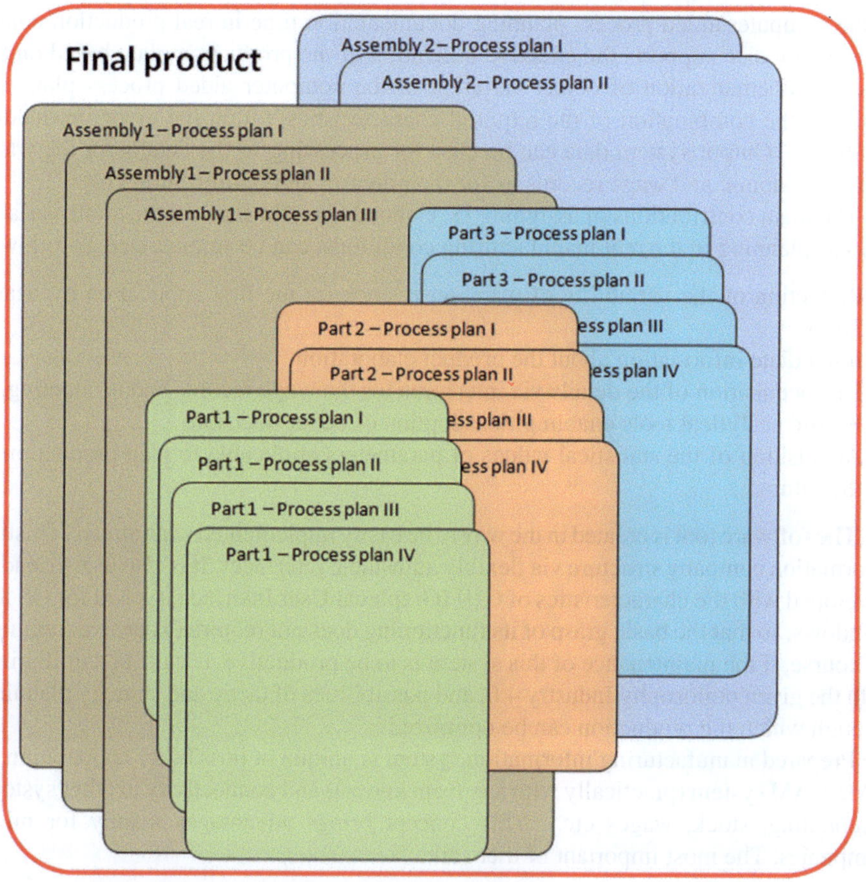

Fig. 9. Structure of produced documentation obtained by multi-variant process planning

4 Conclusion

The analysis of requirements for appropriate software application and based on possibilities to find the information system, which is adapted to East Europe market conditions, brought the idea to create new system. On the basis of the aforementioned theory characteristics, new information system was created by authors and it was applied into real production conditions in the computer aided process planning consisting of approximately 6,000 components. The given product was a result of the cooperation between a German company, providing investments and cooperation of the activities, and Slovak companies providing a technical process planning and the production of a final product.

From the very beginning of the project the established IS served for a suitable analysing of individual real database objects (components, substructures, structures, finished product), i.e. new analytical tools were created when required. Established solution serves the purpose of easier and faster assigning of the process parameters, shortening of the computer aided process planning documentation time in real production conditions, and it also supports the effective utilization of the production plant based on the model mathematization of object variation of the computer aided process planning, fulfilling the combination of the required characteristics within the given production conditions. Output system data can be used for processing of the details for the warehouse, economic and wage records as for their control and optimization [14].

The main contributions of assigning IS, elaborated on the basis of the multi-variable process planning in the real manufacturing conditions, can be summarized as follows:

- Reduction of the variability of warehouse stock (at the first application by nearly 30%).
- Immediate information about the product elaboration
- Fast acquisition of the details via interfaces for the wage records and accounting.
- Elastic analytical tools enabling the adoption of better decisions.
- Acquisition of the statistical values of parameters applicable to plan production in the future.

The software tool is created in the way to be easily implemented to an already existing information company structure via flexibly adjustable interfaces. It is also user-friendly, developed with the characteristics of GUI (Graphical User Interface) typical for OS MS Windows, so that the basic grasp of its functioning does not require expensive trainings. Of course, if the maintenance of this system is to be productive, it must be familiarized with the given philosophy Industry 4.0. and possibilities of tactic and strategy planning, through which the production can be optimized.

Prepared manufacturing information system is unique in possibility cooperation by CAD/CAM system (practically with any from known) and connectivity to other systems (accounting, stock, wages etc.). This concept brings advantages mainly for micro companies. The most important of them are:

- modular conception,
- flexible interconnections to partners,
- possibility of cooperation with wide variety of external software,
- convenient price level.

The ground tasks for near future in focus of authors are:

- research for general format of process plan data,
- investigation of production environs in other European countries,
- study of graphical features uses for process planning.

Acknowledgments. The paper originates with the direct support of Ministry of Education of Slovak republic by grant KEGA 007TUKE-4/2018.

References

1. Panda, A., et al.: Progressive technology – diagnostic and factors affecting to machinability. Appl. Mech. Mater. **616**, 183–190 (2014)
2. Ungureanu, M., et al.: Innovation and technology transfer for business development. Procedia Eng. **149**, 495–500 (2016)
3. Kadarova, J., et al.: Proposal of performance assessment by integration of two management tools. Qual. Innov. Prosperity **17**(1), 88–103 (2013)
4. Schwab, K., et al.: The Europe 2020 Competitiveness Report: Building a More Competitive Europe, World Economic forum, Insight report, Geneva (2014)
5. Markovic, J., Mihok, J.: Legal metrology and system for calibration and verification of the radar level sensors. Qual. Innov. Prosperity **20**(1), 95–103 (2016)
6. OECD Science: Technology and Innovation Outlook 2016, OECD 2016 (2016). ISBN 978-92-64-263062
7. Baron, P., et al.: Proposal of the knowledge application environment of calculating operational parameters for conventional machining technology. Key Eng. Mater. **669**, 95–102 (2016)
8. Lubyova, M., et al.: Labour market in Slovakia, Center of Social and Psychological Sciences, Slovak Academy of Sciences (2016). ISBN 978-80-970850-4-9
9. Stoicovici, D.I., et al.: An experimental approach to optimize the screening in the real operating conditions. Manufacturing Engineering, Issue 2, 2008, Technical University of Kosice, 2008, pp. 75–78 (2008)
10. Ackerman, J.: Presentation of contribution integrative planning of logistics structures and production plants in competence cell-based networks. In: Proceedings of Conference Advances in Production Engineering, Warsaw (2007)
11. Krehel, R., et al.: Mathematical model of technological processes with prediction of operating determining value. Acta Tech. Corviniensis Bull. Eng. **2**(4), 39–42 (2009)
12. Stojadinovic, S.M., Majstorovic, V.D.: Developing engineering ontology for domain coordinate metrology. FME Trans. **42**(3), 249–255 (2014)
13. Filippov, A.V., et al.: Mathematical support for automated geometry analysis of lathe machining of oblique peakless round-nose tools. J. Phys. Conf. Ser. **803**, Article number 012041 (2017)
14. Monkova, K., Monka, P.: Newly developed software application for multiple access process planning. Adv. Mech Eng. **2014**, article no. 539071 (2014)

Strategic Note for a Digital Innovation Policy

Augusta Maria Paci(✉)

Department of Chemical Sciences and Materials Technologies,
National Research Council of Italy (DSCTM-CNR), P.le Aldo Moro, 7, Rome, Italy
augustamaria.paci@cnr.it

Abstract. Future-oriented production methods, systems and tools driven by industry 4.0 generate fundamental transformative processes at local and global level. The "rise of Industry" vision addresses the entire field of advanced manufacturing. Within the high relevant trend of personalization, these thoughts regard people-oriented horizontal policy aspects with related middle-course tasks for moving value chain processes and addressing societal impacts. To enable a society paradigm shift, these thoughts propose a "people-centered" model for realizing high-value production and societal innovation. Trends and emerging popular expectations require the strategic interplay of techno-economic objectives and people-oriented values and life-styles. In this direction, forward-looking and sense-making approach can strengthen advanced manufacturing 4.0 initiatives within the large undergoing societal transformation. Horizontal policy aspects are briefly reported as middle-course tasks towards 2030. The focus goes to social connectivity as a key enabling factor also for manufacturing success to engage demand driven people creativity for value creation for a sustainable progress.

Keywords: Advanced manufacturing · Industry 4.0 · Foresight · R&I policies
Society · People · Personalization · Strategic studies

1 The Rise of Industry 4.0

Future-oriented production methods, systems and tools driven by industry 4.0 generate fundamental transformative processes at local and global level. The "rise of Industry" vision addresses the entire field of advanced manufacturing and the related value chain processes.

Within the high relevant trend of personalization, these thoughts regard people-oriented horizontal policy aspects with related middle-course tasks for moving value chain processes and addressing societal impacts. During systemic change, innovation policy aspects aim at preparing the readiness of contexts and at providing cross-cutting conditions, opportunities for all industries entering the digital age.

These thoughts follow forward-looking "sense-making" approach- as referred by Strategic Foresight – for wider and time-framed impacts (SFRI 2017). Sense-making serves as a basis "to understand and activate relations and interactions between people, places and events in order to create mental models which enable the anticipation of behaviour and

J. Ni et al. (Eds.): AMP 2018, LNME, pp. 156–163, 2018.
https://doi.org/10.1007/978-3-319-89563-5_12

actions". Mental models can be useful for understanding a paradigm shift and for capturing as early as possible weak signals of issues requiring demand-driven responses.

The new linkage between mental models, technology, economy and horizontal policy aspects can introduce a change of perspective in scenarios and the related value added process chains for advanced manufacturing (Kochan 2017). The new relationship, indeed, looks to people-centered policy aspects - beyond SMEs innovation within traditional sectors, which aim at maximizing the return of investment in innovative key enabling technologies.

With a change of perspective, the large society of individuals with new qualities, roles, values and needs will determine and activate the scope of Manufacturing 4.0 - the industrial kernel based on Industry 4.0 technologies and advanced materials.

Industry 4.0 is producing a cross-cutting technology transformation impacting the economy of existing manufacturing sectors and promoting the rise of new industries with related services.

Through Industry 4.0, industries achieve digitally- enhanced objectives such as improved time-to-market, efficiency, quality and flexibility. SMEs are also able to introduce new objectives such as sustainability aspects, traceability of products, dematerialization, enhanced supply chains, extended business relations, new consumer's strategies. Many companies are starting to roll out cloud platforms and gateway hardware. These allow access to multiple users and enable to take advantage by data transformation and from confidential "Know How".

Such big Industry 4.0 innovation model - in which technology and economy represent the main drivers of change for accessing world markets - requires therefore to include societal and environmental perspectives for future-oriented scenarios and actions.

Recently, future scenarios for manufacturing relate current megatrends and drivers (Caldeira 2017). Among the Megatrends, appear very important urbanization and less important new work methodologies and among Drivers (see Fig. 1) appear very important as well as uncertain the human-technology relationship. These aspects highlight the competitive difference for leading new markets, value creation and big data driven strategy with related new capitalization processes. Within this high-level vision for the future of manufacturing and the related reference model (Jovane et al. 2008), strategic studies have contributed leveraging through foresight new ideas and actions (Paci 2009, 2013, 2014).

In the present personalization trend, smart digital devices are part of the transformation of society. They are essential in private life and feature high connectivity, light portability and easiness of use and a lot of amusement, services and social media. These characteristics can shape new people-oriented working modes providing persons not only with new functions and applications but also new ways to reveal unique person qualities adding specific advantage to economic, social and environmental objectives. Humans, as resilient agents, can enhance the tasks of organizations and accelerate digital transformation processes. In a foresight perspective, smart people will cover hybrid roles in the emerging virtual organizational systems: digital workers, digital citizens and consumers and also digital friends so on. Innovative SMEs are following this trend, social media officers and influencers are engaged for market development.

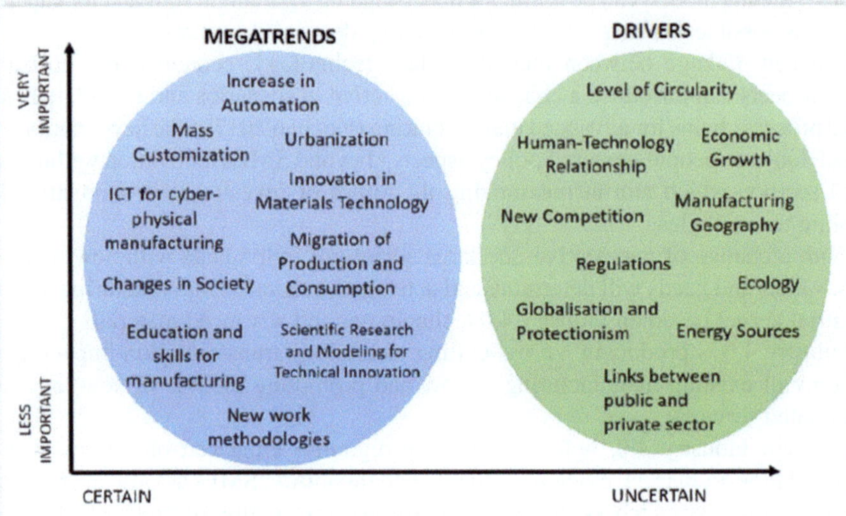

Fig. 1. Megatrends and drivers (source: Caldeira 2017)

Digital trajectories for 2030 can explore how to respond to new values, emotions, personal amusement, personal motivations and design cross-cutting horizontal aspects in order to increase people hope, popular feelings and societal awareness and also reduce failures, business and market uncertainties.

In the People-centered Model – Fig. 2 - the piano keyboard is used as an example of sense-making: black and white are single notes to play for delivering one music.

The people-centered model frames horizontal policy aspects next to digital technologies and foster pro-active and constructive roles of people. The work transformation with new suggestions also for engineering education (Rolstadas 2012) can benefit from well-orchestrated policies including societal paths for socio-economic values.

In the suggested model, a single policy aspect refer to changes related to the digital relationship to explore principles relevant to society for the strategic change as well as the potential barriers.

This model aims to stimulate a reflection on person and society ability in the discovery of new roles needed in manufacturing 4.0. Personal emerging qualities will cover mind agility, flexibility, mobility, relaxed styles, sympathy, empathy, free-thinking, inclusive approaches etc. Initiatives for connecting and integrating people at regional and global level envisage new roles and foreseen societal progress. These initiatives reduce the uncertainty of innovation and build trust in new views, values and visions and can also compensate inertia and fragmentation, providing new flows of ideas into action.

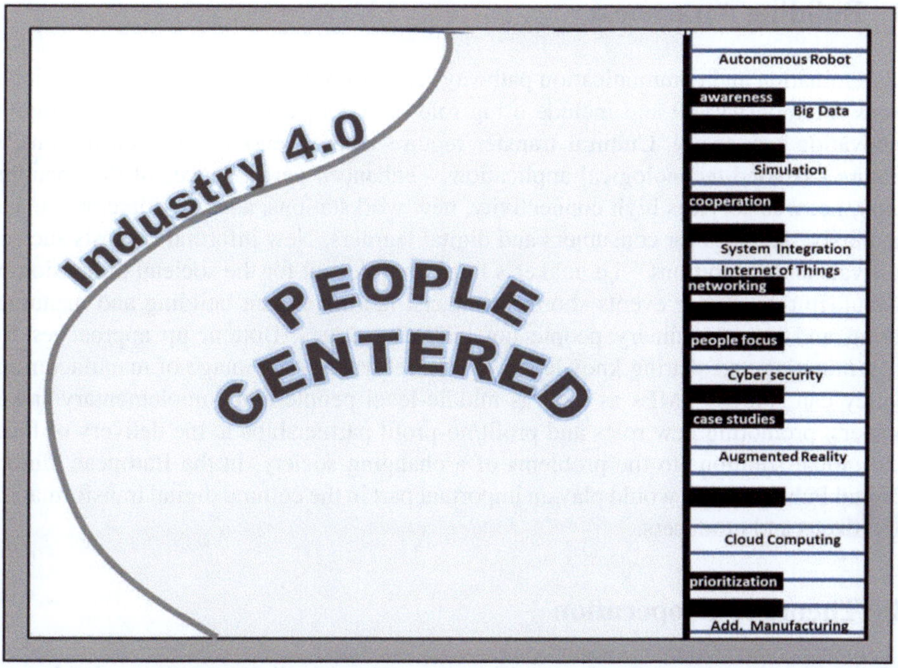

Fig. 2. People and technology: the high value sound

2 Policy Aspects vs Middle-Course Tasks

In forward-looking socio-technical systems, the exponential growth of interactions reflect persons' "social media activism". The challenge for production and manufacturing systems is to leverage these energies and to identify opportunities for firms, supply chains and citizens. Big industrial players communicate vision are based on a complex mix of top-down and bottom-up approaches, while SMEs need to better understand their role in the digital transformation. With Industry 4.0, concepts like co-design or co-creation of value can get new momentum renewing collaboration between research and business environments. Exploring a new mentality, openness and new routines can accelerate the world around the techno-economic kernel of Industry 4.0 Advanced Manufacturing.

People-centered policy aspects can be addressed by middle-course tasks related to following six policy aspects briefly described below:

1. building awareness
2. thematic cooperation
3. international networking and value-adding,
4. people focus
5. case studies
6. prioritization

3 Building Awareness

Dissemination and communication pathways are part of the key trajectories of national plans for Industry 4.0 and include a big role of school education to improve digital innovation propensity. Cultural transfer require talks, events and demonstrations to feature - beyond technological applications - enhanced performances of the phenomenon, i.e. web services high connectivity, new workstations, advanced interactive and immersive services for consumers and digital learners. New informal contests such as festivals and exhibitions – i.e. maker's fair - are relevant for the societal dimension of manufacturing. These events shorten contacts, facilitate team building and manufacturing and attract ordinary people not only consumers. Bottom up approaches for disseminating and sharing knowledge will be relevant to the image of manufacturing: policy can address SMEs as well as middle-level people and complementary stakeholders, promoting new roles and profit/no-profit partnerships in the delivery of local and global solutions to the problems of a changing society. In the European Union, Digital Public Clouds would play an important part in the cultural digital transformation of citizens and producers.

4 Thematic Cooperation

Cooperation fosters building innovative partnership and alliances with new implementation agreements and programmes. Thematic domains can connect types of stakeholders (cities, non-governmental organizations, school, academy and enterprises small, mid-size and start-up) to bridge the cultural gap between generations, between territories and also society, industry and R&I structures. Policy tasks for partnership schemes have the aim to identify and discover mutual benefits in order to renew existing membership and to develop value relationships. Emerging Industrial Partnership for Academy driven by Industry 4.0 for Advanced Manufacturing will renew industry-led partnership involving industry, academy and students. From last year, the Singapore Governmental Alliance for industry 4.0 launched a new cooperation for Research and Innovation with related test labs. A similar cooperation is foreseen in the European Union with the sustainable Innovation Hubs within regional territories.

5 Synergies and Complementarities

Synergies regard the value process chains with related approaches in collaboration and competition. Middle-course tasks are important to promote new approaches for enhancing digital collaboration between specific value partners, to develop SMEs innovation across countries. Policies need also assess social and cultural differences, security and trust, diplomacy and solidarity and promote bilateral or multilateral collaboration initiatives, for profit and for solidarity. Digitally enhanced systems can optimize the implementation of synergies and complementarities enabling value creation in strategic sectors. Digital systems and organizations have to respond to the need of strategic global data, standards and information sharing. In a future perspective, people across countries and cultures will increase unexplored new value relationships.

6 People Focus

Industry 4.0 figures out a landscape responding to industry needs and demand. Currently economic resources - fiscal incentives and competence development –are delivered through complex legal, legislative and economic Country procedures. In middle-course digital citizens will be involved to work according to new methodologies, digital skills, system dynamics and according to new approaches for safety, security, privacy, data harvesting and in a multicultural society dimension. The above envisages pro and cons of incoming hybrid in distribution of work, crossing technical, administrative and managerial activities.

In long-term, education at all level will work in this direction. Mid-term aspects can feature personalized and smart working conditions and personal engagement, full and equal role of women, multicultural workers, company sustainability objectives. Businesses, sharing new people values, aim at either profit and no profit objectives to improve quality of life of families and of large and small cities and territories. They contribute to advance health services, to reinstate qualities of nature, of cultural places and art works. Manufacturing 4.0 could improve the observation of social needs to respond with new processes, products and organizations. The open innovation floor with society can turn ideas into demand-driven paths for successful products and results.

7 Case Studies

Case studies collect and analyse examples of good practices to address real problems and related successful solutions. Case study are often reported in Industry 4.0 current dissemination campaigns. They show concrete industrial efficiency and competitive gains. Case studies attract and grow the interest of recipients as real life stories summarize people role in facing processes with uncertainties, failures, decision-making and recommendations. Successful case studies generally highlight the specific ability of persons and organizations to react quickly and to respond with innovative dynamics to the limitation of constraints (e.g. Crowd-funding). In the digital transformation, people are innovators and case studies show clearly the leadership role at all level for the digital innovation. A middle-course task for case studies could develop a library of "exempla of the digital transformation age" reporting industry cases and ordinary people case studies showing adaptation and survival qualities.

8 Priorities

Prioritization processes are part of the digital transformation and drive decision-levels for improving the systems efficiency. Priorities come from the process of ranking and selected priorities become top-level. Priorities regard decisions, assignments of roles, actions, resources, with relevant implications for the involved communities. Priorities create constraints, controls and a limitation to individual choices.

Favouring people active behaviours to share their motivations, to participate in ranking processes is very relevant in digital systems. This dimension gives space to personal contribution regarded as new value for productivity, business and society. Active and easy acceptance of priorities develop a human dimension that stimulates and accelerates people capacity to respond and solve complex problems and unlock unexpected perspectives, similarly in private life.

In this, mind-sets become capable of fast turning heterogeneous sets of data into valuable information and more sensible to science and innovation breakthroughs.

9 Concluding Remarks

Each policy aspect represent a note of a living sound in the incoming digital age for the whole society. They aim to support value process chains, looking to maximize complementarity among different drivers of change and among stakeholders, workforce and society groups and all people in the projected landscape (The Innovation Group 2017). The interplay of drivers of change (ESET) can contribute to the complex digital transformation of manufacturing processes, businesses and society.

Manufacturing strategies together with research and innovation policies can ensure competitiveness and sustainability through wealth creation. Societies are rapidly changing life styles and habits, therefore it is highly relevant to develop and encourage new people values and conditions for high-value manufacturing.

Acknowledgments. The author would like to thank ing. Detlef Kochan of the Zentrum für angewandte Forschung und Technologie e. V. an der HTW Dresden for his kind support and for the interesting comments. Thanks to prof. Vidosav Majstorovic of the University of Belgrade for the continuity in inspiring and promoting collaborations for the future of manufacturing. A special recognition to the Manufuture Initiative that continues to develop the overall framework for a common vision.

References

SFRI Group: Strategic foresight in EU R&I policy. Wider use, more impact, EC Report 2017. https://ec.europa.eu/research/foresight/index.cfm?pg=fb_policy

Kochan, D.: Advanced manufacturing and Industrie 4.0 for SME. In: Newtech 2017 5th International Conference on Advanced Manufacturing Engineering and Technologies, Belgrade (2017). http://cent.mas.bg.ac.rs/newtech/

Caldeira, J.: Manufuture vision 2030. In: Manufuture Conference (2017). http://manufuture2017.eu/wp-content/uploads/2017/10/pdf-Jos%C3%A9-Carlos-Caldeira.pdf

Jovane, F., Yoshikawa, H., Alting, L., Boër, C.R., Westkamper, E., Williams, D., Tseng, M., Seliger, G., Paci, A.M.: The incoming global technological and industrial revolution towards competitive sustainable manufacturing. CIRP Ann. Manuf. Technol. **57**(2), 641–659 (2008)

Paci, A.M.: Strategic studies towards industrial transformation. Int. J. Total Qual. Manage. Excell. **37**(3), 249–255 (2009). Special Issue, ISSN 1452-0680

Rolstadas, A., Paci, A.M.: Editorial: special issue on engineering education. J. Intell. Manuf. **24**, 421–422 (2012). https://doi.org/10.1007/s10845-012-0649-2

Paci, A.M.: Future oriented activities: mutual benefits towards grand challenges. In: Workshop 2.5, Manufuture Conference, Vilnius, Lithuania, 6–8 October 2013

Paci, A.M., Ritchkoff, A.C.: Practices of focus groups for horizon scanning in research organizations. In: Future Oriented-Technology Analysis Conference, Brussels, 27–28 November 2014

The Innovation Group: Digital Italy 2017: Il Digitale e la forza del nostro Paese, delle sue Imprese e dei suoi Territori (2017). https://www.theinnovationgroup.it/events/digital-italy-summit-2017/?lang=it

Machining Process Improvement Through Process Twins

Erhan Budak(✉)

Manufacturing Research Laboratory, Sabancİ University, Istanbul, Turkey
ebudak@sabanciuniv.edu

Abstract. Selection of cutting conditions is of great importance for high productivity in metal cutting industry. Process twins which make use of physical models considering both mechanics and dynamics of metal cutting are useful tools in the context of Industry 4.0. In this paper, the use of science integrated machining approaches is demonstrated for selection of high performance cutting conditions. The results obtained from simulations of several industrial applications are utilized in selection of cutting parameters and the results are discussed through applications and experiments.

Keywords: Machining process twins · Process modeling · Cutting forces Chatter vibrations

1 Introduction

Production of high quality parts at low cost and shorter time requires more analysis and work at the manufacturing development phase. It has been shown and well known that process simulation tools are very useful in order to decrease number of trials and to improve the machining operations prior to shop floor. For instance, cutting forces cause deflections which may result in quality problems such as form errors. In extreme cases very high cutting forces may cause spindle overloads and tool breakage. Chatter vibrations, on the other hand, deteriorate the surface quality leaving chatter marks on the surface and may even result in damage to the machine tool. Therefore, process planners need to be very careful in selection of the machining parameters.

Obviously, high cutting forces and chatter vibrations can be eliminated by using conservative cutting parameters at the cost of decreased productivity. However, machining process twins based on process models help selection of cutting parameters in such a way that high cutting forces and chatter vibrations are eliminated keeping the productivity at the highest possible level as Industry 4.0 approach.

The objective of this paper is not to give a detailed literature review; however some common references will be discussed. There has been considerable amount of research on cutting force and stability predictions in machining [1] which can be used in development of process twins in the context of Industry 4.0. In addition to the mechanics of the process, geometric calculations are also necessary in creating the process twin. For the modeling of the tool path, due to the variation in tool position and orientation, the

J. Ni et al. (Eds.): AMP 2018, LNME, pp. 164–179, 2018.
https://doi.org/10.1007/978-3-319-89563-5_13

models have to consider in-process workpiece geometry. One of the most challenging parts in creation of the process twin is to predict the cutter engagement boundaries for which both analytical and numerical methods have been used. For 3 axis milling of sculptured surfaces, Lazoglu and Liang [2] presented an analytical approach to predict the engagement boundaries. Z-mapping and Octree methods are among the most popular and robust numerical models proposed in the literature. In Z-mapping method [3, 4] the workpiece is represented with arbitrary vectors in the Z direction and the height of these vectors is updated by the intersection of the swept volume of the cutter body during cutting. In Octree method, the workpiece is divided into cubic elements and as in the Z-mapping methodology whereas arbitrary cubes engaging with the swept volume of the cutter are further decreased in size to finally find an accurate solution. There are also a few analytical models developed for specific conditions. For 5-axis milling, an analytical engagement model considering the orientation of the cutting tool was developed by Ozturk and Budak [5]. They also predicted the cutting forces and form errors during multi-axis milling operations. Another case specific semi-analytical model for taper ball end mills was developed by Ferry and Altintas [6]. These models are limited to specific cases for designated milling conditions.

In this paper, after a brief overview of the process twins, their use in real industrial applications is shown and discussed. Example applications are demonstrated and the paper is finalized with conclusions.

2 Modeling of Machining

Process twins based on process models is one of the most important steps in order to conduct physical simulation of the cutting operation in computer environment. The basic approach is modeling of cutting mechanics by calculating cutting forces, power and torque. The second step is used to model dynamics of cutting, i.e. chatter vibrations, and process stability.

These two process twins are usually developed based on orthogonal and oblique cutting models. In industrial applications, on the other hand, local edge geometry on the cutter used and process conditions may vary in the cutting zone where basic models cannot be directly applicable. For instance, the uncut chip thickness is constant in turning operations where it varies per one rotation of the cutter in milling. Therefore, usually several transformations are needed in order to develop process twins for real-life industrial applications.

2.1 Cutting Models Used for Process Twins

In this approach, the basic approach used to develop process twins for machining operations is based on orthogonal and oblique cutting models. The two-zone contact thermo-mechanical model includes the mechanical behaviors of workpiece material and chip at the primary and secondary deformation zones. The third deformation zone can also be modeled using this approach [7]. The dual contact model used in all 3 deformation zones provides significant improvements with respect to previous models, cutting models which neglected the third deformation zone, and either the sliding contact such as in [8] or the sticking contact as in [9]. It should be noted that the analytical nature of the model makes the computation very fast. In addition, the model calibration needs very limited number of tests compared to the mechanistic models commonly used for machining. The details of these models can be found in [8, 10, 11] whereas a brief overview is given below.

For the development of the process twins, cutting process models based on thermo-mechanical material behavior are used. Figure 1, shows deformation zones, two-zone contact model on the second zone and the third zone contact forces. As for the primary shear zone the thermo-mechanical model is adapted using the Johnson-Cook constitutive relationship.

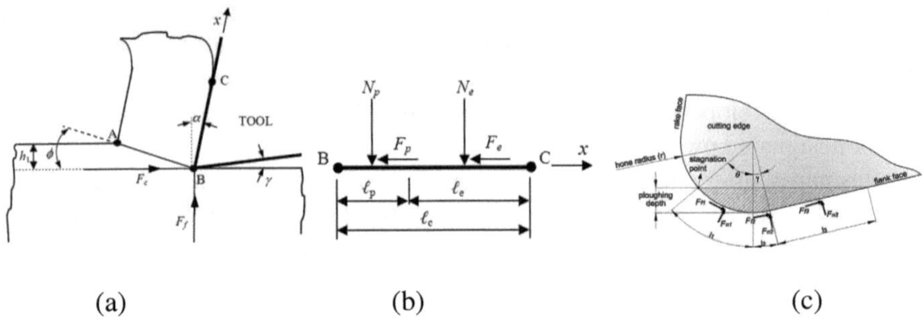

Fig. 1. (a) Sketch of orthogonal cutting (b) sticking and sliding friction regions on the rake face (c) contact forces in the third zone (edge forces)

For high speed machining, the primary shear zone is represented by a thin band where the shear plastic deformation occurs generating the chip. Afterwards, the material exiting from the primary shear zone enters the rake contact with a high normal pressure that creates sticking, i.e. plastic contact, between the tool and the material. After a short distance, the contact state transforms to complete sliding due to the decreasing normal pressure. The sliding friction coefficient on the rake face identified from orthogonal tube cutting tests where the material model coefficients are also updated. After this calibration, the generated process twin can be used for different machining operations involving the same tool and workpiece couple. The outputs of the process twin are the cutting forces, the stress distributions on the rake face and cutting temperatures [12]. The relationship between the sticking, ℓ_p, and the total contact, ℓ_c, lengths of the chip can be calculated by the shear and normal stress distributions on the rake face. Other important parameters include the calculation of Johnson-Cook material constants, sliding friction coefficient and the shear angle. These parameters except shear angle are identified and calibrated using orthogonal cutting tests experimentally. From the experimental results the shear stress is used to calibrate the JC parameters by non-linear regression. Once these parameters are calibrated for different cutting conditions, the developed model can be used to calculate the cutting forces by using the force equilibrium on the chip. The shear angle calculation is conducted analytically by using minimum energy principle [11].

2.2 Machining Stability and Process Twins

Chatter is one of the most important problems in machining as it results in several unwanted effects such as high cutting forces, poor surface finish and shortened tool life. Stability models for machining operations provide stability diagrams which can be used for selection of chatter-free conditions. The analytical chatter stability solution for milling processes by Budak and Altintas [13] has been extended to multi-axis ball end milling in [14] and to generalized milling applications in [15]. Stability solution for turning and boring operations including the effects of insert geometry can be included in the process twin by the model presented in [16].

Although stability diagrams are useful for high speed machining, some difficult-to-machine materials such as aerospace alloys can only be machined at slow cutting speeds due to their low machinability. In previous studies on modeling of process stability at low cutting speeds it is experimentally shown that the process damping is an important source for increased stability limits. The additional process damping effect mainly depends on the indentation volume between the flank face of cutting edge and the work material, which is a function of the cutting parameters and tool geometry as illustrated in Fig. 2(a and b), respectively. Thus, the process damping coefficients are related to the indentation volume and an indentation coefficient to generalize the results based on the energy balance analysis as detailed in [17, 18]. Stability diagrams with the effect of process damping are plotted in Fig. 3 where it is clearly seen that the stability limits significantly increase at slow speeds and with hone radius on the cutting edge.

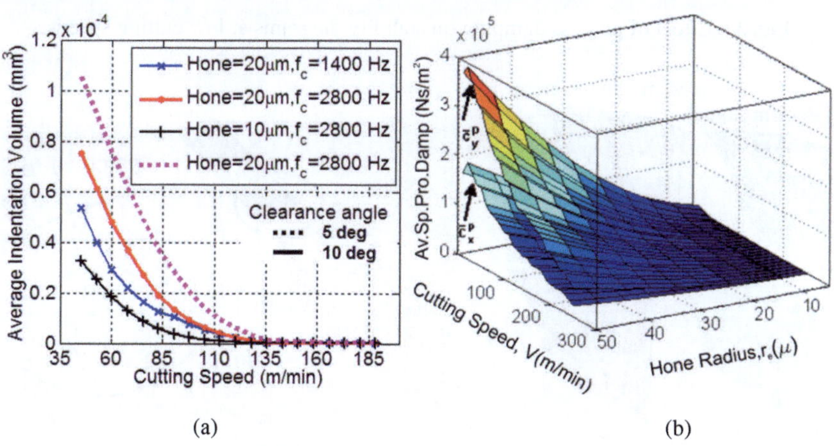

(a) (b)

Fig. 2. Effect of process parameters on (a) indentation volume and (b) process damping.

In general, the transfer function used in the stability analysis is the sum of the tool and workpiece FRFs in feed and cross feed directions, respectively. However, the in-process workpiece (IPW) geometry and hence dynamics of thin walled geometries vary significantly during machining. Therefore, it is not straight forward to measure or identify the varying structural dynamics of the workpiece. The FRF due to the structure of the IPW can be estimated through three steps as detailed in [19]: Mesh generation as shown in

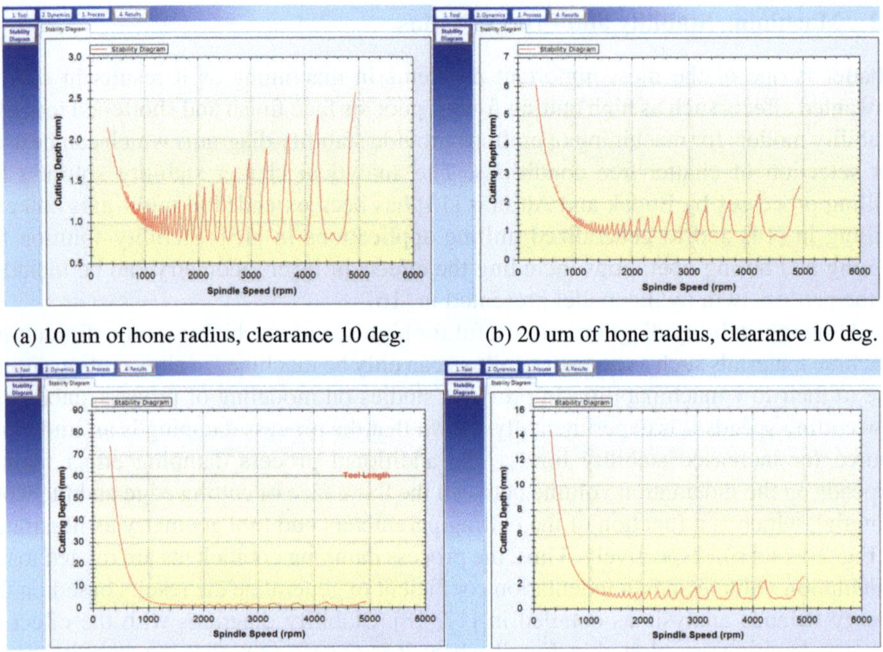

(a) 10 um of hone radius, clearance 10 deg.
(b) 20 um of hone radius, clearance 10 deg.
(c) 40 um of hone radius, clearance 10 deg.
(d) 40 um of hone radius, clearance 20 deg.

Fig. 3. Effect of process damping on stability diagrams at low cutting speeds.

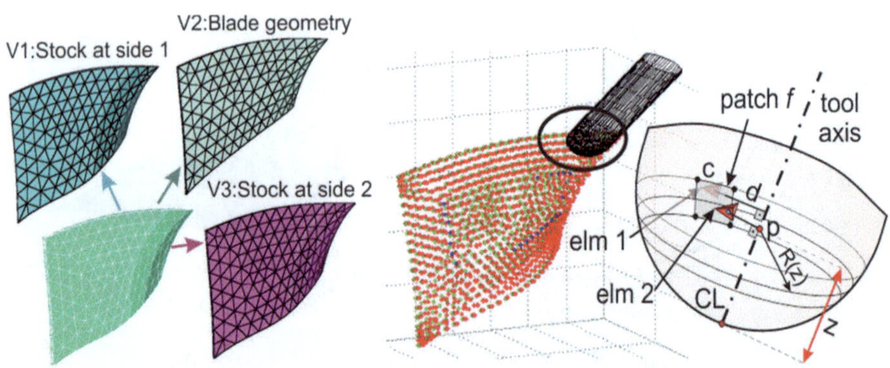

(a) Mesh of final part and the stock geometry.
(b) Determination of removed elements.

Fig. 4. Dynamic cutting system with flexible workpiece.

Fig. 4a, determination of the removed elements at each tool location as shown in Fig. 4b and FRF updating by the matrix inversion method. In step 3, the elements are added to the mesh generated on the finished geometry in the reverse order, i.e. the element machined first is added last. This procedure was used since the FEA results are usually available or easier to generate for the finished part geometry. The application of the structural modification method in such analysis is detailed in [20].

Unified Cutting Model for Process Twins

Cutting mechanics and dynamics models are instrumental for development of process twins. However, there are several challenges in direct application of these models into industrial machining operations. One common case is the applications involving complex cutting tool geometries. Therefore, there is a need for a unified cutting model which can be used for development of process twins to simulate different processes and cutting tools.

In order to implement the aforementioned mechanic and dynamic models into the process twins several parameters must be known: the engagement boundaries between the tool and the workpiece, local cutting edge geometry, the material deformation behavior and the friction between tool and workpiece. For the proposed approach here, the material and friction behaviors are identified from the orthogonal cutting tests [11].

The immersion and the local cutting angles are calculated analytically where a generalized elemental division model is used. In Fig. 5 the application of the approach to several cutting tools can be seen where the cutting tool is divided into many elements which affect the solution time and the accuracy. The immersion between the tool and the workpiece is calculated by the Z-mapping which will be discussed in next section.

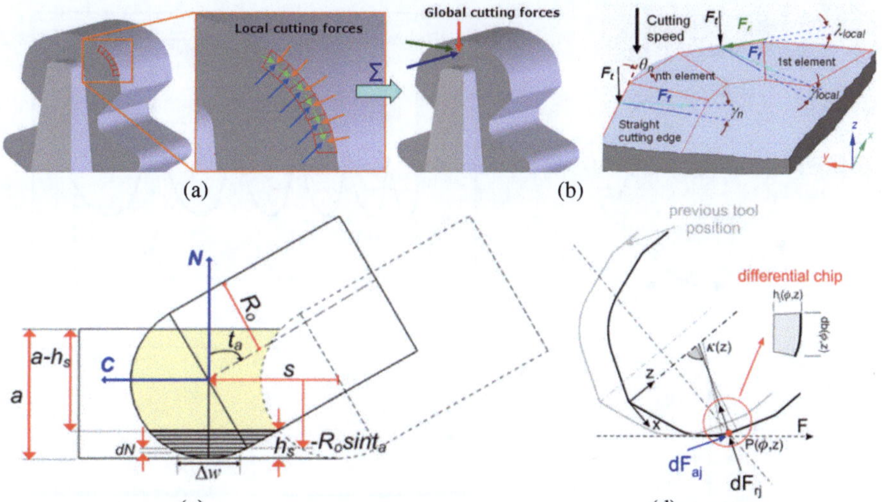

Fig. 5. Differential elements on (a) turning insert nose, (b) broaching tool, (c, d) milling tool.

The last step involves calculation of the local cutting angles for each element. Once these parameters are determined, the local forces can found and transformed into cutting tool or machine tool coordinates for further investigation. The unified model is applied on a multi segmented cutting tool which has 18 cutting edges (Fig. 6a). The measurements and simulations are compared in Fig. 6b, where a good agreement is seen between the simulations and measurements. Another objective in development of process twins is to generate stability diagrams. The proposed solution for the generalized tool geometry shown in Fig. 7a is used to calculate dynamic uncut chip thickness and forces for each axial disc where the problem reduces to an eigenvalue equation [15]. The limiting axial depth of cut is calculated for each disc, compared with the total disc height and the iteration is continued until convergence is satisfied. The stability diagram developed and experimentally verified for a bull-nose end mill as can be seen in Fig. 7b [15].

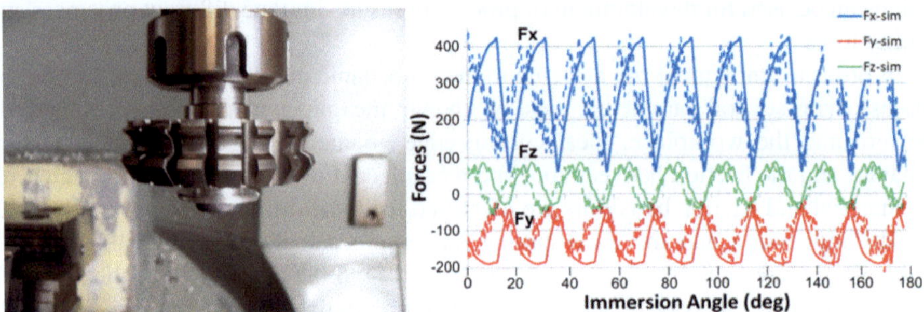

(a) Profiling tool used in the test. (b) Simulated (-) and measured (---) cutting forces

Fig. 6. Tool geometry used in case 2 and experimental results.

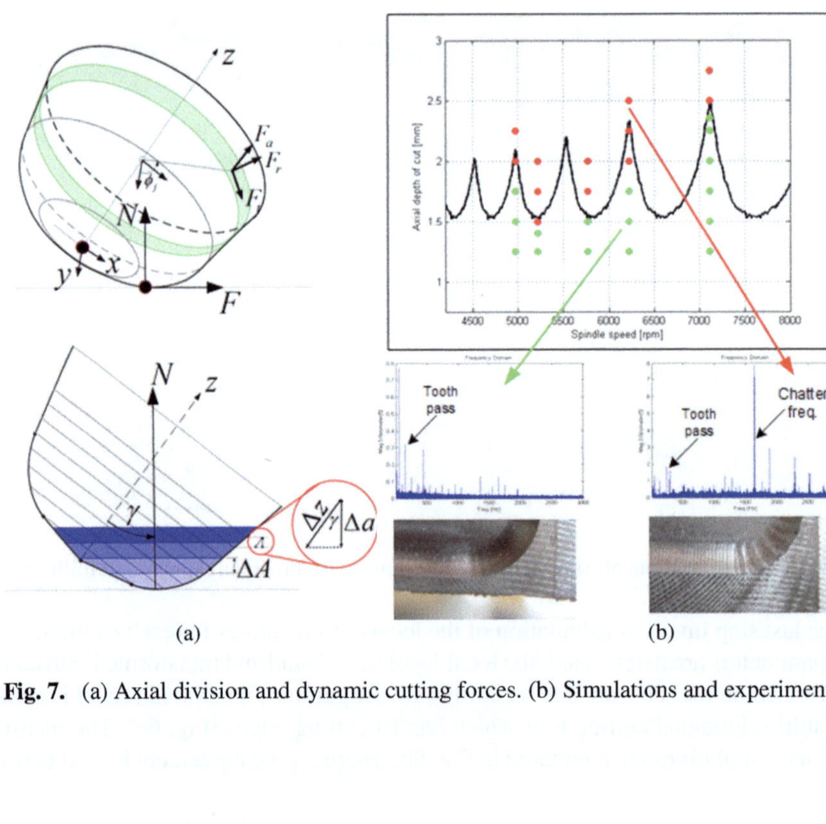

(a) (b)

Fig. 7. (a) Axial division and dynamic cutting forces. (b) Simulations and experiments.

3 Machining Cycle Simulation Using the Process Twin

3.1 Modeling of Tool Geometry and Tool Motion

The position and orientation of the cutting tool is parsed from the CL file, where x, y, and z coordinates of the tool tip and i, j, k components of the unit tool axis vector are provided as shown in Fig. 8. It is required to translate and orient the points according to the tool position and tool axis at each tool move. First the points on the tool envelope are calculated at the initial position (0, 0, 0), with an initial tool axis vector of [0, 0, 1]. Then, the rotations of the initial tool axis vector around i and j vectors are calculated through the inverse solution of the rotation transformation at each tool position. Finally, the points representing the tool envelope are calculated. An illustration of this approach is given in Fig. 8.

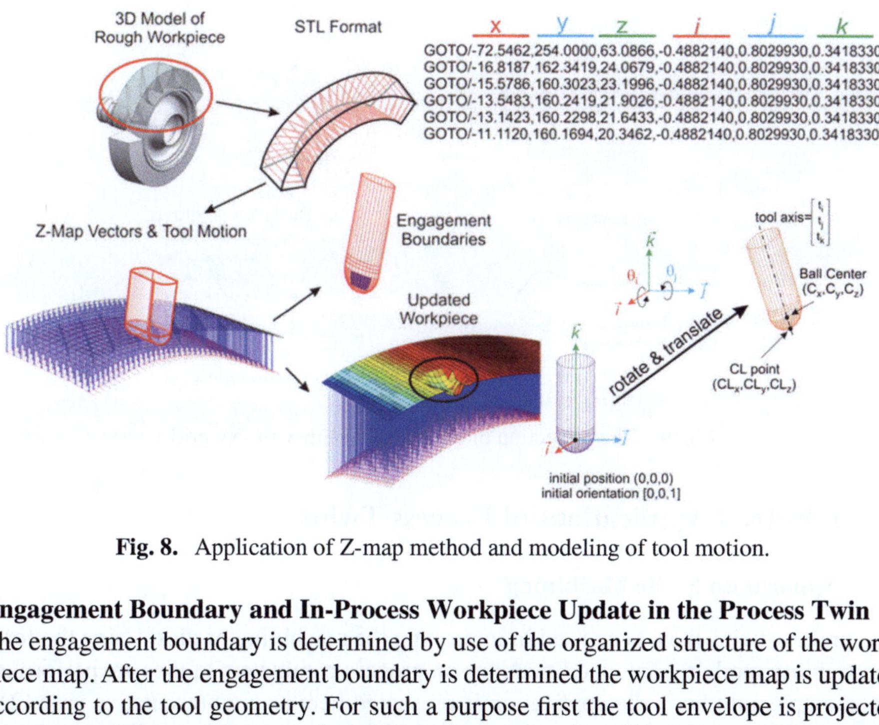

Fig. 8. Application of Z-map method and modeling of tool motion.

Engagement Boundary and In-Process Workpiece Update in the Process Twin
The engagement boundary is determined by use of the organized structure of the workpiece map. After the engagement boundary is determined the workpiece map is updated according to the tool geometry. For such a purpose first the tool envelope is projected onto the workpiece grid on –xy plane. Then, the intersection of the corresponding vectors within the covered area by each tool patch is determined. Finally, the starting and ending z values of the corresponding z-map dexels are updated.

In Fig. 9, the red points represent the locations of the nails covered by the projected polygon area. Identifying the points enclosed by the polygon formed by points p1, p2, p3 and p4 becomes a problem of determining whether a series of points lie within the bounds of a polygon in 2D plane. A simple implementation of this method requires each wall intersection be checked for each point, resulting in an O (N * M) complexity, where M is the number of edges and N is the number of points to be tested. There are two advantages of the implemented algorithm. First, the test points are sorted by the y-value and a binary search is used to find the first point in the list that has a chance of intersecting with a given wall. The sorted list is also used to determine when we have reached the last point in the list that has a chance of intersection. This means that in general only a small portion of points are checked for each wall, rather than the whole set. Secondly, the intersection test is simplified by first checking against the bounding box for a given wall segment. Checking against the bounding box is an inexpensive alternative to the full intersection test and allows us to take a number of shortcuts, minimizing the number of times the full test needs to be done. Finally, the engagement boundaries, i.e. the patches of the cutting tool engaged with the workpiece, are provided to the force model where they are converted to engagement maps in immersion angle domain for simulation of cutting forces.

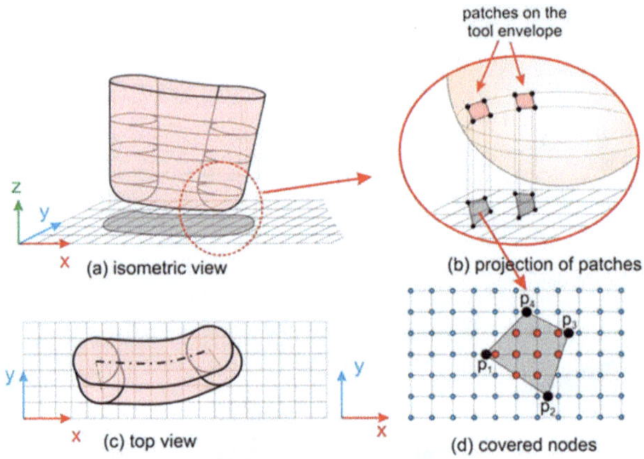

Fig. 9. The projection of tool swept volume on –xy grid.

4 Industrial Applications of Process Twins

4.1 Application to Die Machining

The process twins have been used in improving the machining of dies. Using the force and stability models, some modifications on process parameters of dies are performed. An example auto body die application is presented here as a representative case [21]. The example is a roughing operation on 39NiCrMo3 steel using a milling tool with 66 mm diameter that has 5 circular-16 mm diameter inserts. Radial depth of cut is variable during the process where it changes between 70% and 100% of the tool diameter. Moreover, cutting type is also variable, i.e. process is an up-milling operation at some parts of the tool path and down-milling at other parts. Since cutting forces are higher and stability limits are lower in slotting, the process parameters are selected considering the slotting case to stay on the safe side during the whole process.

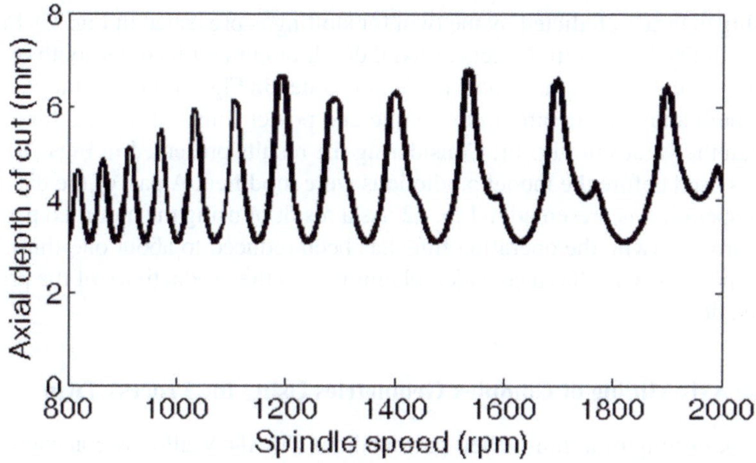

Fig. 10. Stability diagram for the slotting of die.

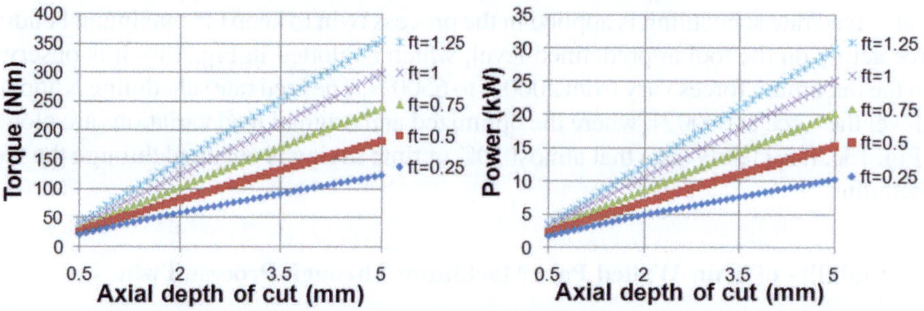

Fig. 11. Predicted torque and power values for different axial depth of cut and feed per tooth values (Spindle speed = 1000 rpm).

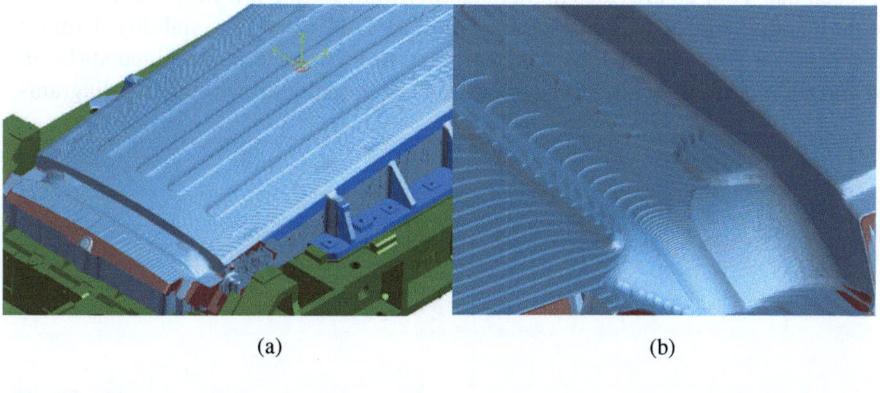

(a) (b)

Fig. 12. The example die after the roughing operation (a) Part of the die (b) Detail view.

Stability diagram predicted by the twin for slotting is presented in Fig. 10. Predicted absolute stability is 3.1 mm. Effects of axial depth of cut and feed per tooth on cutting torque and power are calculated, and are demonstrated in Fig. 11. For the range of usable feed per tooth and axial depth of cut, torque and power limits of the machine tool are higher than the values in Fig. 11. Considering the results presented in Figs. 10 and 11, parameters used before the model predictions were modified. A part of the die after the roughing operation is presented in Fig. 12. As a result of using the modified parameters from the process twin, the operation time has been reduced to about one-third. Hence, use of the process twins has a considerable impact on the productivity of the process as well as quality.

4.2 Five Axis Milling of Complex Geometries Using the Process Twin

5-axis flank milling of a thin-walled part made of Ti6Al4 V alloy is considered as the next application (see Fig. 13). In the simulated process the total depth varies along the tool path and reaches up to 70 mm. There are approximately 150 CL points along the cutting pass and the forces are simulated per 4 CL points. According to the simulation results, feed rate scheduling is applied in the process twin to keep the maximum bending force acting on the tool at predefined level, which is plotted in Fig. 13b. It is observed that the maximum forces vary from 3000 N to 6500 N. The feed rate scheduling is applied to level the force at 6000 N, where the optimized and original feed variations are plotted in Fig. 13c. It is clearly seen that almost 20% of time saving is obtained through the feed scheduling.

4.3 Stability of Thin-Walled Part Machining Through Process Twin

In this application, semi finishing and finishing cycles of a thin-walled Ti6Al4 V part are given. The measured FRF's for three cutting tools are given in Fig. 14.

The generated stability diagrams based on the process twin are illustrated in Fig. 15a. It is seen that Tool 26 provides higher stability limit for the finishing pass. The desired cutting depth is higher than the stability limit of Tool 22 and the stability diagram for Tool 25 has a very narrow lobe which is risky to be used. The machined surfaces are shown in Fig. 15b, where a good agreement is observed with the stability diagrams.

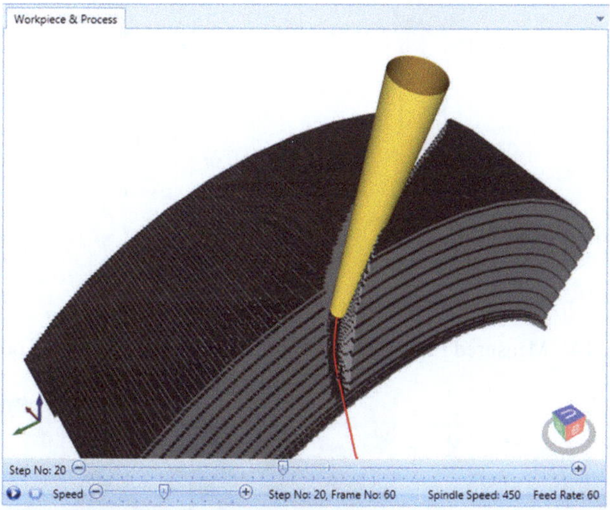

(a) The simulated cycle.

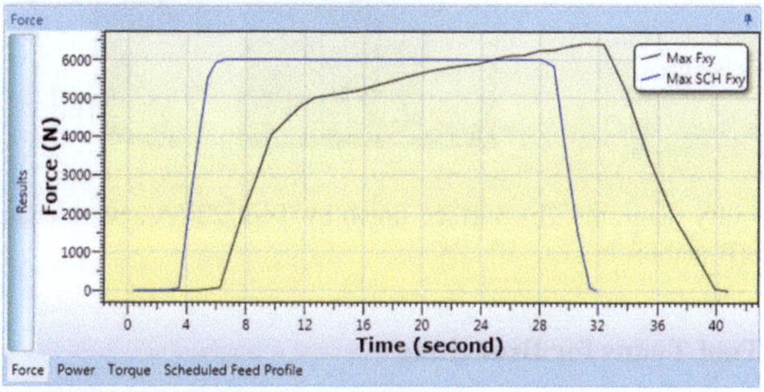

(b) Maximum bending force, Original & Simulated

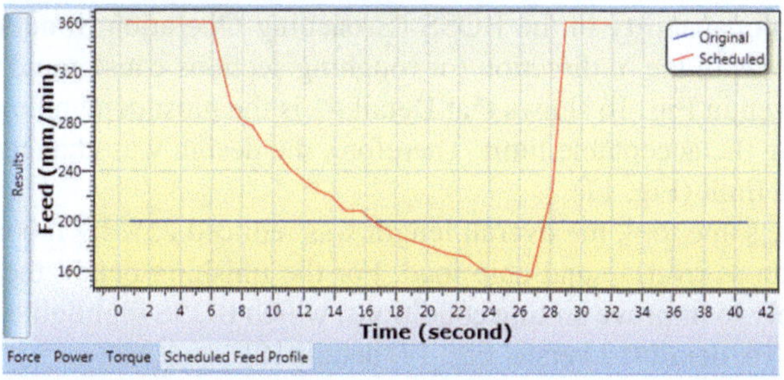

(c) Feed rate profiles, original & scheduled

Fig. 13. The simulated flank milling process and simulation results.

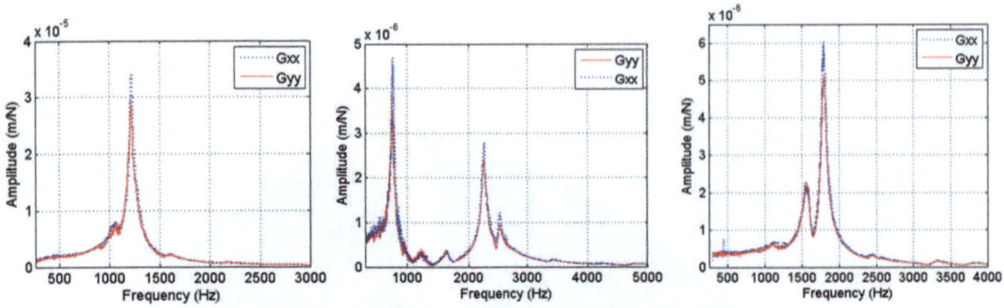

(a) (b) (c)

Fig. 14. Measured FRF's for Tool (a) 22, (b) Tool 25 and (c) Tool 26

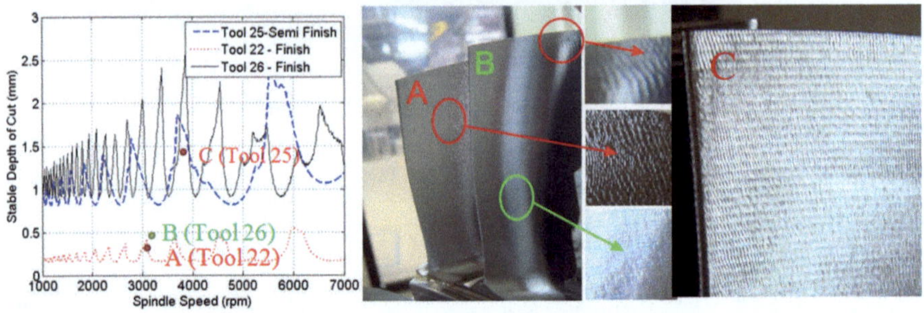

Fig. 15. (a) Stability diagrams, (b) Surface finish by Tool 22 (A-severe chatter), Tool 25 (C-chatter), Tool 26 (B-stable)

4.4 Process-Tool Twins for Broaching

As another application is a fir-tree shaped broach tool which has been optimized using force simulation capability in the BOSS (Broaching Operation Simulation Software) [22]. Large forces in the Y-direction for roughing sections cause rapid tool wear. The force graph given in Fig. 16 shows that Detail #1 is the most demanding detail with its force exceeding the acceptable limit. Therefore, the design was changed to reduce the force under the limit (Fig. 17).

Also for the same tool, the overall length was reduced 25% by modifying different parameters such as overlap and chip load. For the finishing detail, the chip load was increased mainly to improve cutting conditions, which BOSS highlighted as an increase in forces (Fig. 16, detail#12 versus Fig. 17, detail#10).

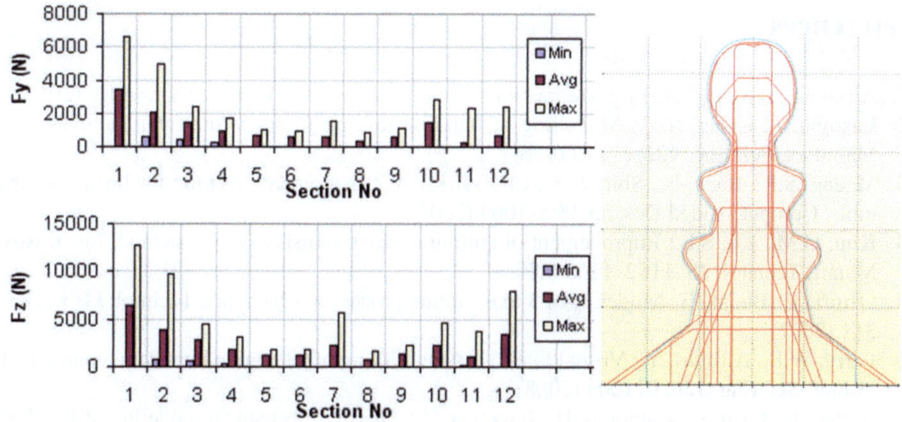

Fig. 16. Original tool.

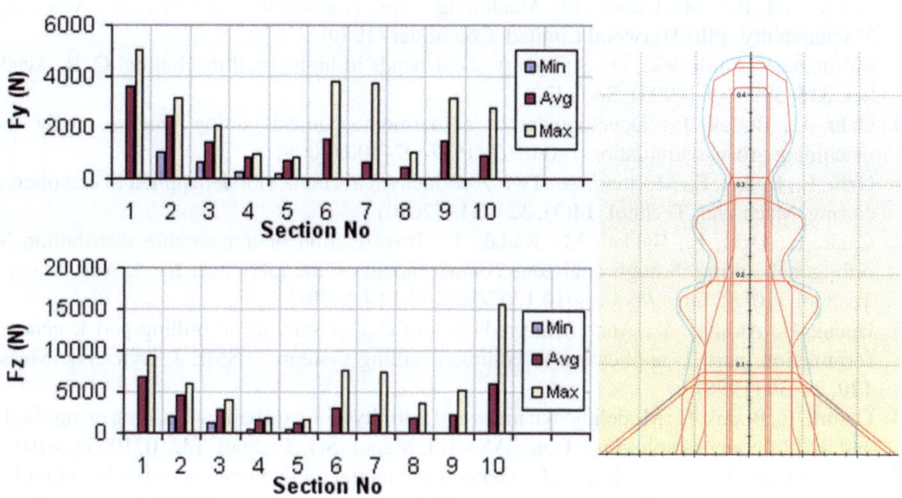

Fig. 17. Optimized tool.

5 Conclusions

Several aspects of implementation of process twins have been presented and discussed in this paper. Substantial improvements and savings can be achieved by using the twins in machining processes. This is demonstrated by several examples.

References

1. Altintas, Y.: Manufacturing Automation. Cambridge University Press, Cambridge (2012)
2. Lazoglu, I., Liang, S.Y.: Modelling of ball-end milling forces with cutter inclination. J. Manufact. Sci. Eng. **122**, 3–11 (1996)
3. Maeng, S.R., Baek, N., Shin, S.Y., Choi, B.K.: A Z-map update method for linear moving tools. Comput.-Aided Des. **35**, 995–1009 (2003)
4. Kim, G.M., Ko, S.L.: Improvement of cutting simulation using octree method. Int. J. Adv. Manuf. Technol. **28**, 1152–1160 (2005)
5. Ozturk, E., Budak, E.: Modeling of 5-axis milling processes. Mach. Sci. Technol. **11**(3), 287–311 (2007)
6. Ferry, W.B., Altintas, Y.: Virtual five-axis flank milling of jet engine impellers – part I-II. J. Manuf. Sci. Eng. **130**, 011005 (2008)
7. Budak, E., Ozlu, E., Bakioglu, H., Barzegar, Z.: Thermo-mechanical modeling of the third deformation zone in machining for prediction of cutting forces. CIRP Ann.-Manuf. Technol. **65**(1), 121–124 (2016)
8. Oxley, P.L.B.: Mechanics of Machining: An Analytical Approach to Assessing Machinability. Ellis Horwood Limited, Chichester (1989)
9. Molinari, A., Dudzinski, D.: Stationary shear bands in high speed machining. C. R. Acad. Sci. **315**, 399–405 (1992). Série Ii
10. Ozlu, E., Budak, E.: Development of a thermomechanical cutting process model for machining process simulations. Ann. CIRP **57**, 97–100 (2008)
11. Ozlu, E., Budak, E., Molinari, A.: Two-zone analytical contact model applied to orthogonal cutting. Mach. Sci. Technol. **14**(3), 323–343 (2010)
12. Cakir, E., Ozlu, E., Bakkal, M., Budak, E.: Investigation of temperature distribution in orthogonal cutting through dual-zone contact model on the rake face. Int. J. Adv. Manuf. Technol. (2018). https://doi.org/10.1007/S00170-017-1479-3
13. Budak, E., Altintas, Y.: Analytical prediction of chatter stability in milling-part I: general formulation, part II: application to common milling systems. ASME J. Dyn. Sys. Meas. **120**, 22–36 (1998)
14. Ozturk, E., Budak, E.: Modeling dynamics and stability of 5-axis ball-end milling using single and multi frequency solutions. Trans. ASME J. Manuf. Sci. Technol. **132**, 021003 (2010)
15. Ozkirimli, O., Tunc, L.T., Budak, E.: Generalized model for dynamics and stability of multi-axis milling with complex tool geometries. J. Mater. Process. Technol. **238**, 446–458 (2016)
16. Ozlu, E., Budak, E.: Analytical modeling of chatter stability in turning and boring operations part I: model development. Trans. ASME J. Manuf. Sci. Technol. **129**, 733–739 (2007)
17. Budak, E., Tunc, L.T.: A new method for identification and modeling of process damping in machining. Trans. ASME, J. Manuf. Sci. Eng. **131**(051019), 1–10 (2009)
18. Budak, E., Tunc, L.T.: Identification and modeling of process damping in turning and milling using a new approach. CIRP Ann.-Manuf. Technol. **59**, 403–408 (2010)
19. Budak, E., Tunc, L.T., Alan, S., Özgüven, H.N.: Prediction of workpiece dynamics and its effects on chatter stability. Milling CIRP Ann. - Manuf. Technol. **69**, 339–342 (2012)
20. Alan, S., Budak, E., Ozguven, H.N.: Analytical Prediction Of Part Dynamics For Machining Stability Analysis. Int. J. Automation Tech. (Spec. Issue Mod. Sim. Cut. Process) **4**(3), 259–267 (2010)

21. Ozturk, E., Ozkirimli, O., Suren, G., Aksu, B., Budak, E.: Increasing productivity in die machining through process modeling. In: Proceedings of the 5th International Conference and Exhibition on Design and Production of Machines and Dies/Molds, Kusadasi, Aydin, Turkey, 18–21 June 2009 (2009)
22. Ozlu, E., Engin, S., Cook, C., El-Wardany, T., Budak, E.: Simulation of broaching operations for tool design optimization. In: 2nd International CIRP Conference on Process Machine Interactions (PMI 2010), Vancouver, Canada,10–11 June 2010 (2010)

Connecting Humans to the Loop of Digitized Factories' Automation Systems

Emanuele Carpanzano[1(✉)], Andrea Bettoni[1], Simon Julier[2], Joao C. Costa[3], and Manuel Oliveira[4]

[1] Department of Innovative Technologies, University of Applied Sciences and Arts of Southern Switzerland, Manno, Switzerland
emanuele.carpanzano@supsi.ch
[2] Department of Computer Science, University College London, London, UK
s.julier@ucl.ac.uk
[3] HighSkillz, Lisbon, Portugal
joao.costa@highskillz.com
[4] Stiftelsen for Industriell og Teknisk Forskning, Trondheim, Norway
manuel.oliveira@sintef.no

Abstract. Emerging factory digitization, along with the increased automation levels it promotes, represents a unique opportunity that manufacturing enterprises must seize. By distributing once centralized decision-making through an ecosystem of smart factory objects, enterprises will be able to increase their productivity, responsiveness and quality levels. However, for continued effective management, humans must adapt to production systems whose behavior is defined by the interactions that take place between these smart objects and the overall automation layers and automatic control functions. These interactions can occur in different ways across many levels of abstraction and complexity, and across many timescales. As a result, it is extremely hard for humans to preserve reliable mental models and this raises the risk of the out-of-the-loop condition. This paper proposes a human-centered automation framework for improved workers' well-being, safety and psychological health. Particularly, the dynamic real time interactions among closed loop control functions and human workers are addressed so as to properly include humans in the feedback loops. Experimental cases of the proposed framework application are presented in the white-goods and in the furniture sectors.

Keywords: Workers well-being · Digital factory · Adaptive automation
Human in the loop

1 Introduction

Today's speculation in the manufacturing domain is in ferment over the concepts emerging along the fourth manufacturing revolution. Such a transformation aims at leveraging on industry digitization to deliver smart factories in which the decision-making burden is distributed among a net of interoperable and interconnected cyber physical systems. Baur and Wee (2015) list four drivers of this paradigm shift: (i) the

J. Ni et al. (Eds.): AMP 2018, LNME, pp. 180–193, 2018.
https://doi.org/10.1007/978-3-319-89563-5_14

astonishing rise in data volumes, computational power, and connectivity, (ii) the emergence of analytics and business-intelligence capabilities, (iii) improvements in transferring digital instructions to the physical world and (iv) new forms of human-machine interaction. However, only the fourth driver is concerned with human-system interaction and, even then, it is only regarded in terms of how new technologies, such as touch interfaces and augmented-reality, can be used to improve worker productivity.

Despite some authors (Bauer and Horváth 2015) focus on the real-time connection between humans, machines and objects, most of the effort being spent, as of today, in this fast evolving innovation race is aimed at developing those enablers capable of making the Industrie 4.0 vision viable from the factory automation side leaving aside its human counterpart. This makes topics like (i) interoperability of machines, devices and sensors for plug & produce, (ii) plant virtualization creating digital twins of the cyber physical systems of systems and (iii) the deconstruction of the automation pyramid for decentralized decision-making climb on top of academia and industry's agenda (Hermann et al. 2016).

However, the intelligibility of the resulting smart factories' behavior challenges a workers' ability to gain situation awareness, to properly supervise and manage those systems. Assistance is necessary to support informed decision-making by aggregating and visualizing information comprehensibly, yet it cannot be regarded as a holistic approach, acting as a passe-partout which is capable of solving any cognitive issue. Indeed, the introduction of new technologies at the shop-floor level, as well as the growing levels of automation the human operator has to deal with, make more and more crucial that the management of manufacturing systems focuses on the human factors and, specifically, on the cognitive workload related to manufacturing operations carried out at different levels of decision.

The increasing burden that the progressive adoption of the new fabrication paradigms is loading on workers' shoulders is already traceable looking at the outcome of the European Working Condition Survey (Eurofound 2016) that shows how the one of plant and machine operators is the group most subjected to work intensity. With 45% of responders reporting that their pace of work is dictated by three or more pace determinants (out of a list of five), operators hold, together with craft workers, an unenviable leadership. The foremost impact on pace of work is caused by the interaction with automatic machines that determine speed of operators' actions in more than 40% of cases.

The risk of estranging people of this group, that are already experiencing the lowest level of creativity and task variety, is thus very high.

Moreover, such operators experience the lowest levels of decision latitude with only 38% being allowed to change the order tasks are executed and only 43% having a say in the way work is organized. In most of the cases (almost 80%) they are neither involved in work improvements actions nor consulted before personal targets are set, contributing to the lowest level of engagement among all groups (score of 63 on a 100-scale) and making the group the worst in terms of satisfaction about working conditions. The effects on well-being of people belonging to the plant and machine operator groups are also evident: 38% feel that work affects their health negatively and they are also the first in the statistics of sickness leave.

All the situations mentioned above paints in gloomy colors the future of industry workers if they are left as just a complementary resource in the consideration of managers and organization leaders. The transformation and adaptation process that is now implementing the decentralized automation paradigm is a great chance to introduce worker-centered design and deployment of production systems. Such an approach will on the one hand increase well being and health of workers, and on the other hand improve the overall reliability and resiliency of production systems, which can count on an effective integration of both human and digital capabilities.

To this end, this paper proposes a methodology to integrate cognitive workload into the design of workplaces to match the human safety and well-being necessities and tasks' cognitive requirements. Specifically, it introduces a model to include in the automation loop worker-related instances and methods to classify the fabrication tasks involved in the production process according to their cognitive complexity and the required capacity, thus enabling an anthropocentric management and optimization of the manufacturing activities.

2 Human-Automation Interaction Challenges

The origins of the attention towards human-machine interaction can be traced back to the 1960s (Hancock et al. 2013) and has been characterized as the need to understand, predict and counteract man-machine mismatching by focusing on the interaction among three mutually constrained factors, namely the world to be acted on, the agents who act on the world and the external representation through which agents experience that world (Sarter 1994). This principle is still valid today, but the world as well as the agents interacting with it have evolved over time to feature different challenges from the past. Most interaction nowadays takes place in complex systems controlled or supervised by human operators where some of the main issues are their ability to effectively oversee and interact with automation when needed (Endsley 2017) and the difficulty in anticipating all potential interactions within and between system components (Bolton et al. 2013).

Over time, the most important approach aimed at addressing these challenges has been so-called *adaptive automation* (Carpanzano and Jovane 2007) - the idea of having machines and automation that adapt to the cognitive and physical demands of humans in a momentary and dynamic manner (Romero et al. 2015), starting from the knowledge of the collective human-machine system states. Hancock et al. (2013) report that recent trends related to methods of adaptation to operator's cognitive and physical needs draw inspiration from physiologically-based approaches, neurophysiological imaging and neuroergonomics. Also, the same authors point out that there is a shift from progressively greater individual- and task-related diagnosticity for adaptive function allocation towards a fuller agenda of human-machine intimacy. Bolton et al. (2013) also mention cognitive work analysis and cognitive tasks analysis as techniques used to design the interaction.

Expected benefits of a careful management of this interaction include reducing out-of-the-loop performance problems and related issues mainly concerning loss of situational awareness and high mental workload (Kaber et al. 2001).

Wickens et al. (2010) provide a synthesis of the traditional conceptual model depicting the link between the degree of automation and the effect on the variables impacting the human-machine interaction. On the one hand, a greater degree of automation in man-in-the-loop systems brings benefits in performance routine circumstances, but causes a reduction of workload as humans are left with less to do. On the other hand, costs rise when automation fails, humans must intervene and a degraded failure response can occur due to loss of situational awareness. In turn, this latter can be generated by different causes, the main ones cited in literature (for example Endsley 2017) contributing to this field of research being: complacency (operator overreliance on automation); generation effect (operators are not involved in generating actions); changes in information presentation with automation; automation over trust and operator engagement. Some mitigation actions (e.g. effective intuitive displays or effective training) can be implemented to overcome the implicit trade-offs brought by this conceptual model including the variables impacted by the degree of automation (routine and failure performance, situational awareness and workload).

Bolton et al. (2013) and Endsley (2017) identify another possible cause of the out-of-the-loop condition referred to as brittle automation: operators may encounter problems in situations outside the scope of the automation's design parameters since human intervention is required when programming does not cover the needed functionalities.

Mode confusion (Bolton et al. 2013) is a further form the out-of-the-loop condition can take. This could be due to insufficient feedback about the state of the automation from the human-machine interface or to the lack of a correct mental model preventing human operators to properly understand how the automation is behaving. Mode confusion can lead to automation surprise effect (mental models and expectations could not be updated and that will lead to a surprise in facing unexpected system states); errors of commission (inappropriate actions are carried out) or errors of omission (appropriate actions are not carried out).

Moreover, automation risks to be perceived as a threat to workers professional profile, especially when it is introduced without a proper transition and a management care for reskilling workers, thus leading to low job satisfaction levels. As a matter of fact, the automation of a task may dramatically reshape its skill demand set moving it from features such as observation, reasoning, problem-solving to less proactive ones thus requiring future job reallocation.

3 A Reference Framework for Human-Automation Interaction

The systemic nature of the mentioned challenges and the pervasiveness of automation in dictating the behavior of whole production processes calls for all-encompassing interpretive, design-oriented and operative models. Those models must be holistic. Rather than using a piecemeal approach of mitigating each undesirable effect on its own, these models put human-automation interaction at the center of the system representation and incorporate this holistic perspective in the design and operation of automated-driven manufacturing systems.

It is only by conceiving the automation system as an extension of human capabilities that negative gaps can be avoided thus reintroducing the man in the loop of factory automation. The framework that embodies this vision is presented in Fig. 1 and, in a conceptual way, it aims at pointing out the additional blocks that need to be added to the traditional production system control loop in order to address the dual nature of human-automation and to unify them in a single view, integrating effectively digital and human capabilities in a systematic way.

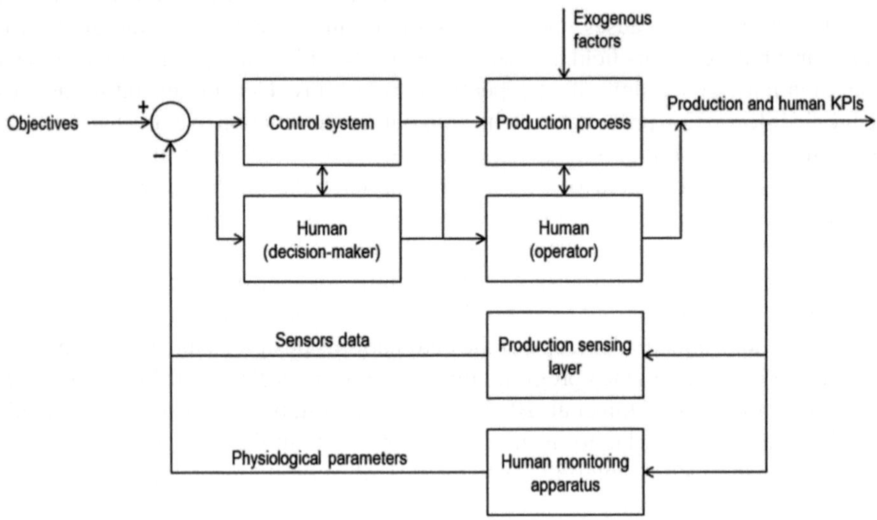

Fig. 1. Framework for human-in-the-loop factory adaptive automation systems.

The human dimension is thus added, and not only as a controlled variable of the system, but becoming integrated part of the decision-making loop. In doing this, the proposed framework aims at making explicit those tacit assumptions about the nature of the interactions taking place at decision-making and operative levels between automation and its human counterpart so as to abate cognitive barriers and build up a synergistic cooperation. To this end it recognizes that only by means of a complete characterization of workforce capabilities it becomes possible to free up its constrained potential and improve both production system productivity, reliability and resiliency as well as workers' health and well-being.

This reflects in an extension of the objectives of the whole system that, beyond typical productivity targets (i.e. line throughput), has to take care of the working conditions experienced within the monitored loop. Management of quasi real-time physiological data flows becomes a critical system requirement to rely on a continuous monitoring of deviations from personalized safe patterns. Leveraging on the availability of such collected big data sets, self-learning algorithms operate to identify risky conditions and to trigger interventions that reduce the cognitive burden experienced by the worker. This calls for dynamically reconfigurable automation and organizational policies (Carpanzano et al. 2014), defined at the many levels of the production system (Brusaferri et al. 2011), to identify and, hopefully, to remove the systematic causes of the encountered cognitive gaps. Among these it is worth mentioning low levels of job rotation, skill mismatch, insufficient job instructions and unfeasible automation pace.

Moreover coordination strategies need to couple automation's efficiency with the flexible human mind set. The proposed approach distinguish two kinds of human figures based on the level of interaction they present with the automation system: on the one hand, decision-makers are referred to as those human who oversee, cooperate and, potentially, may overtake the role and functions of the available automation; on the other hand, shop floor operators interact with the production process and see automation mostly in terms of dictated pace and operative support in task completion.

Both these kinds of cooperation have to be differently treated, tailoring their functions to the peculiar features each one presents and to the expected goals for each category.

The application of this framework requires that two distinct phases are considered to fully achieve the sought human-automation harmonization:

(1) system design phase: where an automated environment that synergistically cooperates with human decision-makers is setup;
(2) system operation phase: where real-time monitoring of the system leads to changes and adjustments applied to improve production system productivity and reliability as well as workers' well-being at the same time.

3.1 System Design

Design of human-automation interaction aims at introducing cognitive considerations in the definition of the processes that regulate conjoint behavior of humans and automation systems. However, most of the times, it is focused mainly on modulating the level of automation that humans will have to deal with, aiming to minimize the risks associated to mode confusion and, consequently, to surprise effect (Eom and Lee 2015).

The design can be based on the scope of automation with Parusaraman et al. (2000) identifying four classes of functions where automation can play a role, namely (i) information acquisition, (ii) information analysis, (iii) decision and action selection, and (iv) action implementation. Alternatively it can focus on the extent of the support that automation is meant to provide as Endsley and Kiris (1995) point out building a continuum of shared responsibility between human and automation that ranges from (i) no support up to (ii) decision support, (iii) consensual decision, (iv) monitored decision and, eventually, (v) full automation.

In order to overcome the current limitations, the proposed approach starts from a characterization of the working environment and of the agents that will operate in it, namely the operators. In so doing, it is recognized that, without proper models and the availability of personal and specific information, it is impossible to successfully integrate a human-centered perspective into the automation control loop. This makes it possible to reshape the resulting system, putting the human dimension as key cornerstone by acting on the framework building blocks as follows (labels are added at the end of each bulleted point for reference in the experimental scenarios):

- Human (decision-maker): by developing processes and support tools meant to aid the decision-makers in delivering the human-centered approach at the operational level and by categorizing the mental models and behaviors in their interaction with automation (DM-D).
- Human (operator): by knowing the personal attitudes and preferences of operators and by formalizing them through a multi-level classification applied to each operator in terms of skills, functional capabilities (physical, intellectual and sensorial), physiological and psychological baseline characteristics, anthropometric measures, personal needs (O-D).
- Control system: by enlarging its scope to include the human dimension as a monitored domain and operators' well-being as a target variable and by defining the interfaces for a closed loop dynamic and situation-aware coordination with the human decision-makers (CS-D).
- Production process: by including support tools (e.g. augmented reality) aimed at enhancing situation awareness of operators and by modularizing and parameterizing both tasks and workstations so that they can cope with single operators' characteristics and so become an extension of those (PP-D).
- Production sensing layer: by setting up a middleware that can collect, in a robust and formalized way, big data flows coming from the underlying IoT devices installed in the shop-floor and that can interpret and aggregate them in meaningful added value parameters (SL-D).
- Human monitoring apparatus: by defining the parameters of interest for the detection of risky conditions (e.g. stress, loss of attention, mental fatigue) and by connecting physiological sensing devices (e.g. wristbands) with well-trained detection algorithms (HA-D).

This is reflected in the reorganization of control loop objectives to include human-related targets such as well-being, work satisfaction, ergonomics, motivation. As a consequence, metrics that can support their monitoring have to be arranged drawing from different areas of the human factors field.

3.2 System Operation

While the design of the human-automation interaction is focused on properly configuring an adjustable worker-aware environment, the system operation deals with its real time dynamic adaptation to cope with the evolving conditions of production situations and human conditions.

With reference to the framework of the man-in-the-loop automation system presented in Fig. 1, there are two main consuming modes of the information gathered at the level of workplace sensing and human monitoring: the first one is the quasi real-time monitoring of aggregated KPIs for detecting the insurgence of cognitive demanding or risky conditions aimed at triggering relieving interventions (from a simple take-a-break suggestion to the introduction of personalized task instructions or the provision of physical support); the second one is a feedback loop directed to the system design aimed at solving those issues that cannot be addressed by the available interventions.

When it comes to the building blocks of the man-in-the-loop framework, their behavior in the system operation phase is as follows (labels are added at the end of each bulleted point for reference in the experimental scenarios):

- Human (decision-maker): has to consider a multi-dimensional objective function whose scope includes the human well-being and safety aspects and has to learn to interact with the adaptable automation interfaces maintaining conscience of whether his/her mental models are deteriorating (specific training is considered as a mandatory precondition) (DM-O).
- Human (operator): interacts with an adaptive production environment and is supported by personalized information-providing interfaces and psychophysical complementing devices. Acceptance of those support tools is key to maintain the operator in the loop as well as dedicated actions (conveyable also through the support tools) to avoid skill loss, passive mode and over trust in automation (O-O).
- Control system: balances the multiple objectives for the optimization of the control variables (including human-related parameters) aiming at an overall maximum making trade-offs between system productivity and workers' well-being. It distributes part of its decisional burden to the underlying factory smart objects. In the same way it shares decision-making with its human counterpart dynamically adjusting to his/her needs the provided level of support (CS-O).
- Production process: at the organizational level (e.g. production scheduling, job allocation, etc.), it adapts to the static and dynamic features of the available workforce while, at workstation level, it becomes the extension of the human capabilities guided by the human condition detection algorithms in choosing the appropriate nature and level of support it has to provide to the operator (PP-O).
- Production sensing layer: constantly produces environmental raw data, stores them for performing analytics in the future, and aggregates them in more synthetic measures and indicators (SL-O).
- Human monitoring apparatus: is responsible for the continuous monitoring of workers' condition and couples sensing devices with reasoning systems capable to infer non-obvious anomalous conditions detection triggers (HA-O).

4 Framework Application in Industrial Cases

In the following two sections partial implementations of the proposed framework in real world industrial environments are presented. A mapping to the element of the framework is provided by referencing to the labels previously defined.

4.1 Design for Human-Centered Lines in the White-Goods Industry

White goods production environments, due to a combination of high variability of production lots and traditionally low sectorial margins, feature work-intensive assembly lines that exploit human flexibility to deliver a highly diversified range of stock keeping units (SKU).

Companies in this sector typically employ the principles of lean manufacturing achieving, among other targets, the elimination of non-value added activities including setup times. Production then flows without interruptions and at an intensive throughput rate, paced by the factory takt time (i.e. the rate at which a finished product needs to be completed in order to meet customer demand).

Automation plays the metronome role within the working environment and is dedicated to preassembly operations (e.g. stamping of oven shells) and to management of in-line materials flow and handling of parts, including parts presentation. Operators' time management and autonomy in sequencing the production tasks is thus highly dependent on factory automation that, combined with segmentation of operations in micro tasks (cycle time may be as low as 30 s) lead to high cognitive demands and the risk of alienation.

The experimental campaign carried out at one of the market leaders in the production of white goods, and home appliances in general, was aimed at making workplaces, and production environments at large, adaptive to the peculiar skills, expertise and characteristics of each worker with the ultimate goal to make the most out of worker's knowledge and potentials across all age groups and different roles (Bettoni et al. 2014).

To this end, the experimental activities focused on the system design phase envisioned within the framework proposed in this paper and particularly on:

- Workers profiling, by systematically characterizing a set of volunteer individual workers with information belonging to different domains (May et al. 2015). This has been achieved through: (i) a scanning campaign to acquire anthropometric measurements thus allowing ergonomic adjustment of workplace dimensions; (ii) an assessment of functional capacities (physical, intellectual and sensorial) by means of a three-step methodology and tool; (iii) a survey-based assessment of skills, knowledge and personal needs to promote matching of available skills and job demand (O-D). This allowed to go beyond previous attempts made by the value stream management to assess the workforce capabilities along four dimensions (experience on the job, safety management, quality aspects and ergonomics) in order to establish a personal skill matrix for each worker-workstation couple. The lack, experienced in the past, of a systemic approach to the human-automation interaction and of an abstraction layer to reuse the categorization independently of the specific production line prevented any substantial improvement until the proposed framework was introduced.
- Knowledge organization, by introducing a shared repository, named KNOW Platform, based on a semantic representation of factory demands and workforce capabilities in terms of knowledge, skill, personal needs (KSN) and physical, intellectual and sensorial (PIS) capacities with the software infrastructure to support concurrent access of both data provider and data consumer tools (CS-D).
- Production line balancing, by complementing worker-centered dimensions to the typical cycle time balancing problem, to include demand of skills and functional capacities and the ergonomic impacts. This allowed to distribute tasks in feasible job

contents and to validate the resulting line, in a single environment during the design phase, against the characteristics of a real pool of workers (PP-D).

- Worker-aware job allocation, by envisioning a mathematical model based on linear optimization that has been implemented as a decision support system that team leaders can use at the beginning of shift to obtain a starting base for assigning workers to workstations in a way that maximizes use of available capacities (based on skills and capacities matching), promotes job satisfaction (as it considers workers' allocation preferences) and reduces exposure to health issues (through ergonomic impact evaluation) (DM-D).
- Training paths definition, by extracting analytics out of the allocation outcomes to support human resources management identifying existing gaps and in realigning available skills in the workforce with those actually required in production (Pinzone et al. 2016). Creation of relevant job profiles is also included in this training needs detection in order to define meaningful training paths that are customized on workers' career evolution (DM-D) (O-D).

The above-mentioned experimental activities have been carried out in two assembly lines, one for microwave ovens and another for fridge production. Relying on the data used to preliminarily populate the KNOW platform, jobs have been designed and assigned to workstations pursuing multiple objectives: (i) technical objective (maximum saturation set at a level equal to 90% of the takt time); (ii) skill matching objective (by balancing the cognitive burden considering the presence of critical requirements increasing the possibility that a pool of workers can effectively and completely match the skill demand); (iii) ergonomics objective (reducing the risk of musculoskeletal disorders). Results of the line optimization have been compared against the choices that team leaders would have done with the same set of jobs.

Experimental campaign confirmed the potential of the framework in achieving several benefits out of the simultaneous changes applied to the building blocks during the design phase as indicated in Table 1.

Table 1. Benefits coming, at design level, from the introduction of man-in-the-loop framework application in a white-goods plant.

Introduced feature	Experienced benefit
Workplace adaptability	Reduction of cumulative trauma disorders and psychological stress
Characterization of both workers and production lines from a human-centric perspective	Integration of all necessary human-related information in the decisional-process
Multi-objective approach in line balancing	Increased awareness of all the human-related aspects at a glance and reducing the several design iterations that were necessary before
Training paths definition	Improvement of fit between workers' profiles and automation needs

4.2 Real-Time Support for Operators in the Furniture Industry

The furniture industry is composed by multiple subsectors, characterized according to the function (storage, sitting, surfaces and spaces) and materials of the furniture (e.g. wood, plastic, metal, glass, plywood). In the companies adopting large scale production, targeting economies of scale, the production lines are characterized by different levels of automation and the human operator has a pivotal role throughout, particularly in the assembly and packaging lines. There is a high diversification of products, with high customization driven by market factors such as seasonality and trends. Consequently, the success of a company relies heavily on creative capacity in combining raw materials, technology and design to address the consumer needs. The high variability of products, with relatively small sized lots in a fast paced production line, implies a strenuous work in high-production environments where operators are driven to achieve demanding production targets.

The involved wood-furniture manufacturing company is one of the few European companies equipped for economies of scale and recognized as a trendsetter on the global level in terms of applying innovation in their products and manufacturing processes. The company considers training strategic to their business, but the wide range of contexts requiring training such as launch of a new products, infrequent product demand, high product variability, makes personalized training tailored to the individual needs difficult to achieve and entailing significant costs. In addition, the access to information on demand when needed remained a challenge.

To address the aforementioned challenges and to create human-centered workplace, an augmented reality-based solution was devised consisting of:

- the HUMAN platform, which consisted of a solution for the operational phase of production that captures signals from the environment, including the operator, to analyse the human-centered work environment. With an understanding of the operator's work situation at any instance, the platform determines whether a deviation is taking place, its severity and how best to support the operator to achieve the desired outcomes (SL-O) (HA-O);
- a device to detect the levels of stress by measuring variances in the breathing patterns of the operator using a thermographic imaging camera (Cho et al. 2017) and machine learning (HA-O);
- knowledge and activity models, to support the situational awareness of the platform concerning the operator's work environment, (i) a task model to capture the current state of assembly task progression; (ii) a worker model to capture the operator's competence, past experience and performance baselines; (iii) a product model, (iv) a stress model to effectively make decisions concerning the impact of stress levels on the operator's performance and the ability to reach targets (CS-O) (PP-O) (O-O);
- the augmented reality service providing different levels of just-in-time knowledge as required by the operator taking into account the situation as mapped out in the knowledge and activity models (PP-O) (O-O);
- the process-mining tool to support the system design phase, so different stakeholders within the organization (e.g. team leader, manager, operator) can do exploratory

analysis of work sessions with the aim of improving product and/or process design (PP-D).

For purposes of validation, a prototype of the solution has been assessed at the company in a production cell setup for training purposes for the pre-assembly of a particular furniture unit. Both experienced and novice operators took part of the evaluation study. The service provided different levels of support for the operator depending on their level of expertise, with novice operators demonstrating the greatest benefit to achieve an established performance rhythm in reduced amount of time when compared to traditional training processes. The solution demonstrated the potential to support line balancing by complementing the operators with the required knowledge just-in-time, thereby increasing the pool of operators that meet the job requirements.

5 Conclusion

The diffusion of the Industry 4.0 paradigm advocating higher level of digitization for factories (Valente et al. 2015) is a further trigger to pay more and more attention towards the human-automation interaction to safeguard the human health and well-being whilst pursuing production systems performance optimization.

In this paper this topic is explored by proposing a conceptual framework depicting how the traditional automation control loop should evolve making explicit, in the adaptive automation realm, the role of humans and the main interaction needs to be synergistically managed to successfully connect humans in the loop.

The framework is analyzed at two levels: the design level highlighting how each constituting block should be designed to take simultaneously into condensation both productivity and well-being targets and the operational level where the functioning of the system is illustrated to show how the interaction takes place.

Partial implementation of the framework in real contexts is illustrated through two application cases in industrial sectors, where the diffusion of automation is actually rising challenges when it comes to manage the operators roles transition required by higher levels of digitization, namely the white-goods sector and the furniture sector.

The white-goods experience shows how the design of the system components has changed to achieve a real human-automation integration and coordination. Worth of mention is the great preliminary effort needed to collect, structure and cross-link data related to both operators characteristics and production processes parameters. This evidences the need to revolutionize the way systems are thought about since their conception in the automation designers' minds.

The second experiment is focused on the operative side of the framework and presents some tools that have been put in place to collect and analyze real-time data from operators, including stress-related parameters. Also situational models are developed so to make sure that working environment, execution of tasks and operators' conditions are aligned to prevent human's burden and low satisfaction. In case such monitoring system detects problems related to lack of instruction or information hindering operators, augmented reality is used to provide them with the needed piece of knowledge. This set of tools supporting the working life ensures that the human-automation interaction is constantly monitored and any possible deviation is timely approached putting operators in the condition to always generate value for them and for the factory.

The presented framework, proposing a holistic approach, potentially supports factories in their journey towards digitalization, yet a fundamental prerequisite is that a cultural change pervasively interests any organizational level of the company to achieve its final goal: a harmonious and synergic interaction between humans and automation. Once the cultural change is accepted, the experiments have shown that the exploitation of technological advances as well as the ability to manage big data allows to create a complex system where the continuous monitoring of both human- and production-related parameters ensures their simultaneous optimization.

Results achieved so far are promising, but the way forward for a complete implementation of the conceptual framework still requires both researchers and practitioners attention for the development of tools to cover all the interactions needs and for their experimentation towards wider industrial application.

Acknowledgments. This work has been partly funded by the European Commission through Man-Made (MANufacturing through ergonoMic and safe Anthropocentric aDaptiveworkplacEs for context aware factories in EUROPE) project (Grant Agreement No: 609073) and through HUmanMANufacturing project (Grant Agreement No: 723737). The authors wish to acknowledge the Commission for its support. The authors also wish to acknowledge their gratitude and appreciation to all the Man-Made and HUMAN partners for their contribution during the development of various ideas and concepts presented in this work.

References

Baur, C., Wee, D.: Manufacturing's next act. McKinsey Quarterly (2015)

Bauer, W., Horváth, P.: Industrie 4.0-Volkswirtschaftliches Potenzialfür Deutschland. Controlling **27**(8–9), 515–517 (2015)

Bettoni, A., Cinus, M., Sorlini, M., May, G., Taisch, M., Pedrazzoli, P.: Anthropocentric workplaces of the future approached through a new holistic vision. In: IFIP International Conference on Advances in Production Management Systems, pp. 398–405. Springer, Heidelberg (2014)

Bolton, M.L., Bass, E.J., Siminiceanu, R.I.: Using formal verification to evaluate human-automation interaction: A review. IEEE Trans. Syst. Man Cybern. Syst. **43**(3), 488–503 (2013)

Brusaferri, A., Ballarino, A., Carpanzano, E.: Reconfigurable knowledge-based control solutions for responsive manufacturing systems. Stud. Inform. Control **20**(1), 31 (2011)

Carpanzano, E., Jovane, F.: Advanced automation solutions for future adaptive factories. CIRP Ann. Manuf. Technol. **56**(1), 435–438 (2007)

Carpanzano, E., Cesta, A., Orlandini, A., Rasconi, R., Valente, A.: Intelligent dynamic part routing policies in plug&produce reconfigurable transportation systems. CIRP Ann. Manuf. Technol. **63**(1), 425–428 (2014)

Cho, Y., Bianchi-Berthouze, N., Julier, S.J.: DeepBreath: Deep learning of breathing patterns for automatic stress recognition using low-cost thermal imaging in unconstrained settings. In: 7th International Conference on Affective Computing and Intelligent Interaction (Submitted). arXiv preprint arXiv:1708.06026

Endsley, M.R., Kiris, E.O.: The out-of-the-loop performance problem and level of control in automation. Hum. Factors **37**(2), 381–394 (1995)

Endsley, M.R.: From here to autonomy: lessons learned from human–automation research. Hum. Factors **59**(1), 5–27 (2017)

Eurofound: Sixth European Working Conditions Survey – Overview report, Publications Office of the European Union, Luxembourg (2016)

Eom, H., Lee, S.H.: Human-automation interaction design for adaptive cruise control systems of ground vehicles. Sensors **15**(6), 13916–13944 (2015)

Hancock, P.A., Jagacinski, R.J., Parasuraman, R., Wickens, C.D., Wilson, G.F., Kaber, D.B.: Human-automation interaction research: past, present, and future. Ergonomics in Des. **21**(2), 9–14 (2013)

Hermann, M., Pentek, T., Otto, B.: Design principles for industrie 4.0 scenarios. In: 2016 49th Hawaii International Conference on System Sciences (HICSS), pp. 3928–3937. IEEE (2016)

Kaber, D.B., Riley, J.M., Tan, K.W., Endsley, M.R.: On the design of adaptive automation for complex systems. Int. J. Cogn. Ergonomics **5**(1), 37–57 (2001)

May, G., Taisch, M., Bettoni, A., Maghazei, O., Matarazzo, A., Stahl, B.: A new human-centric factory model. Procedia CIRP **26**, 103–108 (2015)

Parasuraman, R., Sheridan, T.B., Wickens, C.D.: A model for types and levels of human interaction with automation. IEEE Trans. Syst. Man Cybern. Part A Syst. Hum. **30**(3), 286–297 (2000)

Pinzone, M., Fantini, P., Fiasché, M., Taisch, M.: A multi-horizon, multi-objective training planner: building the skills for manufacturing. In: Advances in Neural Networks, pp. 517–526. Springer, Cham (2016)

Romero, D., Noran, O., Stahre, J., Bernus, P., Fast-Berglund, Å.: Towards a human-centred reference architecture for next generation balanced automation systems: human-automation symbiosis. In: IFIP International Conference on Advances in Production Management Systems, pp. 556–566. Springer, Cham (2015)

Sarter, N.B.: Strong silent and "out-of-the-loop": properties of advanced (cockpit) automation and their impact on human-automation interaction (Doctoral dissertation). The Ohio State University (1994). Retrieved from OhioLINK. (Order Number 9517075)

Valente, A., Mazzolini, M., Carpanzano, E.: An approach to design and develop reconfigurable control software for highly automated production systems. Int. J. Comput. Integr. Manuf. **28**(3), 321–336 (2015)

Wickens, C.D., Li, H., Santamaria, A., Sebok, A., Sarter, N.B.: Stages and levels of automation: an integrated meta-analysis. In: Proceedings of the Human Factors and Ergonomics Society Annual Meeting, vol. 54, no. 4, pp. 389–393. Sage Publications, Los Angeles, September 2010

Development of Skills and Competences in Manufacturing Towards Education 4.0: A Teaching Factory Approach

Dimitris Mourtzis(✉)

Laboratory for Manufacturing Systems and Automation,
Department of Mechanical Engineering and Aeronautics, University of Patras, Patras, Greece
mourtzis@lms.mech.upatras.gr

Abstract. Industry 4.0 manufacturing paradigm, apart from the technological revolution requires also a shift from the traditional education to an advanced set of methods for developing skills and building digital competences, summarized in the term Education 4.0. Human skills undoubtedly require upgrading to handle the key enabling technologies sufficiently, including machines as cyber-physical systems, augmented reality, human-robot collaboration, and smart devices. In the present work, these new requirements for the enhanced manufacturing workflow are presented. Furthermore, existing education 4.0 approaches are investigated, and finally, a Teaching Factory concept adapted to the Industry 4.0 paradigm needs is proposed.

Keywords: Manufacturing · Industry 4.0 · Education 4.0 · Teaching factory

1 Introduction

Due to the recent advancements in information technology (IT), manufacturing research has entered a new era of product planning, analysis and troubleshooting, moving forward to the Industry 4.0 paradigm [1]. Digital manufacturing has been considered, over the last decade, as a highly promising set of technologies for reducing product development times and cost as well as for addressing the need for customization, increased product quality, and faster response to the market requirements [2]. Digital manufacturing has become a common trend worldwide as computer-integrated manufacturing systems have eliminated data handling errors and enhanced decision making. Virtual Reality (VR) technology, Computer simulation using CAD modelling tools, and finite element analysis have assisted manufacturing engineers to reach decisions faster, while reducing errors [3].

This would ensure that activities, such as design, planning, machining, etc., are done first-time-right without the need for subsequent re-work and modifications. On the other hand, cloud computing has already managed to spread across many different industries and numerous users around the world, despite the fact that it is clearly a philosophy born in the 21st century as a natural outcome of globalization [4] and as the shift of industrialized countries towards a more service-oriented society model [5]. A massive

J. Ni et al. (Eds.): AMP 2018, LNME, pp. 194–210, 2018.
https://doi.org/10.1007/978-3-319-89563-5_15

increase in the amount of collected data, leading to "datasets whose size is beyond the ability of typical database software tools to capture, store, manage and analyze", known as Big Data, can be observed in the last years [6]. However, the effective exploitation of Big Data encloses great challenges and requires novel approaches and tools in order to maximize its impact [7, 8]. As a result, the challenges concerning the integration of distributed manufacturing resources as well as the cooperation and information exchange among different manufacturing sites, departments and organizations can be tackled in an efficient way [9]. Human skills undoubtedly require upgrade to handle the key enabling technologies sufficiently, including machines as cyber-physical systems, cloud platforms, human-robot collaboration, and smart devices. The present paper contributes in this direction by presenting the transformation of manufacturing systems towards Industry 4.0 and the concurrently change of the required skills through the transition of traditional education practices to Education 4.0. This shifting from the traditional Teaching courses to Education 4.0 is realised through this paper by presenting a Teaching factory.

Regarding the structure of the present paper, after this introductory section, the transformation of the manufacturing systems to Industry 4.0 is presented in detail in Sect. 2. Section 3 presents the expected change of the education system, while Sect. 4 presented the Teaching Factory concept. Finally, Sect. 5 compiles the conclusions.

2 Manufacturing Transformation to Industry 4.0

2.1 Traditional Manufacturing

During the previous decades a number of technologies have emerged to support all manufacturing phases. For most of them, an extended testing period was required until these technologies were adopted into the common practice.

In the traditional practice, after communicating with the customer and receiving the requirements for a new product, the design phase begins.

In this phase the design of the product is realized by the responsible engineers using Computer Aided Design (CAD) software which belongs to the standard procedures of manufacturing companies, while it is being constantly evolved [10]. Using CAD, the engineers are able to perform product designs faster compared to the handmade designs and therefore drastically increasing not only their productivity, but also the creativity and the job satisfaction [11]. The analysis of the CAD geometries to extract knowledge on the physical properties (mechanical, electrical, etc.) of the product is performed using the so-called Computer Aided Engineering (CAE) tools to determine the robustness of components and assemblies [12]. Through simulation and optimization methods, the CAE tools analyse the physical properties of the designed parts, providing meaningful insights that support effective design.

Before the production phase, the production planning engineering team creates the schedule for the resources of the shop-floor [13]. Scheduling can be performed using various methods and criteria based on the objectives that need to be fulfilled [14]. Meanwhile, the manufacturing enterprise has to ensure that the required resources will be available at the time of production, through inventory tracking and effective supplier management [15].

After the production schedule is created and the manufacturing tasks are assigned to the resources, the Computer Aided Process Planning (CAPP) and Computer Aided Manufacturing (CAM) tools process the CAD information and generate code for computer numerical control machine tools and robots. CAM may also support adaptation in the case of an unexpected availability of a resource, providing the capability to generate a new numerical code for a part, through the machine specific post-processors via unified data modelling [16].

Two major parts of the production are the manufacturing processes and the assembly. During the manufacturing phase, on-board supervision mechanisms on the manufacturing resources provide real-time information about the status of the resource and the progress of the process [17]. Nevertheless, this information is usually consumed at the time of its generation and it is scarcely assessed to conclude into actual knowledge of the production. Regarding the assembly phase, the research interest has generated mature results that are already applied in production. Robot automation in assembly affects positively the increase of productivity, especially in the assembly of heavy parts [18].

In the subject of the maintenance of machine-tools and equipment, preventive maintenance strategy is the one that is most commonly used in industry. Preventive maintenance, as a maintenance approach, is introduced to inhibit the equipment failure before it actually occurs [19]. Preventive maintenance actions are usually performed without considering the condition of the resource to be maintained. Therefore, unnecessary maintenance actions are performed increasing the operational costs of the industries.

Not all manufacturing companies employ all the aforementioned technologies. Especially small SMEs usually aim to strengthen some specific phases of their production, or they gradually proceed to the digital era, avoiding risks from the implementation and the investment cost of cutting-edge technology [20].

Despite the increase in productivity that these software systems provide, they still lack the capabilities for seamless information transfer and collaboration.

2.2 Industry 4.0 Transition and Requirements for New Skills

The transition to the Industry 4.0 is facilitated through the advances of the technologies related to Augmented Reality (AR), Information and Communication Technologies (e.g. Cloud and networking), smart sensors and others.

2.2.1 Key Enabling Technologies for Industry 4.0

AR is a novel human–computer interaction tool that overlays computer-generated information over the operator's field of view. The information that is provided to the user of the AR technology is context sensitive as it is usually depended on the location of the user and the object that they observe, providing an additional level of perception [21]. AR has already been demonstrated as a solution in many applications, in manufacturing and other fields [3].

Cyber-Physical Systems (CPS) are open, linked-up systems that operate flexibly, cooperatively (system-to-system cooperation), and interactively (human-to-system cooperation). They link the physical world seamlessly with the virtual world of information technology and software [22], and to do so they use various types of available data, digital communication facilities, and services [23]. CPS are closely related to the Internet of Things, where seamless communication among physical objects is achieved through embedded systems and communication networks [8].

A step forward is the integration of the digital tools related to product development into Cloud platforms, in order to enhance collaboration among the various actors at this stage of the product lifecycle providing ubiquitous access to information [24]. It can be concluded that the latest advances in ICT technologies are the means for the interconnection of the different elements of a manufacturing system towards a digitalized ecosystem. Integration standards with semantic representation of information, such as OPC-UA, along with Web-services and wireless sensor networks enable the seamless communication among tangible resources and humans. Especially, with the use of mobile and wearable devices humans enter the cyber world and communicate with machines.

2.2.2 Introduction of Enabling Technologies to Industry 4.0 Workflow and Required Skills

The Industry 4.0 paradigm introduces a new workflow for production systems by integrating the latest advances in Information and Communication technologies. From the very first step of the requirements collection, the social media enable customers opinion retrieval in a crowd sourcing manner [25]. As far as product design is concerned, AR is becoming a major part of the prototyping process. In collaborative VR/AR systems, the 3D models are displayed in a collaborative design session. In this phase, the AR technologies mainly involve engineers whose existing technical skills do not require radical enhancement, but may benefit from the advanced product visualization. Nevertheless, they need to familiarize with the concept of AR and understand its capabilities. Furthermore, the potential of cloud technologies in collaborative design is well-discussed in [26]. However, the design for X philosophy is required in order to achieve the goals set by the manufacturing company [27].

In an AR-based machining simulation environment, the users can gain awareness on the process information overlay, e.g. cutting parameters, and axes movement. The advantages of applying AR in these applications are that the user can accumulate knowledge and information when operating on real machines. The use of IoT devices for the real-time monitoring systems for the manufacturing resources, towards a CPS approach, may generate meaningful information that can be delivered via AR to production line operators [28, 29]. However, this novel information provision is intended also for employees with technical education. Therefore, the technical workforce should be capable of processing main operating characteristics and performance indicators related to the process and the resource and decide on potential changes. Specifically, this capability can support the identification of disturbances in production prior to their occurrence, supporting agile reaction [30].

Maintenance activities, e.g., preventive and corrective maintenance, are performed according to pre-defined procedures of the maintenance tasks. The CPS can support the condition-based preventive maintenance through the monitoring of the actual status of the manufacturing resources [31]. Maintenance technicians need to be trained in the respective procedures, and they sometimes need to seek help from supporting systems and experts [32]. AR technology shows merit in the maintenance applications in two aspects. Firstly, user interfaces can be rendered in a ubiquitous manner so that the operator perceives the instructions with less effort.

Secondly, user interactions in the AR environment can facilitate maintenance data exchange and allow remote collaboration to be achieved intuitively, while the on-spot technician performs the task [33]. Moreover, the collaboration through mobile devices in a social media manner facilitates the problem-solving potential, especially when dealing with unexpected events in the production [34]. The use of AR in actual production environments requires a workforce that is highly familiarized with it. This is firstly a safety issue, due to the limitations that AR projections introduce to the visibility of the user, while also it may lead to dizziness and increased cognitive load if used for extended periods of time.

During assembly, except for AR supported assembly [35], human-robot collaboration is a major enabler for increasing production rates while maintaining high quality in operations that involve humans. Hence, the operators need to be able to work inside a hybrid work cell and be familiar with the technology that is exploited to facilitate this collaboration [36]. In these situations, high alertness of the operator is critical while also following the strict task sequencing that the production planning has generated to make human-robot collaboration effective. The use of sensors, mobile devices, wearables, and context sensitive data provision transforms the operators of a digitalised manufacturing system into Operators 4.0. Thus, advanced learning models for machines are important so that humans and machines could develop skills that complement each other under any working conditions. One future research direction is an approach for "human-in-the-loop" machine learning, which enables humans to interact efficiently and effectively with decision-making models [37].

The integration of the Industry 4.0 technologies can be performed gradually without necessarily introducing radical changes in the production. This transformation is depicted in (Fig. 1). Nevertheless, the way of thinking about managerial issues and the approaches to decision making, need to adapt to the Industry 4.0 philosophy. This mostly affects the approaches for the data handling and the related software selection. Therefore, an important skill for the future engineers is to be able to distinguish the information related to the current task that can actually give value and support decision making. And based on this information, the selection of the appropriate software systems that will provide integration capabilities and process the data captured, towards improving the production processes through the accurate definition of the corresponding KPIs [38]. Furthermore, an open-minded systemic thinking will allow future engineers to adopt emerging technologies without delays.

Summarising this section, it is possible to aggregate and categorize competencies into four main groups – Technical, Methodological, Social and Personal competencies. [39] As it can be observed, the roles that are close to the production processes begin to require organizational skills to be able to function in the emerging requirements of the Industry 4.0. In addition, the executive roles require some technical knowledge on the

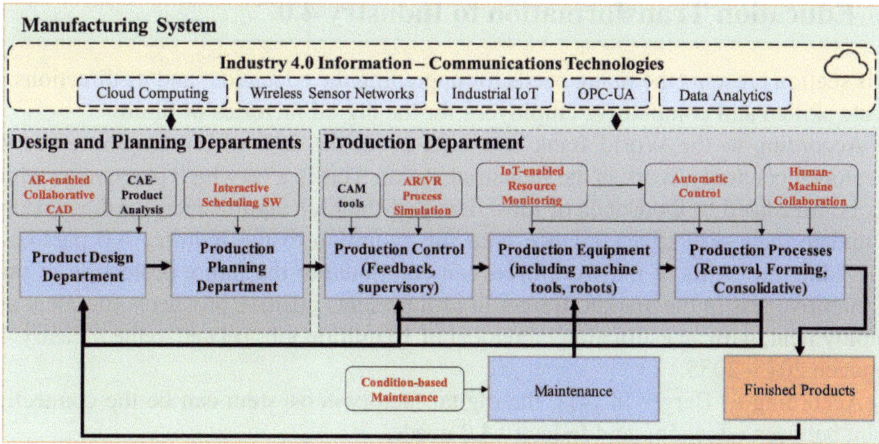

Fig. 1. Manufacturing Industry 4.0

production processes and the applied technologies, in order to exploit the value that is generated in the lower levels of the production. The skills and competences that are emerging from the introduction of the Industry 4.0 technologies, along with their importance to the roles of the employees are depicted in (Fig. 2).

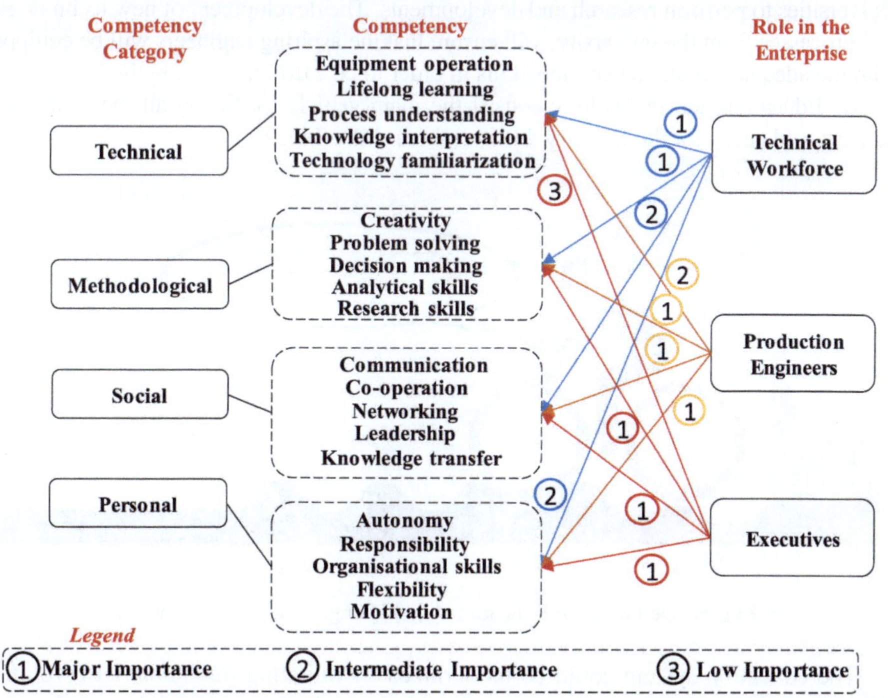

Fig. 2. Competencies for Industry 4.0 and importance to the roles in the enterprise.

3 Education Transformation to Industry 4.0

This section presents the rising needs for upgrading the education, and its directions, to create skilled and competitive employees for the digital factories of future.

According to the World Bank database, about the 20% of the employees of the developed countries, work in the Industrial sector. This is a very high percentage which can be translated to thousands of jobs, denoting the high significance for these people to update their skill in accordance with the evolution to the Industry 4.0 paradigm. Upgrading the skills of the workforce, can contribute to the future protection of their competitiveness in the market. Berger in [40] presents positive prospects for talent and employment, with an estimation increment of 10 million jobs related to the industry 4.0 between 2015–2035.

According to Berger in [41], the digital factory ecosystem can be the connecting point between education and Industry 4.0 needs.

Based on this ecosystem there is a chain which connects: (i) Government, (ii) Manufacturers, (iii) technology companies, (iv) Venture capital, and (v) University. In detail, the Government should provide grants and tax incentives to the research and innovation. On the other hand, the manufacturers should have the will to implement new technologies and new processes on a global market scale, while the technology provision companies should transform the new ideas and technologies into new business models and products. Finally, the venture capital should invest on testing and proving new ideas, while also supporting Universities to perform research and developments. The development of new technologies and processes from the university, will ensure that the aspiring engineers will be equipped with the adequate competences and skills in order to be a driving force in the industries of future. Education and technology consist the main vehicle for facing all the challenges, barriers, and needs for the Industry 4.0 workforce skills (Fig. 3).

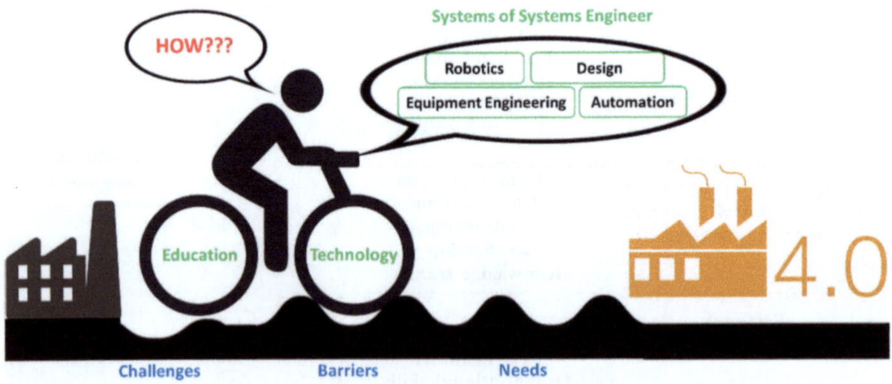

Fig. 3. Definition of Education 4.0 challenges, barriers, and needs

The education system could be modernized by imitating the industrial revolution (Fig. 4). Particularly, from the apprenticeships which were the dominant education practice in the decades of 50'–60', education was transformed into mass education through the universities, and nowadays, the universities have started to make online and digital Teaching towards personalized Teaching content.

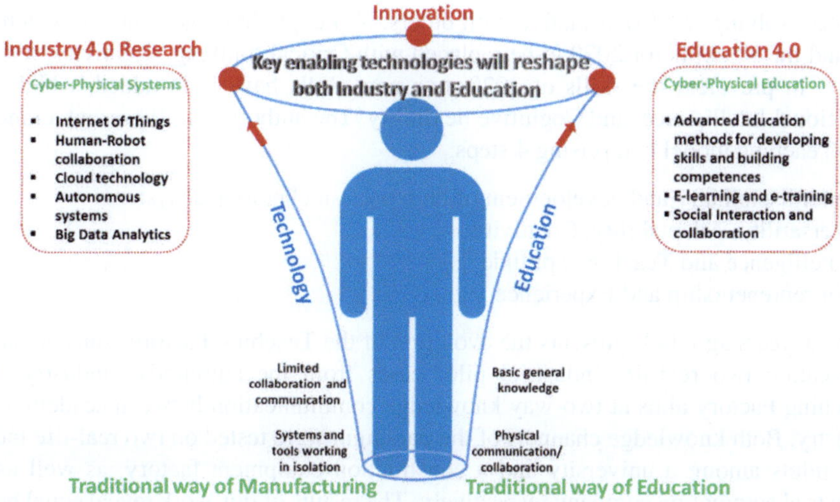

Fig. 4. Definition of Education 4.0

Table 1 shows the dominant knowledge requirements for the modern manufacturing systems, versus the knowledge requirements for the digital factory. As we can see in Table 1, the basic general knowledge around engineering has been transformed into Intelligence, while their capability to perceive their environment, a complex ecosystem with human operators and CPS, plays now a very important role, especially in dynamic production systems which change their workflow according to the production requirements. Moreover, the knowledge of using specific IT tools has been generalized to a capability to easily adapt to various IT tools, with the Teaching aptitude of the employees to be fast adaptive to every new tool or a new version of existing tools. Finally, the social interaction has been added to the digital factory skills requirements, since the good communication of manufacturing networking throughout the supply chain is nowadays the core competitive advantage.

Table 1. Skills for Engineering Education 4.0

Modern Factory knowledge	Digital Factory knowledge
Basic general knowledge	Intelligence
-	Senses to perception
Knowledge on using manufacturing equipment and specific IT tools	Advance knowledge on using various IT tools
	Teaching aptitude
-	Experience
Experience	Ability to improve
Ability to improve	Creativity
Creativity	Social Interaction
-	

According to a recent study of WEF (2017), about the 35% of core skills will change between the years 2015–2020. In the same research, a ranking is performed between the top 10 skills in 2015 and the 2010, with the first 3 skills for the 2015 to be: 1st Complex problem solving, 2nd Coordinative with others, 3rd People Management, for which the 2nd and the 3rd skills for 2020 to be replaced with Critical thinking, Creativity, respectively. In prevalent the skills of 2020, two new skills have been added, which are: Emotional Intelligence, and Cognitive flexibility. The authors in [42] present an industrial Teaching model comprising 4 steps:

- Interdisciplinary and development of high socio-technological systems
- Versatility, Adaptability, Creativity
- Intelligence and Teaching aptitude
- Entrepreneurship and Experience

Few years ago, [43] presents the evolution of the Teaching Factory concept and its application two real-life industrial pilot cases from the automotive industry. The Teaching Factory aims at two-way knowledge communication between academia and industry. Both knowledge channels of the paradigm were tested on two real-life industrial pilots among a university and a construction equipment factory, as well as an industrial automation company respectively. The results of this work reveal equal benefits for both sides. From the industrial side, they secure a connection to the ever-changing technologies and the on-going research, while providing new ideas and solutions beyond the traditional company's process. For the academia, Teaching factories offer a useful connection between the theory and the generated knowledge and their integration under real cases, especially considering that the participants lack related experience. Finally, Teaching Factory encourages entrepreneurship in universities and innovation within companies. The actual benefit of the economic system by adopting new techniques towards Education 4.0 is illustrated in Fig. 5.

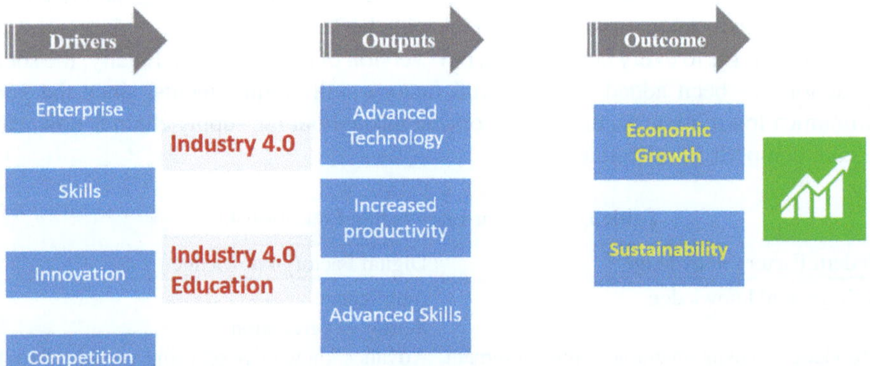

Fig. 5. Teaching factory 4.0 benefits

4 Innovative Teaching Factory Approach

Following the findings presented in the previous sections, related to the requirements for new skills and the adaptation of education to fulfill the demands of Industry 4.0, it is apparent that an innovative Teaching Factory approach is essential for the academia and the industry. The proposed Teaching Factory approach upgrades the human operator skills and competencies needed for the 4th Industrial evolution paradigm as presented in detail in the previous chapter.

This innovative Teaching Factory approach refers to the under-graduate students of engineering science university departments. Therefore, it contributes to the skills and competences required mainly for production Engineers. The proposed approach is already applied to the undergraduate curriculum and includes 4 steps, where the participants, separated in groups, go through the requirements collection to the design and the production of a complex product [44].

It is of high importance for students to be familiar with a structured way of thinking for solving an industrial problem, starting with the collection of requirements, followed by the specification of the environment that the product will be developed, and after that the actual design and the implementation. The previous structured way of thinking is very important for their professional career later. The proposed Teaching Factory Paradigm is illustrated in Fig. 6.

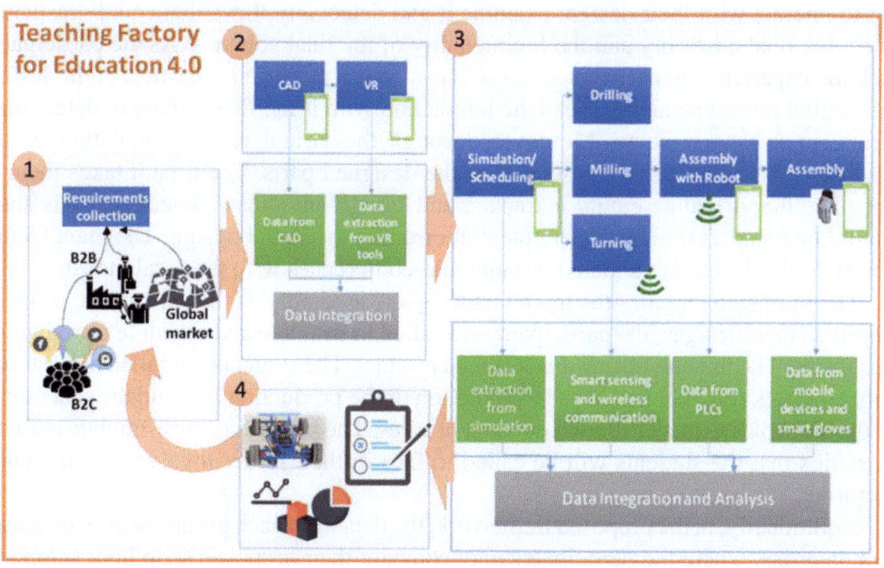

Fig. 6. Teaching Factory for Education 4.0

This teaching factory paradigm aims to educate young Engineers in the Industry 4.0 workflow that is presented in Sect. 2.

The Teaching Factory concept is based on the knowledge triangle notion (teaching, research, and technology transfer) aiming to become a new paradigm of both academic and industrial learning [42, 45]. The first step is the requirements collection. The students should initially recognize the target group of market, whether the product will be provided in the end-consumers, business companies, or the regional market; all these aspects affect the way that the requirements collection will be performed. For example, if the target group of products is other industries in a business-to-business manner, then the students will have the opportunity to organize a number of meetings (web-meetings and on-site) in order to collect all the appropriate information, to start a design that meets the industrial needs. This is achieved through the close collaboration of the university with the local industry. If this product is a regional product, then the frugal innovation aspects should be considered during the design, which will be based more in the culture of the specific area, and the affordability of the product. [46] Finally, if the students design a product that will be provided in the end-consumers, then, the exploitation of the social media and the comments in several forums should be considered in order to collect the appropriate requirements. In this case, several open source software tools could be considered to facilitate the process.

The second step is the design of the product. In this stage, the students are called to prepare the initial design of their product in a CAD software. The part designs from each group are stored in a central database. The product designs are then used for an augmented/virtual reality evaluation workshop. The participants are called to validate their designs in real scale using Augmented Reality and then to simulate the assembly process using Virtual Reality [47]. In the first phase, the Teaching Factory participants get to interact with their design, examine it and detect any flaws that could drastically affect the final assembly and the functionality of the final solution. As the participants lack the experience to early detect these flaws and to produce a first-time-right design, this digital prototype and the realistic interaction with it facilitates them to detect their errors, in this early stage. Moreover, through the virtual reality simulation of the assembly, they may test the assembly of the designed parts. The time it takes them to complete the virtual assembly is tracked and stored; any delays detected at this stage should be noted as it may reflect potential errors in the initial design. The manufacturability of the design is validated through tele-conferences with the local industry [48].

The third step includes the manufacturing and assembly of the parts of the product. Based on their designs, the participants are called to simulate and schedule a production line that will be manufacturing the designed product. The students are called to use novel software packages that support the simulation of the production line and the scheduling of the task so as to meet different demand profiles. The provided cases simulate realistic scenarios that the students will be called to face in their future involvement in manufacturing.

Additionally, in the proposed framework, the data from each group are directly stored in a database, where the participants may compare their proposed shop floor setup and the achieved performance with the other groups, evaluating their own proposals. This procedure aims to involve the group in a self- evaluating improvement, during which they familiarize with a simulation software and eventually the requirements of a production line. After simulating the production and familiarizing with the theoretical background of each process required, the participants are called to manufacture the parts for their product. The Teaching Factory offers them three processes to utilize: milling, drilling and turning. Each station/machinery is upgraded in a Cyber- Physical System (CPS) equipped with sensors that monitor the process and is connected with a wireless network that collects all the recorded data and stores them in the database. Each group uses the given near field communication (NFC) card to state their group number and the starting time of each process, and under the supervision of an experienced operator of the machinery, manufacture the required parts. The data from the installed sensors are recorded and sent to the central database and also to a mobile device application from which the participants may read the recorded data in real time, monitor the status of the machine and notice any deviations. Through this part of the Teaching Factory 4.0, the participants familiarize themselves with the Industry 4.0 technologies that shape manufacturing and start to be integrated in production lines. The assembly of the final product is executed in two parts. The first part is performed with the aid of a robotic arm, also under the human-robot collaboration framework. Since the majority of the assembly processes in modern manufacturing is performed by automations, it is important that Education 4.0 integrates these approaches in the courses and Teaching workshops. The aspiring engineers are called to collaborate with the robot in an assembly process, using augmented reality goggles to see the instruction what to do and the trajectory of the robot when it is moving, and also communicate with it through a smart watch. Through the smart watch, the operator may interact wirelessly with the robot, confirming the completion of an action, seeing useful notifications and validating the completion of the whole task. During the task, each group collects data about the required time and also from the PLC of the robot, thus creating a useful dataset about the task, from which they may draw some information. Following the directives of digitalization introduced by Industry 4.0, the participants wear a smart glove during the assembly process. The smart glove hosts sensors that track the movements, grasps and assemblies of the operator creating a map of its movements. Using the NFC card, each group indicates the beginning and competing time of each assembly step. Moreover, to support this process, the operator may use a mobile device application that utilizes augmented reality to visualize assembly instructions for the standardized parts of the designed product.

Through the recorded data, each group may evaluate their performance in the assembly process, by monitoring the required time and detecting redundant movements that could be a result of an error in the original design or manufacturing.

After the completion of the third step, each group tests the developed product, in a sequence of trials. Completing the Teaching factory 4.0, the participants acquire a set of highly useful skills that will support their future careers in manufacturing. Apart from familiarizing themselves with the traditional steps of product design and manufacturing tools, the aspiring engineers become educated in the technologies dictated by Industry 4.0 and will shape future manufacturing. Teaching Factories, following the technological breakthroughs, need to evolve so as to provide the workshops that will support the Education 4.0, making the aspiring engineers used to working in the industrial environments, in realistic conditions and considering resources and design limitations, as they will be called to do in their future careers.

All the steps of the process were upgraded to provide digital data on the task: mobile devices apps were used in the design phase steps while embedded sensors provided data coming from the machines. Making the students familiar with these features that are forming the cyber-physical systems of future production lines, prepares the aspiring engineers for the manufacturing environments that they will work in. Data integration and analysis of the gathered data provides an insight for the students to better understand the manufacturing phases and evaluate their performance through the task. The major aspects from the proposed Teaching Factory are presented in Fig. 7.

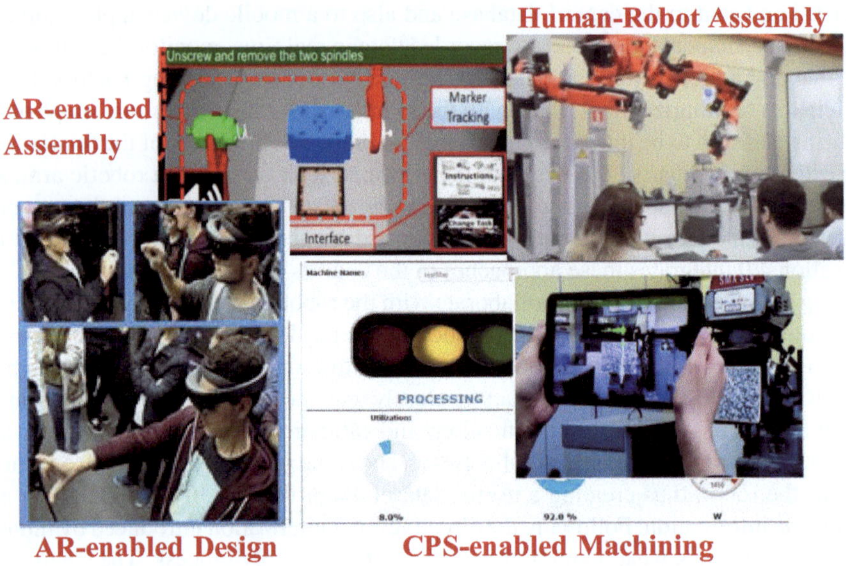

Fig. 7. Major aspects of the proposed Teaching Factory

5 Conclusions

The present study investigates the new requirements for employee's skills raised from the new manufacturing paradigm of Industry 4.0. The new manufacturing workflow is presented along with the needs for new skills and competences. The model of Education 4.0 is presented based on these needs. Finally, a Teaching Factory concept is proposed following the requirements of Industry 4.0 and Education 4.0. The proposed framework aims to facilitate the training and to aspire engineers in better envisioning and evaluation of their product designs, raising their interest in product design and manufacturing, and eventually increasing the overall effectiveness of the Teaching Factory. Furthermore, young engineers are familiarized with the novel digital technologies that are expected to shape manufacturing in the years to come and broaden the horizon of thought of the participants towards Industry 4.0 paradigm. The integration of innovative technologies has the potential to improve the insight of the designer on the final product thus increasing error detection in the early stages of design and the confidence of the participants for their design. A major outcome of the proposed Teaching Factory is the training of young engineers of being able to work in a group and deliver a tangible result in near-real manufacturing conditions [49].

Future enhancement of the proposed Teaching Factory aims to complement the curriculum with workshops on the development of the specific Industry 4.0 applications. Hence, the young engineers will learn in depth the technologies for the development of the AR applications, CPS monitoring systems, human robot collaboration etc. This knowledge will make them capable of proposing the adoption of these technologies in their future work in industry. Therefore, major steps towards the adoption of Industry 4.0 practices in real manufacturing systems will be performed.

References

1. Deloitte: Challenges and Solutions of the Digital Transformation (2014). https://www2.deloitte.com/content/dam/Deloitte/ch/Documents/manufacturing/ch-en-manufacturing-industry-4-0-24102014.pdf. Accessed 30 Jan 2018
2. Chryssolouris, G., Mavrikios, D., Papakostas, N., Mourtzis, D., Michalos, G., Georgoulias, K.: Digital manufacturing: history, perspectives, and outlook. Proc. IMECHE Part B J. Eng. Manuf. **222**, 451–462 (2008)
3. Nee, A.Y.C., Ong, S.K., Chryssolouris, G., Mourtzis, D.: Augmented reality applications in design and manufacturing. CIRP Ann. Manuf. Technol. **61**, 657–679 (2012)
4. Wu, D., Greer, M.J., Rosen, D.W., Schaefer, D.: Cloud manufacturing: strategic vision and state-of-the-art. J. Manuf. Syst. **32**(4), 564–579 (2013)
5. Meier, H., Roy, R., Seliger, G.: Industrial product-service systems—IPS2. CIRP Ann. Manuf. Technol. **59**(2), 607–627 (2010)
6. SAS: Big Data Meets Big Data Analytics (2012). https://eric.univ-lyon2.fr/~ricco/cours/slides/sources/big-data-meets-big-data-analytics-105777.pdf. Accessed 30 Jan 2018
7. Court, D.: Getting Big Impact from Big Data (2015). http://www.mckinsey.com/insights/business_technology/getting_big_impact_from_big_data. Accessed 30 Jan 2018
8. Mourtzis, D., Vlachou, E., Milas, N.: Industrial Big Data as a result of IoT adoption in Manufacturing. In: CIRPe2016, Procedia CIRP, 5th CIRP Global Web Conference-Research and Innovation for Future Production, vol. 55, pp. 290–295 (2016)
9. He, W., Xu, L.: A state-of-the-art survey of cloud manufacturing. Int. J. Comput. Integr. Manuf. **28**(3), 239–250 (2015)
10. Horváth, I., Vroom, R.W.: Ubiquitous computer aided design: a broken promise or a Sleeping Beauty? Comput. Aid. Des. **59**, 161–175 (2015). https://doi.org/10.1016/j.cad.2014.10.006
11. Barfield, W., Shieldst, R., Cooper, S.: A survey of computer-aided design: implications for creativity, productivity, decision making, and job satisfaction. Int. J. Hum. Factors Manuf. **3**(2), 153–167 (1993). https://doi.org/10.1002/hfm.4530030205
12. Kochhar, N.K., Kelkar, S.G.: The role of CAE in product development at Ford Motor Company. In: Computational Fluid and Solid Mechanics, pp. 15–18 (2003). https://doi.org/10.1016/b978-008044046-0.50007-5
13. Efthymiou, K., Pagoropoulos, A., Mourtzis, D.: Intelligent scheduling for manufacturing systems: a case study. Lecture Notes in Mechanical Engineering, pp. 1153–1164 (2013). https://doi.org/10.1007/978-3-319-00557-7_94
14. Mourtzis, D., Doukas, M., Vlachou, E.: A mobile application for knowledge-enriched short-term scheduling of complex products. Log. Res. **9**(1), 3 (2016). https://doi.org/10.1007/s12159-015-0130-7
15. Atieh, A.M., et al.: Performance improvement of inventory management system processes by an automated warehouse management system. Proc. CIRP **41**, 568–572 (2016). https://doi.org/10.1016/j.procir.2015.12.122

16. Newman, S.T., Nassehi, A.: Universal manufacturing platform for CNC machining. CIRP Ann. **56**(1), 459–462 (2007). https://doi.org/10.1016/j.cirp.2007.05.110
17. Teti, R., et al.: Advanced monitoring of machining operations. CIRP Ann. **59**(2), 717–739 (2010). https://doi.org/10.1016/j.cirp.2010.05.010
18. Michalos, G., Makris, S., Chryssolouris, G.: The new assembly system paradigm. Int. J. Comput. Integr. Manuf. **28**(12), 1252–1261 (2014). https://doi.org/10.1080/0951192x.2014.964323
19. Mori, M., Fujishima, M.: Remote monitoring and maintenance system for CNC machine tools. Proc. CIRP **12**, 7–12 (2013). https://doi.org/10.1016/j.procir.2013.09.003
20. Kristianto, Y., et al.: A study of technology adoption in manufacturing firms. J. Manuf. Technol. Manag. **23**(2), 198–211 (2012). https://doi.org/10.1108/17410381211202197
21. Azuma, R., Baillot, Y., Behringer, R., Feiner, S., Julier, S., MacIntyre, B.: Recent advances in augmented reality. IEEE Comput. Graph. Appl. **21**(6), 34–47 (2001). https://doi.org/10.1109/38.963459
22. Sztipanovits, J., Ying, S.: Strategic R&D opportunities for 21st century: cyber-physical systems, Connecting computer and information systems with the physical world. Report (2013). http://www.nist.gov/el/upload/12-Cyber-Physical-Systems020113_final.pdf. Accessed 30 Jan 2018
23. Mikusz, M.: Towards an understanding of cyber-physical systems as industrial software-product-service systems. Proc. CIRP **16**, 385–389 (2014). https://doi.org/10.1016/j.procir.2014.02.025
24. Mourtzis, D., Vlachou, E.: Cloud-based cyber-physical systems and quality of services. TQM Emerald J. **28**(5), 704–733 (2016)
25. Forbes, H., Schaefer, D.: Social product development: the democratization of design, manufacture and innovation. Proc. CIRP **60**, 404–409 (2017). https://doi.org/10.1016/j.procir.2017.02.029
26. Valilai, F., Houshmand, M.: A collaborative and integrated platform to support distributed manufacturing system using a service-oriented approach based on cloud computing paradigm. Robot. Comput. Integr. Manuf. **29**(1), 110–127 (2013)
27. Sassanelli, C., et al.: Testing the methodology to generate Design for Product Service Supportability (DfPSS) guidelines and rules: an application case. Proc. CIRP **64**, 265–270 (2017)
28. Mourtzis, D., Vlachou, E., Zogopoulos, V., Xanthi, F.: Integrated production and maintenance scheduling through machine monitoring and augmented reality: an Industry 4.0 approach. In: APMS Conference 2017, Hamburg, Germany, 3–7 September (2017)
29. Segovia, D., Mendoza, M., Mendoza, E., González, E.: Augmented reality as a tool for production and quality monitoring. Proc. Comput. Sci. **75**, 291–300 (2015). https://doi.org/10.1016/j.procs.2015.12.250
30. Wittenberg, C.: Human-CPS interaction - requirements and human-machine interaction methods for the Industry 4.0. IFAC-PapersOnLine **49**(19), 420–425 (2016). https://doi.org/10.1016/j.ifacol.2016.10.602
31. Mourtzis, D., Vlachou, E., Milas, N., Xanthopoulos, N.: A cloud-based approach for maintenance of machine tools and equipment based on shop-floor monitoring. In: CIRP CMS 2015, Procedia CIRP, 48th CIRP Conference on Manufacturing Systems, Naples, Italy, 24–26 June, vol. 41, pp. 655–660 (2015)
32. Mourtzis, D., Vlachou, E., Zogopoulos, V.: Cloud-based augmented reality remote maintenance through shop-floor monitoring: a PPS approach. ASME J. Manuf. Sci. Eng. **139**(6), 061011 (2017)

33. Masoni, R., Ferrise, F., Bordegoni, M., Gattullo, M., Uva, A.E., Fiorentino, M., Carrabba, E., Di Donato, M.: Supporting remote maintenance in Industry 4.0 through augmented reality. Proc. Manuf. **11**, 1296–1302 (2017). https://doi.org/10.1016/j.promfg.2017.07.257
34. Mourtzis, D., Doukas, M., Milas, N.: A knowledge-based social networking app for collaborative problem-solving in manufacturing. Manuf. Lett. **10**, 1–5 (2016). https://doi.org/10.1016/j.mfglet.2016.08.001
35. Makris, S., Pintzos, G., Rentzos, L., Chryssolouris, G.: Assembly support using AR technology based on automatic sequence generation. CIRP Ann. Manuf. Technol. **62**(1), 9–12 (2013)
36. Makris, S., Karagiannis, P., Koukas, S., Matthaiakis, A.S.: Augmented reality system for operator support in human–robot collaborative assembly. CIRP Ann. Manuf. Technol. **65**(1), 61–64 (2016)
37. Zhong, R.Y., Xu, X., Klotz, E., Newman, S.T.: Intelligent manufacturing in the context of Industry 4.0: a review. Engineering **3**(5), 616–630 (2017). https://doi.org/10.1016/J.ENG.2017.05.015
38. Mourtzis, D., Fotia, S., Vlachou, K.: Lean rules extraction methodology for lean PSS design via key performance indicators monitoring. J. Manuf. Syst. **42**, 233–243 (2017)
39. Hecklau, F., Galeitzke, M., Flachs, S., Kohl, H.: Holistic approach for human resource management in Industry 4.0. Procedia CIRP **54**, 1–6 (2016). https://doi.org/10.1016/j.procir.2016.05.102
40. Berger, R.: The Industrie 4.0 transition quantified: how the fourth industrial revolution is reshuffling the economic, social and industrial model. Report (2016). https://www.rolandberger.com/de/Publications/pub_the_industrie_4_0_transition_quantified.html. Accessed 30 Jan 2018
41. Berger, T., Frey, C.: Digitalization, Jobs, and Convergence in Europe: Strategies for Closing the Skills Gap (2016). http://www.oxfordmartin.ox.ac.uk/downloads/reports/SCALE_Digitalisation_Final.pdf. Accessed 30 Jan 2018
42. Mavrikios, D., Papakostas, N., Mourtzis, D., Chryssolouris, G.: On industrial learning & training for the Factories of the Future: a conceptual, cognitive & technology framework. J. Intell. Manuf. **24**(3), 473–485 (2011). Special Issue on Engineering Education
43. Rentzos, L., Mavrikios, D., Chryssolouris, G.: A two-way knowledge interaction in manufacturing education: the teaching factory. In: Procedia CIRP, 5th Conference on Learning Factories, Bochum, Germany, 7–8 July, vol. 32, pp. 31–35 (2015)
44. Chryssolouris, G., Mavrikios, D., Rentzos, L.: The teaching factory: a manufacturing education paradigm. In: CIRP-CMS 2016, 49th CIRP Conference on Manufacturing Systems, Stuttgart, Germany, 25–27 May, vol. 57, pp. 44–48 (2016)
45. Chryssolouris, G., Mavrikios, D., Papakostas, N., Mourtzis, D.: Education in manufacturing technology & science: a view on future challenges & goals. Inaugural keynote. In: Proceedings of the International Conference on Manufacturing Science and Technology, Melaka, Malaysia, August 2006 (2006)
46. Colledani, M., Silipo, L., Yemane, A., Lanza, G., Bórgin, J., Hochdφrffer, J., Georgoulias, K., Mourtzis, D., Bitte, F., Bernard, A., Belkadi, F.: Technology-based product-services for supporting frugal innovation. In: 8th IPSS Conference, Product-Service Systems Across Life Cycle (CIRP IPSS), Procedia CIRP, Bergamo, Italy, 20–21 June, vol. 47, pp. 126–131 (2016)
47. Rentzos, L., Vourtsis, C., Mavrikios, D., Chryssolouris, G.: Using VR for complex product design. In: Virtual, Augmented and Mixed Reality, Applications of Virtual and Augmented Reality. Lecture Notes in Computer Science, vol. 8526, pp. 455–464 (2014)

48. Rentzos, L., Doukas, M., Mavrikios, D., Mourtzis, D., Chryssolouris, G.: Integrating manufacturing education with industrial practice using teaching factory paradigm: a construction equipment application. In: CMS 2014, Procedia CIRP, 47th CIRP Conference on Manufacturing Systems, Ontario, Canada, 28–30 April, vol. 17, pp. 189–194 (2014)
49. Chryssolouris, G., Mourtzis, D., Stavropoulos, P., Mavrikios, D., Pandremenos, J.: Knowledge management in a virtual lab collaborative training project. In: Bernard, A., Tichkiewitch, S. (eds.) Methods and Tools for Effective Knowledge Life-Cycle-Management, vol. 3, pp. 435–446 (2008)

Opportunistic Maintenance for Wind Turbines Considering External Opportunities – A Case Study

H. Truong Ba[1(✉)], M. E. Cholette[1], P. Borghesani[2], and L. Ma[1]

[1] Queensland University of Technology, Brisbane, Australia
h.truongba@qut.edu.au
[2] University of New South Wales, Sydney, Australia

Abstract. This paper aims to develop an opportunistic maintenance (OM) policy for the generator of a hypothetical wind turbine using methods developed recently by the authors. The OM policy considers external opportunities caused by low wind speeds which produce little-to-no electric power. The results show that some cost savings are achievable by taking maximal advantage of these low-speed wind events, particularly when electricity prices are at their peak cycle.

Keywords: Opportunistic maintenance · Maintenance optimization
Partial opportunity · Wind turbine maintenance

Abbreviations:

PM:	Preventive maintenance
OM:	Opportunistic maintenance
CM:	Corrective maintenance
NHPP:	Non-homogeneous Poisson Process

Notation

Throughout, capital letters denote random variables while lower case variables denote deterministic quantities.

t :	age of a system.
t_{PM} :	planned PM maintenance interval.
T :	the length of a renewal cycle
p_{PM} :	probability of cycle stop due to PM
p_{OM} :	probability of cycle stop due to OM.
p_{CM} :	probability of cycle stop due to CM.
D :	duration of opportunity.
$d_{min}(t)$:	decision function of minimum opportunity duration that will be accepted.
d_{PM} :	fixed duration of PM activities.
CR_{PM} :	cost rate relating to PM activities.
CR_{OM} :	cost rate relating to OM activities.
CR_{CM} :	cost rate relating to CM activities.

J. Ni et al. (Eds.): AMP 2018, LNME, pp. 211–226, 2018.
https://doi.org/10.1007/978-3-319-89563-5_16

$c_{CM}(t)$: the cost of one CM instance.
c_b : the direct cost for maintenance activities.
$c_l(t)$: cost rate per unit of downtime
$f(\cdot), F(\cdot), \bar{F}(\cdot)$: The PDF, CDF and survival function of machine.
TC : total maintenance cost
CR : The expected maintenance cost per unit of time (cost rate).

$$[x]^+ \triangleq \begin{cases} x, x \geq 0 \\ 0, x < 0 \end{cases}$$

1 Introduction

Wind energy is one of the fastest growing renewable energy sources for electricity generation in Australia and around the world. As wind energy becomes an increasingly mature field, the O&M portion of wind farm costs is becoming significant. Hence, employment of an efficient maintenance policy is critical to the economical operation of wind farms.

Maintenance strategies applied to wind turbines are similar to other industrial systems and include corrective maintenance (CM) and preventive maintenance (PM) approaches, both time based and condition based [1]. Compared to CM, PM is more beneficial because it can reduce the number of unplanned failures [2]. A lot of research has therefore been focused on optimizing PM policies for wind turbines using both time-based models [3–7] and condition-based models [8–12].

Recent research into opportunistic maintenance (OM) policies could provide significant benefits for wind turbine maintenance practice. OM is a particular type of preventive maintenance, which is conducted when an opportunity for lower maintenance costs occurs. OM is usually considered for the functional components of a system when corrective or preventive maintenance activity occurs on other components (i.e. group maintenance) [13–15]. This type of OM constitutes the vast majority of OM research in the field of wind turbine systems [3–5, 16–19]. Savings on set-up costs have been the focus of most OM studies (i.e. grouping maintenance policies), particularly for wind turbines. The reason is that many wind farms are located in offshore or remote zones where it is challenging to send maintenance crews. Therefore, logistical issues have been the main focus on wind turbine OM policies [1, 5, 20, 21].

This type of opportunity is often referred to as *internal*, in contrast to *external* opportunities which are caused by outside phenomena such as weather events which can stop industrial systems and create the chances for maintenance. Previous studies [22, 23] have discussed the potential benefit of implementation of this type OM policy for industrial systems, particularly production systems which are impacted by weather factors, e.g. agricultural process industries. Wind turbine operation is obviously affected directly by wind speed, and speeds near or below the cut-in speed can significantly decrease (or halt) production and therefore create low-cost maintenance windows, i.e. less maintenance-induced downtime.

This study aims to apply a recently developed OM model [23] considering external opportunities to maintenance activities of a wind turbine. The considered external opportunities consist of time-windows in which wind speed is too low for the operation of the turbine. The proposed policy is illustrated for a hypothetical wind turbine installed in the Mount Emerald windfarm, Queensland – Australia. A sensitivity analysis is conducted to provide a deeper understanding of the benefits of this type of OM.

The remainder of this paper is organized in four sections. Section 2 summarizes the mathematical model developed in [23] and used in this study; Sect. 3 discusses the case study of a hypothetical wind turbine within the Mount Emerald windfarm; and Sect. 4 presents the numerical results and analysis. Section 5 (conclusions) summarizes the main findings and proposes future potential developments.

2 Model Summary

This study is aimed at developing a preventive maintenance (PM) policy with opportunistic maintenance (OM) for a hypothetical wind turbine. The theoretical bases for this task are detailed in [23], and this section will summarize the methodology for this specific application. Wind turbine components are replaced (perfect repair) at a planned time or when an external opportunity arrives, which may yield some cost savings. The external opportunities considered in this model are time-windows (discretized in days) when the average wind speed is less than the cut-in threshold of wind turbine. At this time, the wind turbine is not functioning and this constitutes a chance to do preventive maintenance without incurring in additional downtime costs. Historical wind data at the selected location is used to develop the opportunity model and production loss cost of wind turbine. The developed OM policy for a wind turbine can be considered as a basic starting point to develop a more advanced grouping maintenance policy, thus combining external opportunities to the internal events discussed in previous wind-turbine maintenance literature [9, 17, 18]. It is noted that wind turbines also stop when the wind speed is higher than the cut-out threshold and this also creates an opportunity for maintenance; however, this situation is not considered because of safety considerations.

The OM model in this study is based on model 2 of the study in [23], which assumes perfect corrective maintenance, i.e. the systems/components are replaced with a new asset if a failure occurs before planned or opportunistic maintenance. In this model, the OM component of the maintenance policy is represented by the decision variable $d_{min}(t)$, whereas the planned maintenance component is governed by the planned-maintenance interval t_{PM}. In between two planned maintenance events (separated by t_{PM}), when an opportunity arrives, its suitability for conducting OM is assessed based on the threshold $d_{min}(t)$ (minimum acceptable opportunity duration): if the duration D satisfies $D < d_{min}(t)$, the opportunity is rejected, whereas if $D \geq d_{min}(t)$ opportunistic maintenance is conducted at time t. The limit case where $d_{min}(t) = 0$ represents an OM policy which accepts any opportunity at time t, whereas the case when $d_{min}(t) = \infty$ represents a policy rejecting any opportunity at time t, regardless of its duration D. The objective of the policy optimization will therefore be to minimize the expected cost rate by jointly optimizing the decision variables t_{PM} and $d_{min}(t)$.

The expected total cost per unit time is determined based on a maintenance cycle of duration T, which is defined by the interval between two subsequent maintenance actions (corrective or preventive):

$$\begin{aligned} E\left[\frac{TC}{T}\right] &= E\left[\frac{C_{OM}(T_{OM}, D)}{T_{OM}}\right]p_{OM} + \left[\frac{c_{PM}(t_{PM})}{t_{PM}}\right]p_{PM} + E\left[\frac{c_{CM}(T_{CM})}{T_{CM}}\right]p_{CM} \\ &= E[CR_{OM}] + E[CR_{PM}] + E[CR_{CM}] \end{aligned} \tag{1}$$

where $c_{CM}(T_{CM})$, $c_{PM}(t_{PM})$ and $C_{OM}(t, D)$ are the corresponding costs of CM, PM and OM activities, and p_{CM}, p_{PM} and p_{OM} are the probabilities that the cycle is terminated by CM, PM or OM, respectively. Note that the time variable t is reset to zero at every maintenance operation, thus starting a new cycle.

The CM cost $c_{CM}(T_{CM})$ includes two parts: the fixed cost and the downtime cost $c_l(T_{CM})$ which depends on the time-varying electricity prices and instantaneous turbine power output. The PM-related cost is comprised of two parts: the direct cost c_b(e.g. spare parts and labor) and the production losses due to the PM duration (d_{PM}), which has production loss per unit time $c_l(t)$. The PM cost c_{PM} is then computed as

$$c_{PM} = c_b + c_{loss} = c_b + c_l(t_{PM}) \cdot d_{PM}. \tag{2}$$

The OM cost is divided into two parts: (1) the direct cost c_b which is the same as the direct cost of a PM activity, and (2) the production loss $c_l \cdot [D_{PM} - D]^+$, which in the OM case depends on the length of opportunity duration, which is a random variable. Thus the OM cost is computed as

$$C_{OM}(t, D) = c_b + c_l(t) \cdot [d_{PM} - D]^+. \tag{3}$$

In order to determine the probabilities p_{OM}, p_{CM} and p_{PM}, the timeline will be discretized into N time intervals of length Δt, $(t_{i-1}, t_i]\; i = 1, 2, \ldots, N$ as following the approach detailed in study [23], where $d_{min}(t)$ is considered constant for each interval. These probabilities and the cost rate components in Eq. (1) are computed as the same procedure with assumption that there is only one opportunity in each interval. The next two Sects. 2.1 and 2.2 summarize some theoretical results from [23], adjusted to the specific case to compute these terms.

2.1 Determining p_{OM}, p_{CM} and p_{PM}

Using this discretization, the probability of OM may be calculated as

$$p_{OM} = \sum_{i=1}^{i_{PM}} p_{OM_i} \tag{4}$$

where p_{OM_i} probability of conducting OM in the interval $(t_{i-1}, t_i]$ and $i_{PM} \in \{1, 2, \ldots, N\}$ is the time index where PM occurs. The probability p_{OM_i} may be computed as:

$$p_{OM_i} = p_{AO_i} \prod_{j=0}^{i-1} (1 - p_{SO_j}) \tag{5}$$

where p_{AO_i} is the probability that an accepted opportunity occurs in interval $(t_{i-1}, t_i]$ and p_{SO_i} is the probability of the arrival of at least one suitable opportunity in interval i:

$$p_{SO_i} = \bar{F}_{D,t_i}(d_{min}(t_i)) \cdot F_{O,t_{i-1}}(t_i) \tag{6}$$

where $F_{O,t_{i-1}}(t_i)$ is the probability of opportunity occurrence at time interval t_i, which is computed according to NHPP characteristics [24, 25]:

$$F_{O,t_{i-1}}(t_i) = \frac{\gamma(k, \Lambda(t_{i-1}, t_i))}{(k-1)!} \tag{7}$$

where $\gamma(k, u) = \int_0^u x^{k-1} e^{-x} dx$, is the lower incomplete gamma function and $\Lambda(t_s, t)$ is the mean value function of NHPP considering from the start time t_s:

$$\Lambda(t_s, t) = \int_{t_s}^{t} \lambda_O(\tau) d\tau \qquad t \in [t_s, t_f] \tag{8}$$

The probability p_{AO_i} is computed as follows:

$$p_{AO_i} = \bar{F}_{D,t_i}(d_{min}(t_i)) \int_{t_{i-1}}^{t_i} \bar{F}(t) \cdot f_{O,t_{i-1}}(t) dt \tag{9}$$

PM occurs when neither CM nor OM occur. That is, PM is the joint event that the system survives up to t_{PM} and no suitable opportunity occurs prior

$$p_{PM} = \bar{F}(t_{i_{PM}}) \cdot \prod_{i=1}^{i_{PM}} p_{\overline{SO_i}} \tag{10}$$

The probability of CM is:

$$p_{CM} = 1 - p_{OM} - p_{PM} \tag{11}$$

2.2 Optimization Problem

The expected cost rate may subsequently be computed using Eq. (1). The expected cost rate relating to PM activity is determined:

$$E[CR_{PM}] = \left[\frac{c_b + c_l(t_{PM}) \cdot d_{PM}}{t_{PM}}\right] \cdot p_{PM} \tag{12}$$

The expected OM cost rate $E[CR_{OM}]$ is determined as:

$$E[CR_{OM}] = \sum_{i=1}^{i_{PM}} p_{OM_i} \cdot E[CR_{OM}|OM_i] \tag{13}$$

where the term $E[CR_{OM}|OM_i]$ is computed as:

$$E[CR_{OM}|OM_i] = \begin{cases} 0 & \bar{F}_{D,t_i}(d_{min}(t_i)) = 0 \\ E\left[CR_{OM}^m|OM_i\right] & otherwise \end{cases} \tag{14}$$

where $E\left[CR_{OM}^m|OM_i\right]$ is the expected OM cost rate relating to the maintenance action given that the OM is conducted at interval $(t_{i-1}, t_i]$:

$$\begin{aligned} &E\left[CR_{OM}^m|OM_i\right] \\ &= \frac{1}{p_{AO_i}} \cdot \bar{F}_{D,t_i}(d_{min}(t_i)) \\ &\cdot \int_{t_{i-1}}^{t_i} \left(c_b + c_l(t) \cdot \int_{\min(d_{min}(t_i),d_{PM})}^{d_{PM}} [d_{PM} - D] f_{D,t_i}(D) dD\right) \cdot \frac{\bar{F}(t) \cdot f_{O,t_{i-1}}(t)}{t} dt \end{aligned} \tag{15}$$

The cost rate $E[CR_{CM}]$ relating to CM activity is defined as:

$$E[CR_{CM}] = \int_0^{t_{i_{PM}}} c_{CM}(t) \cdot \frac{P_{\overline{OM}}(t) \cdot f(t)}{t} dt = \sum_{i=1}^{i_{PM}} \int_{t_{i-1}}^{t_i} c_{CM}(t) \cdot \frac{P_{\overline{OM}}(t) \cdot f(t)}{t} dt \tag{16}$$

where $P_{\overline{OM}}(t)$ is computed as:

$$P_{\overline{OM}}(t) = \left(1 - \bar{F}_{D,t_i}(d_{min}(t_i)) \cdot F_{O,t_{i-1}}(t)\right) \prod_{j=0}^{i-1} (1 - p_{SO_j}) \tag{17}$$

where i is the index that satisfies $t \in (t_{i-1}, t_i]$.

In order to utilize the equations above, the distribution of system failure time $\{f(t), \bar{F}(t)\}$, opportunity occurrence $\{\lambda_O(t)\}$ and duration $\{f_{D,t_i}(D), \bar{F}_{D,t_i}(d)\}$ are determined in Sect. 3 associated with case study of the generator of wind turbine at Mount Emerald windfarm, Queensland – Australia.

3 Case Study of Wind Turbine

This section discusses a novel case study which illustrates the effectiveness of the proposed policy for wind turbine maintenance. An optimal TB-OM policy is developed to maintain the generator of a hypothetical wind turbine located at Mount Emerald windfarm, Queensland - Australia. The time frame is discretized into weeks, in which considered external opportunities (in days) are when the average wind speed is less than the cut-in threshold of wind turbine. At this time, the wind turbine is not functioning and it creates the chance to do preventive maintenance. Therefore, the wind data of Mount Emerald and electricity prices of Queensland are used to develop the opportunity model and production loss cost of wind turbine.

3.1 Failure Mechanism

In this illustrative example the degradation path of generator proposed in study of Zhu et al. [19] is used to develop the failure mechanism of wind turbine components. The degradation path in this study was modelled as the Lundberg–Palmgren formula:

$$X(t, \phi_1, \phi_2, \theta) = \phi_1 + \theta \cdot t^{\phi_2} \tag{18}$$

where the lifetime t represents the number of days of system operation. The positive random parameter θ represents the wind load and is assumed to follow a Weibull distribution with two parameters, scale β and shape α, which can be obtained by data fitting. The constant parameter ϕ_1 is the initial degradation level, and ϕ_2 is a given constant that depends on the type of component. The degradation was also limited by a failure threshold H.

According to the formula (18) above, the CDF of failure time is determined as:

$$\begin{aligned} F(t) &= \Pr\{X(t, \phi_1, \phi_2, \theta) \geq H\} \\ &= \Pr\{\phi_1 + \theta \cdot t^{\phi_2} \geq H\} = \Pr\left\{\theta \geq \frac{H - \phi_1}{t^{\phi_2}}\right\} \\ &= 1 - F_\theta\left(\frac{H - \phi_1}{t^{\phi_2}}\right) \end{aligned} \tag{19}$$

Since θ follows Weibull distribution $Weib(\alpha, \beta)$, the PDF of failure time can be determined as:

$$\begin{aligned} f(t) &= \frac{dF(t)}{dt} = \frac{d\left[e^{-\left(\frac{H-\phi_1}{\beta t^{\phi_2}}\right)^{\alpha}}\right]}{dt} \\ &= \left(\frac{H - \phi_1}{\beta}\right)^{\alpha} \cdot \alpha\phi_2 \cdot t^{(-\alpha\phi_2 - 1)} \cdot e^{-\left(\frac{H-\phi_1}{\beta t^{\phi_2}}\right)^{\alpha}} \end{aligned} \tag{20}$$

This study also use the same parameters of generator in [19] as following: θ followed a Weibull distribution with scale $\beta = 1.02$ and shape $\alpha = 6.9$; the initial degradation level $\phi_1 = 3$; the type parameter $\phi_2 = 0.51$; and the maximum (failure) threshold of degradation $H = 15$.

The costs of PM action and CM action are assumed 10,000 AUD and 50,000 AUD, which equivalent to the values used in [19]. It is also assumed that PM takes 3 days and yields some production loss cost.

3.2 Opportunity Model According to Wind Speed

In this case study, the days when the average wind speed is less than cut-in threshold are considered as opportunities. If there are n consecutive days with lower cut-in wind speed, it means that there is an opportunity with duration of $D = n$ days. Hence, in order to develop the opportunity model for this windfarm in Mount Emerald, Queensland - Australia, the wind speed model is first studied.

The data on the wind speed distribution at the Mount Emerald site can be found in a report prepared by Parsons Brinckerhoff Australia [26]. The wind speed at case study location fitted to the Weibull distribution with shape of $\beta = 2.11$ and scale of $\alpha = 10.1$ (m/s). However, the report also mentions that the average wind speed varies monthly within a given year. Therefore, in this case study, it is supposed that the wind speed follows different Weibull distributions according different months of year. The shape parameter $\beta = 2.11$ is maintained and the scale parameters $\alpha_k (k = 1\ldots12)$ are determined again according to different mean wind speeds (Table 1).

Table 1. The scale parameters of Weibull distribution for wind speed at mount Emerald, Queensland, Australia

Month	Average wind speed (m/s)	Scale parameters	Month	Average wind speed (m/s)	Scale parameters
1	6.3	7.113	**7**	11.6	13.098
2	8.0	9.033	**8**	9.2	10.388
3	7.2	8.129	**9**	10.3	11.630
4	11.2	12.646	**10**	9.8	11.065
5	10.3	11.630	**11**	9.1	10.275
6	9.4	10.614	**12**	6.6	7.452

For each time interval (week), an opportunity occurs if there is at least one day with average wind speed lower than the cut-in value. According to [26], the wind turbines used in this study produce a small amount of power when the wind speed is 4 m/s. Hence, it is reasonable to set the cut-in threshold of 3.5 m/s. The probability that a weekday in a certain week k is an opportunity can be computed as $p_k = \Pr(WS \leq 3.5)$, where WS follows the Weibull distribution with parameters according to the month that week k belongs. From this, the probabilities of opportunity occurrence for each week can be determined (i.e. probability that at least one weekday has wind speed lower than cut-in threshold):

$$p_k^O = 1 - (1 - p_k)^7 \quad (21)$$

From the weekly probabilities of opportunities, the associated arrival rates for each week are calculated as:

$$\lambda_O(t) = \lambda_k^O = -\ln\left(1 - p_k^O\right) \quad (22)$$

The probabilities of opportunity occurrences for 52 weeks are presented in Fig. 1. The opportunity is stated as lasting for n days if there are n consecutive days having wind speed lower than the cut-in threshold from the first low wind speed day. With this definition, the discrete distributions of opportunity durations can be determined by enumeration. Figure 2 shows the distribution of possible opportunity durations for each month.

3.3 Downtime Cost Rate

When conducting the required PM activity, the wind turbine must be stopped for the maintenance duration, which causes some production loss. The downtime (production loss) cost per time unit (day) depends on the electricity price and average wind speed which affects the possibly produced electricity output. These factors are stochastic and time-varying.

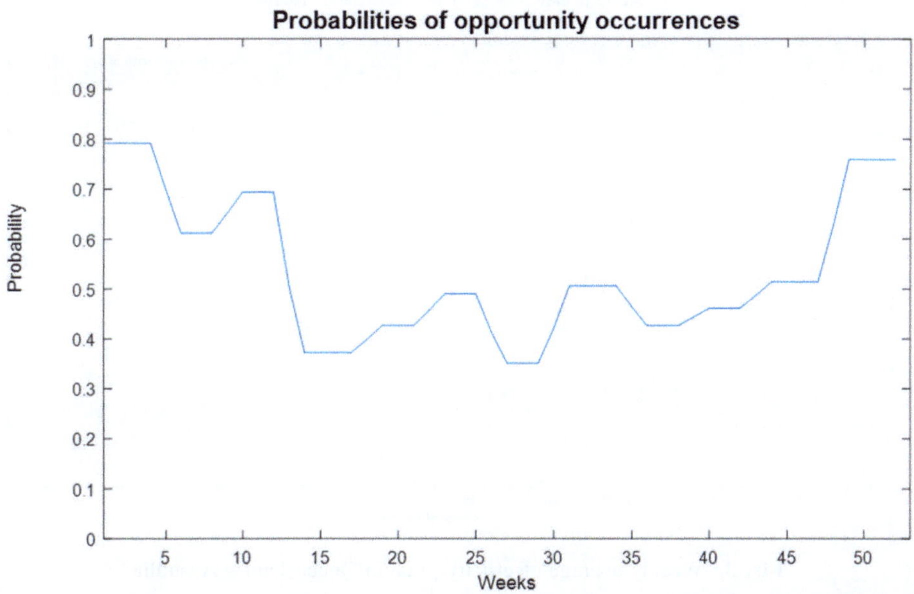

Fig. 1. Annually probability of Opportunity occurrences (at least one weekday has average wind speed lower than 3.5 m/s)

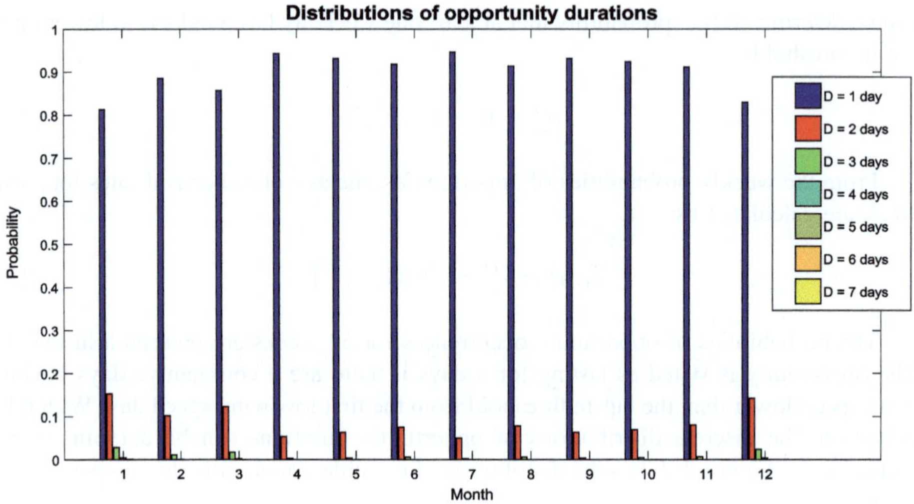

Fig. 2. Discrete distributions of opportunity durations

For the electricity prices, 11 years of prices from Queensland, Australia were synchronously averaged and subsequently smoothed with a 31-day sliding window (see Sect. 3.2 of [27]) to determine the average electricity prices for each week. The resulting weekly average electricity prices are presented in Fig. 3.

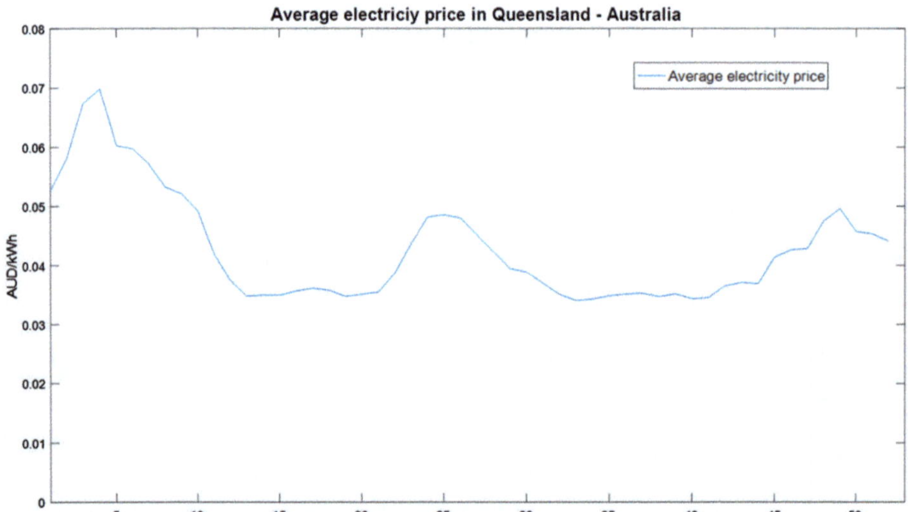

Fig. 3. Weekly average electricity price in Queensland – Australia

Combining the average wind speed discussed above and the power output of the particular wind turbine (REpower 3.4-104 WTG at air density 1.09 kg/m^3 [26]) with the electricity efficiency loss of 4%, the average daily power generation of wind turbine can be calculated and is shown in Fig. 4. Finally, the downtime costs are computed as the multiplication between the electricity price and possible power generation Fig. 5.

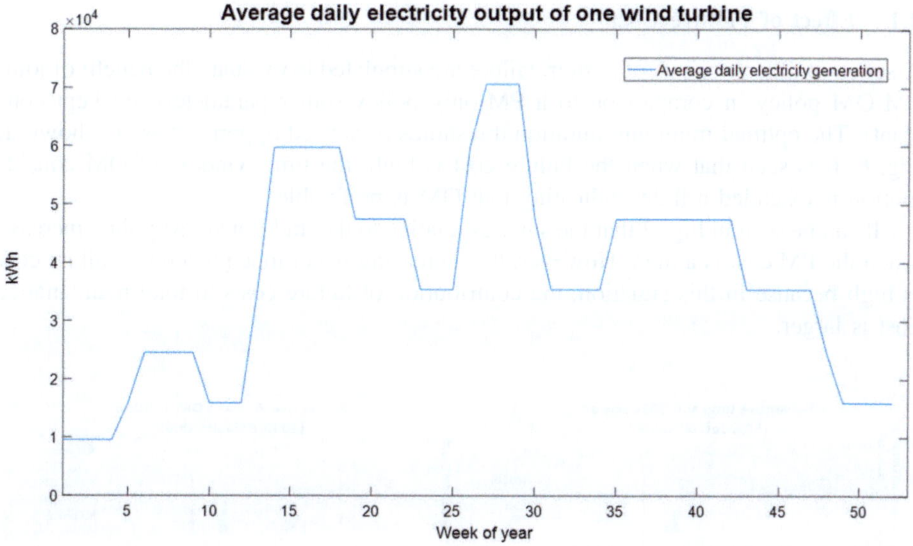

Fig. 4. Average daily possible electricity generation of wind turbine REpower 3.4-104 WTG at air density of 1.09 kg/m3

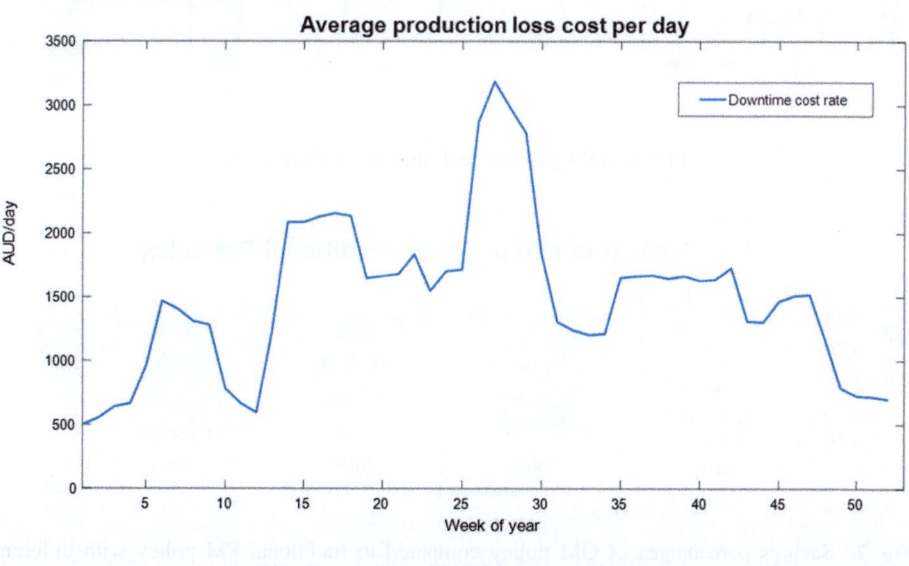

Fig. 5. Average downtime cost per day

4 Results Analysis

The optimization problems are solved approximately using a simple Genetic Algorithm in MATLAB 2016a. The benefits of proposed OM policy are presented for different values of the production loss cost (i.e. electricity price) and opportunity chances.

4.1 Effect of Failure Cost

A series of different costs of system failure are simulated to evaluate the benefit of joint PM-OM policy in comparison to a PM-only policy (other parameters are kept constant). The optimal minimum duration thresholds of arrived opportunities are shown in Fig. 6. It is seen that when the failure cost is high, the time window of OM consideration is extended a little, indicating that OM is preferable.

It can be seen in Fig. 7 that the savings relative to the traditional PM policy increase when the PM cost is a high. However, the increasing is not much when the failure cost is high because in this situation, the contribution of failure costs to total maintenance cost is larger.

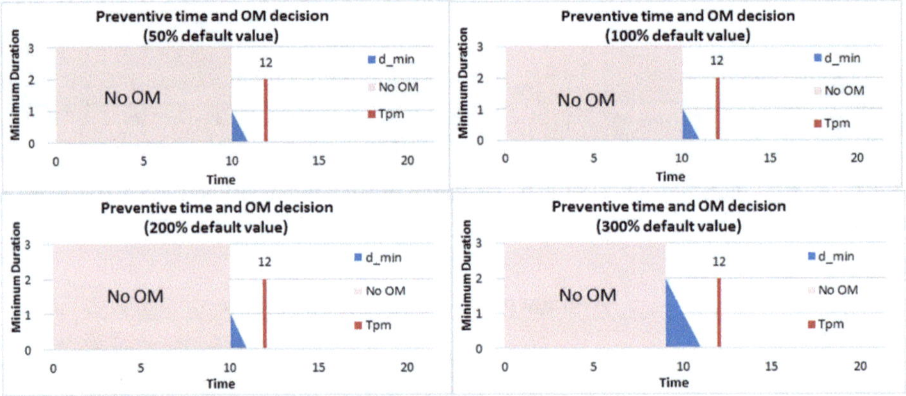

Fig. 6. OM policies with different failure costs

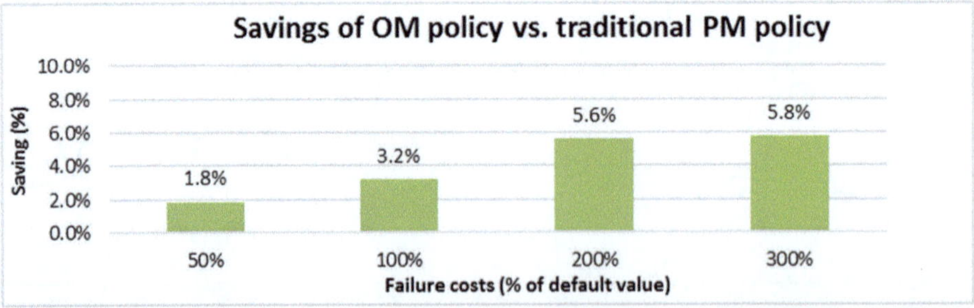

Fig. 7. Savings percentages of OM policy compared to traditional PM policy with different failure costs

4.2 Effect of Production Loss Cost (Electricity Prices)

Similar to previous section, this section discusses the benefit of OM policy in comparison to a PM-only policy with different downtime costs. Clearly, different downtime costs may be induced by the different electricity prices of other regions. Figure 8 shows the optimal OM policies with different downtime costs. When the downtime costs increase, the time window of OM consideration is wider. It makes the chance for doing OM higher and therefore can give better savings in total maintenance cost.

Figure 9 shows the saving percentages of maintenance policy of joint OM and PM to the typical PM policy. The joint OM and PM policies yield some savings compared to the policy with PM only. The percentage of savings increases when the downtime cost is higher, which is reasonable since the aim of the OM policy is to save the production loss cost caused by conducting PM activities.

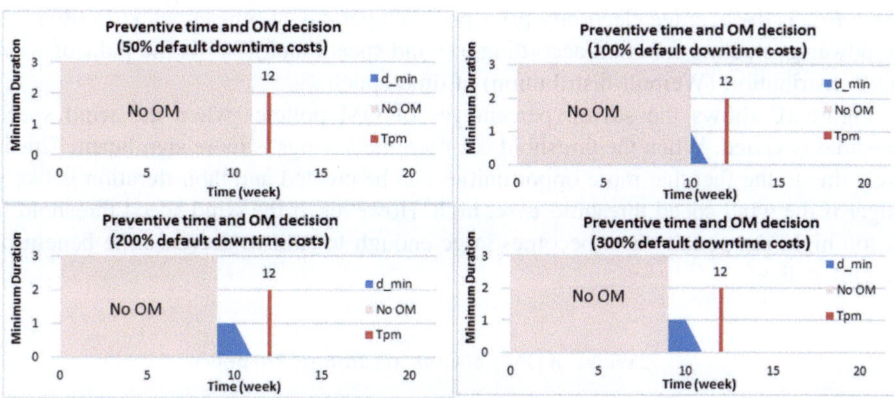

Fig. 8. OM policies with different downtime costs

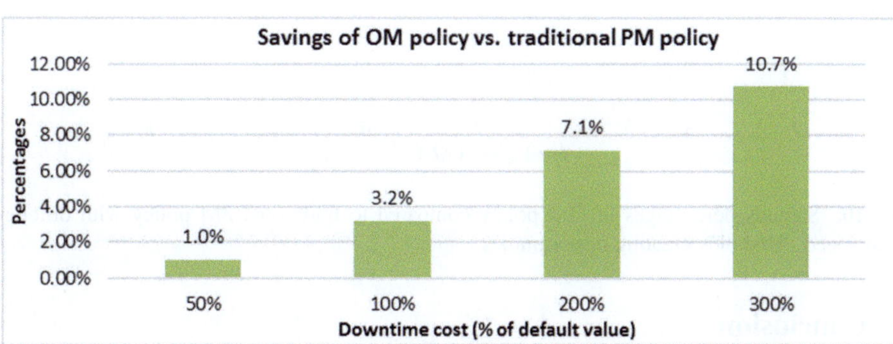

Fig. 9. Savings percentages of OM policy compared to traditional PM policy with different downtime costs

4.3 Effect of Wind Threshold Defining Opportunity Occurrences

In the case study thus far, the cut-in wind speed is considered as the threshold for creating opportunities, i.e. when the wind turbine is completely stopped. However, the opportunity wind speed threshold may be set higher to produce more frequent opportunities. There is certainly some penalty cost for doing OM if the opportunity threshold is higher. This penalty cost is caused by the production loss when the wind speed is less than opportunity threshold but higher than cut-in value. Let w_O and w_{ci} be the wind speed threshold for opportunities and cut-in value, respectively. The average penalty cost per maintenance time unit (day) of doing OM is computed as:

$$cr_{pen,k} = EP_k \cdot \int_{w_{ci}}^{w_O} POW(w) \cdot f_k(w)dw \quad (23)$$

where EP_k is the average electricity price per kWh per day at time epoch k; $POW(w)$ is the power generating function according to wind speed; and $f_k(w)$ is the p.d.f of wind speed distribution (Weibull distribution) at time epoch k.

Figure 10 shows the saving percentages of OM policies when the wind speed threshold is varied. When the threshold is larger, the savings is more significant. This is likely due to the fact that more opportunities can be created and their duration is likely longer if the wind speed threshold is set high. However, if the wind speed threshold is set too high, the penalty cost becomes large enough to start to decrease the benefit of the OM policy.

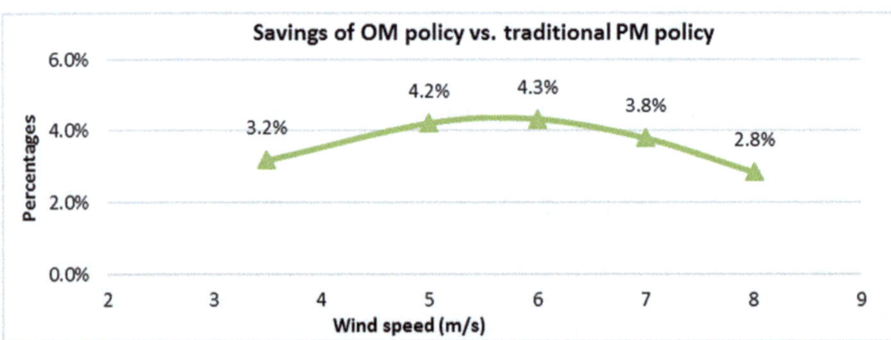

Fig. 10. Savings percentages of OM policy compared to traditional PM policy with different wind speed thresholds creating opportunities

5 Conclusion

This paper has presented an application of an OM policy for wind turbine components considering the external opportunities caused by low wind speed. Conducting maintenance for wind turbines at these periods can save production loss costs. The numerical results show that there are some savings (e.g. 3.2%) incurred by employment the proposed OM policy compared to traditional PM one. The savings is higher when the downtime costs are the major contribution to total costs.

Future work will combine the OM policy with external opportunities and grouping policy for wind turbine systems. During opportunity durations, some components of will be selected to be maintained under the constraint of limited maintenance resources.

References

1. Nguyen, T.A.T., Chou, S.-Y.: Maintenance strategy selection for improving cost-effectiveness of offshore wind systems. Energy Convers. Manag. **157**, 86–95 (2018). https://doi.org/10.1016/j.enconman.2017.11.090
2. Wang, H., Pham, H.: Reliability and optimal maintenance. Springer, London (2006). vol Book, Whole. https://doi.org/10.1007/b138077
3. Gustavsson, E., Patriksson, M., Stromberg, A.B., Wojciechowski, A., Onnheim, M.: Preventive maintenance scheduling of multi-component systems with interval costs. Comput. Ind. Eng. **76**, 390–400 (2014). https://doi.org/10.1016/j.cie.2014.02.009
4. Ding, F., Tian, Z.: Opportunistic maintenance for wind farms considering multi-level imperfect maintenance thresholds. Renew. Energy **45**, 175–182 (2012). https://doi.org/10.1016/j.renene.2012.02.030
5. Abdollahzadeh, H., Atashgar, K., Abbasi, M.: Multi-objective opportunistic maintenance optimization of a wind farm considering limited number of maintenance groups. Renew. Energy **88**, 247–261 (2016). https://doi.org/10.1016/j.renene.2015.11.022
6. Atashgar, K., Abdollahzadeh, H.: Reliability optimization of wind farms considering redundancy and opportunistic maintenance strategy. Energy Convers. Manag. **112**, 445–458 (2016). https://doi.org/10.1016/j.enconman.2016.01.027
7. Shafiee, M., Finkelstein, M.: An optimal age-based group maintenance policy for multi-unit degrading systems. Reliab. Eng. Syst. Saf. **134**, 230–238 (2015). https://doi.org/10.1016/j.ress.2014.09.016
8. Reder, M., Yürüşen, N.Y., Melero, J.J.: Data-driven learning framework for associating weather conditions and wind turbine failures. Reliab. Eng. Syst. Saf. **169**, 554–569 (2018). https://doi.org/10.1016/j.ress.2017.10.004
9. Zhang, X., Zeng, J.: Joint optimization of condition-based opportunistic maintenance and spare parts provisioning policy in multiunit systems. Eur. J. Oper. Res. **262**(2), 479–498 (2017). https://doi.org/10.1016/j.ejor.2017.03.019
10. Ghamlouch, H., Fouladirad, M., Grall, A.: The use of real option in condition-based maintenance scheduling for wind turbines with production and deterioration uncertainties. Reliab. Eng. Syst. Saf. (2017). https://doi.org/10.1016/j.ress.2017.10.001
11. Shafiee, M., Finkelstein, M., Berenguer, C.: An opportunistic condition-based maintenance policy for offshore wind turbine blades subjected to degradation and environmental shocks. Reliab. Eng. Syst. Saf. **142**, 463–471 (2015). https://doi.org/10.1016/j.ress.2015.05.001
12. Tian, Z.G., Jin, T.D., Wu, B.R., Ding, F.F.: Condition based maintenance optimization for wind power generation systems under continuous monitoring. Renew. Energy **36**(5), 1502–1509 (2011). https://doi.org/10.1016/j.renene.2010.10.028
13. Ab-Samat, H., Kamaruddin, S.: Opportunistic maintenance (OM) as a new advancement in maintenance approaches. J. Qual. Maintenance Eng. **20**(2), 98–121 (2014). https://doi.org/10.1108/jqme-04-2013-0018

14. Pham, H., Wang, H.: Optimal (τ, T) opportunistic maintenance of a k-out-of-n:G system with imperfect PM and partial failure. Naval Res. Logistics **47**(3), 223–239 (2000). 10.1002/(SICI)1520-6750(200004)47:3<223::AID-NAV3>3.3.CO;2-1
15. Zhang, X.H., Zeng, J.C.: A general modeling method for opportunistic maintenance modeling of multi-unit systems. Reliab. Eng. Syst. Saf. **140**, 176–190 (2015). https://doi.org/10.1016/j.ress.2015.03.030
16. Zhu, W., Fouladirad, M., Bérenguer, C.: A multi-level maintenance policy for a multi-component and multifailure mode system with two independent failure modes. Reliab. Eng. Syst. Saf. **153**, 50–63 (2016). https://doi.org/10.1016/j.ress.2016.03.020
17. Zhang, C., Gao, W., Guo, S., Li, Y., Yang, T.: Opportunistic maintenance for wind turbines considering imperfect, reliability-based maintenance. Renew. Energy (2016). https://doi.org/10.1016/j.renene.2016.10.072
18. Erguido, A., Crespo Márquez, A., Castellano, E., Gómez Fernández, J.F.: A dynamic opportunistic maintenance model to maximize energy-based availability while reducing the life cycle cost of wind farms. Renew. Energy (2017). https://doi.org/10.1016/j.renene.2017.07.017
19. Zhu, Q., Peng, H., van Houtum, G.-J.: A condition-based maintenance policy for multi-component systems with a high maintenance setup cost. Eindhoven University of Technology, Netherlands, BETA Working Paper 400 (2012)
20. Irawan, C.A., Ouelhadj, D., Jones, D., Stålhane, M., Sperstad, I.B.: Optimisation of maintenance routing and scheduling for offshore wind farms. Eur. J. Oper. Res. **256**(1), 76–89 (2017). https://doi.org/10.1016/j.ejor.2016.05.059
21. Sarker, B.R., Ibn Faiz, T.: Minimizing maintenance cost for offshore wind turbines following multi-level opportunistic preventive strategy. Renew. Energy **85**, 104–113 (2016). https://doi.org/10.1016/j.renene.2015.06.030
22. Ba, H.T., Cholette, M.E., Borghesani, P., Ma, L.: A quantitative study on the impact of opportunistic maintenance in the presence of time-varying costs. In: 2016 IEEE International Conference on Industrial Engineering and Engineering Management (IEEM), 4–7 December 2016, pp. 1360–1364 (2016). https://doi.org/10.1109/ieem.2016.7798100
23. Truong Ba, H., Cholette, M.E., Borghesani, P., Zhou, Y., Ma, L.: Opportunistic maintenance considering non-homogenous opportunity arrivals and stochastic opportunity durations. Reliab. Eng. Syst. Saf. **160**, 151–161 (2017). https://doi.org/10.1016/j.ress.2016.12.011
24. Ross, S.M.: Introduction to Probability Models. 10, 11th edn. Elsevier, Amsterdam (2014). vol Book, Whole
25. Cinlar, E.: Introduction to Stochastic Processes. Courier Corporation (2013)
26. Mount Emerald Wind Resource and Energy Yield Assessment: Parsons Brinckerhoff Archive, RATCH Australia Corporation Limited (2012)
27. Truong Ba, H., Cholette, M.E., Wang, R., Borghesani, P., Ma, L., Steinberg, T.A.: Optimal condition-based cleaning of solar power collectors. Sol. Energy **157**, 762–777 (2017). https://doi.org/10.1016/j.solener.2017.08.076

Cyber-Physical Manufacturing in Context of Industry 4.0 Model

Vidosav D. Majstorović(✉), Mirjana Velimirović, Marija Glišić, Julija Kostić, Elena Đura, Milan Rančić, and Radivoje Mitrović

Faculty of Mechanical Engineering, University of Belgrade, Belgrade, Serbia
vidosav.majstorovic@sbb.rs

Abstract. The Cyber Manufacturing Systems (CMSs) are based on integration of the Cyber Physical Systems (CPSs) and connection between Internet of Things (IoTs) and Cloud technology (CT). These are high-level methodologies for development of new generation factories, which are more intelligent, flexible and self-adaptable. CMSs generates Big Data, horizontally by integration (network of machines, processes and sensors) and vertically by control (usually defined over five levels) which should be analytically processed and managed by the Cyber-Physical Manufacturing (CPM) or CPS. In this way, two functional entities, real and virtual, are established, through which: (a) the advanced connectivity that ensures real-time data acquisition from the physical world and information feedback from the cyber space; and (b) intelligent data management, analytics and computational capability that constructs the cyber space. Industry 4.0 is Program initiated by German Government and industry as a new model of automatization of manufacturing technologies. CPM is the key element of Industry 4.0.

In this paper, a detailed analysis of the current level of development of the Industry 4.0 model has been made in EU and globe, with a special place and role of CPM in current model. A brief overview of the Industry 4.0 concept in Serbia is given as well.

Keywords: CPS · CPM · Industry 4.0

1 Introduction

The industry sector is the most important economic branch of EU economy, which is the basis for employment growth (in other economic activities up to 2, 3 jobs).

Industry (under the term "manufacturing", excluding mining, construction and energy) provides "added value" by creating products through the transformation of materials and raw materials. According to data from 2016, one out of ten organizations in the EU are from the industry (a predefined manufacturing sector), which has 2 million employees considering total amount of 35 million. It generates 80% of EU exports (by value), and also provides 80% of innovations and new technological solutions on annual basis [22]. There are no such detailed data for Serbia, but some preliminary analyzes show that in the last two and a half decades "de-industrialization" has occurred in our country, which was detrimental to the economic development of the country. Purpose

J. Ni et al. (Eds.): AMP 2018, LNME, pp. 227–238, 2018.
https://doi.org/10.1007/978-3-319-89563-5_17

of AMP Conference launched in 2016 by the Faculty of Mechanical Engineering in Belgrade and the Belgrade Chamber of Commerce was to start the "wheel" of the industry in Serbia again, but this time using the sophisticated technological trends - the Industry 4.0 model for Serbia.

However, "de-industrialization" in recent decades had impact in the EU. The first reason is "shifting" production in China over the last two decades, due to cheap labor, and the second is the result of global economic crisis and the decline in business activity. This led to the fact that at the end of 2015, the share of EU industry in total "added value" of economy fell below 15%, considering that the highest level was over 31%. The European Commission has set a target to 20% in 2020.

Germany was and remains the EU industry leader. Global trends and the economic crisis had the least affect to the industry of this country.

Nevertheless, this country has launched a new world cycle of industrial and technological development, defining the Industry 4.0 Program at the CEBIT fair in Hannover, October 2011 [20].

A year later, the EU adopted the Digital Strategy of the Single Market, making Industry 4.0 the official digital production strategy for European industry [21, 23].

2 Industry 4.0 Program – Facts and Figures

One of definition of Industry 4.0 is: *''Industry 4.0 is a German-government-sponsored vision for advanced manufacturing. The underlying concept of Industry 4.0 is to connect embedded systems and smart manufacturing facilities to generate a digital convergence between industry, business and internal functions and processes. Industry 4.0 refers to a fourth industrial revolution (following water/steam power, mass production and automation through IT and robotics) and introduces the concept of "cyber-physical systems" to differentiate this new evolutionary phase from the electronic automation that has gone before''* [20].

This definition contains several key words, and the most important is - advanced manufacturing. This means that advanced manufacturing is the basis for the fourth industrial revolution, with industrial manufacturing being integrated into digital technologies on the Internet. Industry 4.0 is the original German term. In the same context, the following terms are used worldwide: a smart factory, a factory of the future, intelligent manufacturing.

Industry 4.0 model is based on development and application of: (a) Information Communication Technologies (ICT) on a digital basis that integrate all stages of the product, from development to the end of use, (b) cyber-physical systems, which use ICT to monitor and manage physical processes and systems using intelligent sensors, robots and additive production, (c) communication networks (wireless and internet communications), through which processes, machines, people and systems, production, support functions, suppliers and distributors are connected (d) simulation, modeling and virtualization in product design and process management (virtual clones), (e) the collection of vast amounts of data, their analysis, optimization and use, either immediately in production, or through additional analysis and cloud computing, and (f) development of intelligent panels for communicating man-machine, machine-machine and machine-man [20–23].

Today, seven years after development and application of Industry 4.0, the first research in industrialized countries shows that this model contributes to the growth of industrial production from 6 to 8% per year. In Germany, GDP will increase by 1% per year and the number of employees in industry will increase by 400,000 over the next ten years. Also, investments in industrial Internet, which in 2012 reached to $ 20b, will be up to $ 500b in 2020. All this contributes to the growth of the industry's added value, from the current $ 25b to $ 1.4b in 2020 [23].

Significant changes occur in terms of raising the level of knowledge of employees in manufacturing, since it transition from soft programming to soft computing skills, where processes and systems are monitored and intervened in incident situations (tool breakage, etc.). This means that employees in factories where the Industry 4.0 model is applied should have knowledge in the field of ICT technologies. Research shows that by the year 2020, the EU market will require around one million of IT specialists in secondary and higher education (70% : 30%) for Industry 4.0 in field of manufacturing.

Digital manufacturing, as the base of the Industry 4.0 model, has spread from Germany to the EU, North America and the Far East [12, 15]. Throughout these regions, national programs have been established in accordance with the basic concept of Industry 4.0, in this way integrating: university (high technological research and engineering education) - decision makers (governments, ministries, chambers of commerce) - economy (production organizations, start-up companies). In the Far East, by 2020, $ 60b will be invested in the research and development of Internet of Things (IoT). The goal of these activities is to establish a smart, sustainable and inclusive economy in 2020 worldwide.

Digital production has brought a lot of changes in manufacturing processes, business models and interactions between them.

Smart factories have high manufacturing flexibility, using: (a) automation of the manufacturing process through the transfer of product data in the manufacturing chain, the integration of real-time engineering-production and control activities, (b) manufacturing in a chain, completely different products, unique for the individual (c) the production of innovative products, since prototypes or new products can be quickly manufactured without the complicated processing of tools or the installation of new production lines; and (c) digital design (products, technologies) and simulations, reducing time deliveries up to 90%. Smart factories enable drastic increase in product quality and production without write-offs, because the accuracy of processing is managed on-line. Data collection and analysis of Big Data is now explored in the Industry 4.0 concept, which provides enormous savings in production without write-offs. Preliminary analyzes of 100 European manufacturing factories in the field of automotive and aviation industry show that it is possible to make savings in the field of eliminating write-offs at a five-year level of 160b € [2]. Productivity in CPM models increases from 20 to 60% [1], using cloud model diagnostics and maintenance.

Model Industry 4.0 enables the involvement of customers in the process of designing and modifying products, thereby generating on-line new innovative products according to their requirements and ideas. In this way, smart products with specific characteristics are created, and CPM become intelligent business model [2].

All these elements speak of brand new, applied approaches to development of factories that become reality in industrialized countries of the EU, Far East and North America [4]. This approach sets new groundwork for business model relationships in supply chain, especially in the case of companies providing Internet services and telecommunications in the field of defining standards for the transmission and use of Big Data. For these reasons, special investments in the field of industrial Internet development are needed. It is predicted that in the EU by 2020, investments will be 140b € [2]. Investments in CPM for SMEs are also high, and they can be a limiting factor for their application in this area.

Important segments of the Industry 4.0 model are ownership and data security, especially when it comes to innovative products in the chain of CPM business models. These data are particularly important for certain areas of industry (pharmaceutical, military and automotive). One of the solutions for securing ownership and data security can be the ISO 27001: 2013 standard that can be essential to protect data exchanged between machines, systems and databases. International standard for communication protocols, data formats and interfaces provide interoperability in different sectors and different countries, thus encouraging the widespread adoption of Industry 4.0 technology and its application across the globe.

3 Industry 4.0 Program – World Wide Approaches

In 2014, the European Commission introduce a program of "renaissance" of the European industry [21, 23], based on digital technologies (cloud computing, big data analytics, new industrial internet applications, smart factories, robotics and 3D printing). This program provides a fundamental contribution to increasing European competitiveness through redefining business models for CPM and creating new products and services. Also, in 2012, when the EU defined industrial policy, it identified six priorities, where three of them are related to the Industry 4.0 model: advanced manufacturing (CPM), smart grid and digital infrastructure (industrial internet) [23]. In 2013, the EC formed a working group for advanced and clean manufacturing, and a year later the Strategic Policy Forum on Digital Entrepreneurship, with the aim of supporting the digital transformation of European industry. It also supports the development of national centers of excellence for digital production, as well as a range of activities related to education, communication standards, ICT and digital skills, digital economy, etc. [23] Within the Seventh Research Program - Horizon 2020, specific CPM research units have also been defined and supported.

Since 2011, when Industry 4.0 has entered the scene, national programs for its development and application in national industries have been launched around the world. In the meantime, this model has begun to develop in different industries. Today Industry 4.0 models have been applied in the pharmaceutical, military, processing industry and even in services.

The EU is the most developed in application Program Industry 4.0, Fig. 1.

Fig. 1. National initiatives for I4.0 in EU. Source: https://ec.europa.eu/digital-single-market/en/cordination-european-national-regional-initiatives [22]

Fourteen countries have their own national program that are most widely used by industry in Germany, Italy, France and Spain [21].

Each of the countries listed defines the National Strategies to Digitizing European Industries, which is based on: (a) Pillar 1 - Digital Industrial Platforms actions & Research, Development and Innovation actions, (b) Pillar 2 - Standardization actions, regulation and test beds, (c) Pillar 3 - Digital Innovation Hubs actions, and (d) Pillar 4 - Skills development. The defined framework represents the integration of government/ministries - science/research - industry in the concept of Industry 4.0 models and CPM [24].

After 2011, initiatives were initiated for national I4.0 programs around the world, Fig. 2.

Fig. 2. National initiatives for I4.0 across the globe. Source: https://www.dcc-aachen.com/ [25]

UK, USA, Japan, China and Singapore have been the most successful in these Programs. As one of the results of these global activities, launch of a global network of Industry 4.0 Digital Capability Center, Aachen (Germany) [25].

In the UK, 2016, a national I4.0 Program is based on four core components: IIoT (Industrial Internet of Things), CPM, Big Data, Reliable digital infrastructure, Fig. 3.

Industry 4.0 platform, big companies in the USA initiate the start [27]: (a) in March 2014 AT&T, Cisco, General Electric, IBM and Intel founded the Industrial Internet Consortium (IIC) in order to coordinate the priorities for the industrial Internet, and to enable the technical applications required for this, (b) meanwhile 250 companies have joined the movement, including some from Germany, and (c) the aim of the Industrial Internet Consortium is to bring together "operational systems", Fig. 4.

Industrial Internet Consortium and Platform Industry 4.0 are collaboration for Interoperability, based on: (a) RAMI 4.0, the focus is on manufacturing in depth and IIRA crosses multiple application domain, (b) industry as a whole and must work across domains; manufactured goods are one of those domains, and (c) interoperability by V – H approach [27].

Japanese approach on I4.0 program is - from Industry 4.0 to Society 5.0 [28]. In March 2017, Japan announced, a far-reaching program that aims to go far beyond the transformation of manufacturing and any other industry. Society 5.0 includes Industry

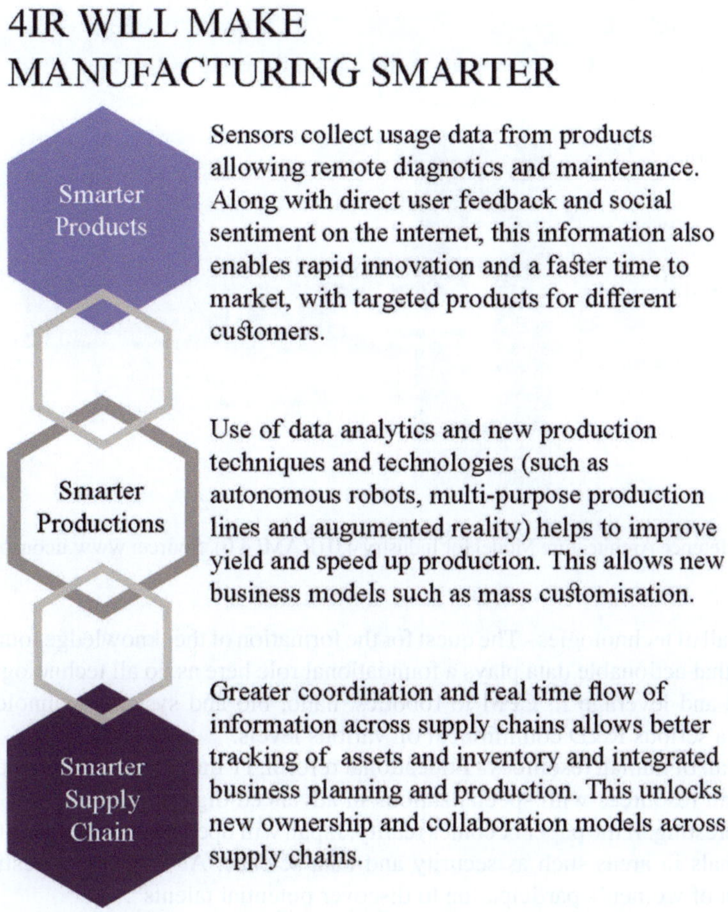

Fig. 3. UK key elements of I4.0 Program. Source: UK Manufacturing and the 4th Industrial Revolution – 4IR Program, 2017 [26].

4.0 but touches upon Japanese society: in the broadest possible sense, including culture, legislation and even philosophy. The Japan Business Federation (Keidanren) has set out a plan for addressing the transformation.

The 5 walls to 'break through' in moving to Society 5.0, and the Keidanren position paper, a "formulation of national strategies and integration of government promotion system". This includes the architecting of a 'handy IoT system' and a think-tank function.

The wall of the legal system - Whereby laws need to be developed to implement advanced techniques.

In practice this would also mean regulatory reforms and a push of administrative digitization (good news for all the document capture and information management people out there).

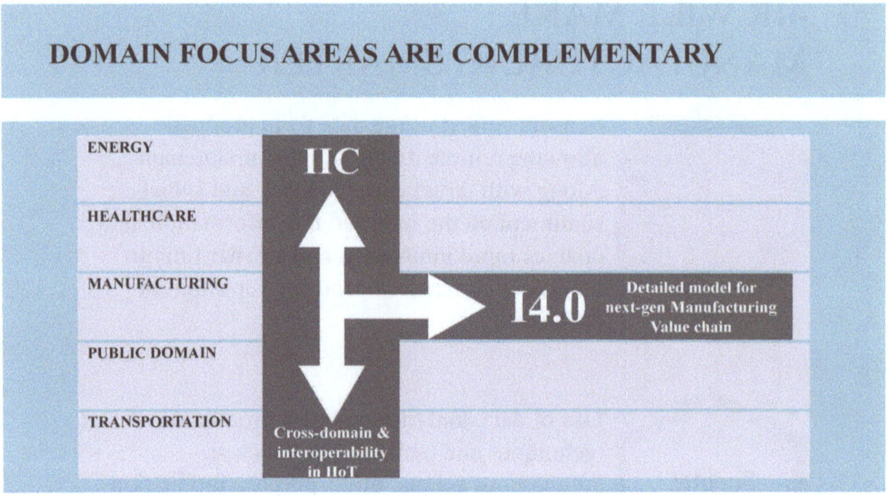

Fig. 4. Reference Architecture Model for Industry 4.0 (RAMI 4.0). Source: www.iiconsortium.org [27]

The wall of technologies - The quest for the formation of the 'knowledge foundation'. It's clear that actionable data plays a foundational role here as do all technologies/areas to protect and leverage it, CPM to robotics, nano, bio and systems technology. The mentions a serious R&D commitment on various levels.

The wall of human resources - Educational reform, IT literacy, broadening the available human resources with specializations in advanced digital skills are just a few of them. Interesting: if the paper becomes reality, Japan will open its doors for highly skilled professionals in areas such as security and data science. At least as interesting: "the promotion of women's participation to discover potential talents".

Social implications, ethics and social acceptance by all stakeholders - "the wall of social acceptance". This is the most society-related aspect of all [27].

Based on Strategic Vision and Outlook of "Made in China 2025", Possibilities and Challenges for Industry 4.0 China Version, China has been the world's factory - in 2015, it produced or assembled: 28% of the world's automobiles; 41% of the world's ships; 80%+ of the world's computers; 90%+ of the world's mobile phones; 60% of the world's color TV sets; 50%+ of the world's refrigerators; 80% of the world's air-conditioners; 24% of the world's power; and Half of the world's steel [28].

The CM2025 initiative also includes ten key sectors that receive special attention:

- Next generation IT
- High-end numerical control machinery and robotics
- Aerospace and aviation equipment
- Maritime engineering equipment and high-tech maritime vessel manufacturing
- Advanced rail equipment
- Energy-saving vehicles and NEVs
- Electrical equipment

- Agricultural machinery and equipment
- New materials
- Biopharmaceuticals and high-performance medical devices.

The key policy tools that China's central and local governments are using in their pursuit of goals related to CM2025 are as follows [28]:

- Forced technology transfers in exchange for market access,
- Market access and government procurement restrictions for FIEs,
- Standards/Subsidies,
- Financial policy, government-backed investment funds, support from local government,
- Technology-seeking investments abroad,
- SOEs: mergers and politicization, public-private partnerships.

4 Industry 4.0 Program in Serbia

In Serbia, since 2006 different kinds of EU initiatives are introduced, related to Advanced Manufacturing, such as:

Manufacture Program. Manufacture Platform in Serbia was established in 2007, and held five International Conference (2009/'11/'13/'15/'17), on same topic. Prof. Dr. Francesco JOVANE, "FATHER'' EU Manufacture Program, participating on Serbian Manufacture Conference 2011, was held Introduction Plenary Presentation. Also we established consortia on national level – 28 members (Faculties, Institutes, SME, ...), since 2008, and participating on EU Manufacture events (Conferences, Panels, ...), more of ten times in this period. Also presenting five papers – case studies from Manufacture Program in Serbia. Establishing REGIONAL Initiative – Manufacture Village (Romania, Hungary, Serbia) – held for Conferences/Panels in Romania and Serbia (2010/'12/'14/'16), and finally three Joint Projects support by EU – Leading by Microelectronica, Bucharest by subject of Manufacture Program.

World Manufacturing Forum – WMF. Those Conference was establishing since 2010 – Conference in Cernobbio, Como, Italy, and Conferences: 2012 – Stuttgart, 2014 – Milano, 2016 – Barcelona, was held. From Serbia we are participating each time – with paper, and representative from Serbia and Member of few Tasks Group. Finally, Prof. Dr. Marco Taisch, Founder and Chairman on WMF, participating with Plenary presentation 2013 on Serbian Manufacture Conference.

Factory of the Future – FoF. Establishing since 2008 – EU Program/support industry excellence, and we are participating from Serbia and member few tasks group. Also representatives this Program participating three time on Serbian Manufacture Conference, 2011 and 2013. Program Industry 4.0 was introduced in Serbia, when the First International Conference was held in Belgrade, on 31st May - 2nd June, (AMP Conference 2016), with the main topic: Advanced Manufacturing Program - INDUSTRY 4.0 model for Serbia. As a result of this Conference, the Program - Advanced

Industrialization of Serbia and Industry Policy, horizon 2020/2030 [29], was created. After this, the Project - Advanced Industrialization of Serbia - Industry 4.0 model for Serbia [30] was also defined. Next year, Second International Conference USA-EU-Japan-Serbia Manufacturing Summit, Belgrade, 7th - 9th June, 2017, Serbia - Smart And Intelligent Products (AMP Conference 2017) was held. So the central theme was - the output from the Industry 4.0 model - Smart And Intelligent Products. As a result of this conference, Position Paper - Smart And Intelligent Products - case study from Serbia. Finally, for 2018, Third International Conference USA-EU-Japan-Serbia Manufacturing Summit, Belgrade, 5th - 7th June, 2018, Serbia - Industry 4.0 model for SMEs (AMP Conference 2018) is planned. A paper on the main topic of the Conference - Industry 4.0 model for SMEs, case study from Serbia [14], is also planned.

After the first AMP Conference, organized by the Faculty of Mechanical Engineering in Belgrade, ten Panels were held on different topics regarding Industry 4.0 model: New Industrial Policy of Serbia, Education of Mechanical Engineers for Industry 4.0 model. Faculty of Mechanical Engineering positioned as a leading higher education and scientific institution for Industry 4.0 model in Serbia and the region [14].

5 Cyber-Physical Manufacturing

High-level methodologies for development of new generation factories, that are more intelligent, flexible and self-adaptable are based on integration of CPS and connection between IoT and Cloud technology. CPM generates Big Data, horizontally by integration and vertically by control which should be analytically processed and managed by CPM/CPS. Two functional entities are established: (a) the advanced connectivity that ensures real-time data acquisition from the physical world and information feedback from the cyber space; and (b) intelligent data management, analytics and computational capability that constructs the cyber space [11]. The five-layer CPS architecture includes features and characteristics [16–19]: (a) smart connection - to collect accurate and reliable data at the lowest level of the sensor, as well as from the controllers from the drives: ERP, SPM, CMM, MES, and their transfer to (b) data-to-information conversion - conversion and analysis of data and information generated at the previous level are carried out at this level, and in order to define the framework for intelligent management, (c) cyber level - this is the central information center (hub) in this architecture, in which a large amount of data is flooded, whose analysis is done immediately, creating databases on the current CPS state, now in the past (twin model, clustering and data mining) [2], (d) cognition level - at this level, knowledge (from the previous notes) is generated, for monitoring, management and decision making, and (e) configuration level - at this level feedback is generated from cyber space, which is used for self-configure and self-adaptive machine in the physical space.

6 Conclusions

The Industry 4.0 program has become a worldwide movement for development and automation of industry on new bases, such as CPM, IoT and Cloud technologies.

This paper has shown detailed development of Industry 4.0 in the most developed regions of the world. We can conclude that this Program Worldwide has become a national priority number one for the industry development.

Also, we can conclude that the Industry 4.0 model has been increasingly used in various fields, making its industrial application, with the results achieved, an impetus for others.

We can proudly say that our country is one of the few countries in the world (30 of them) that are working on developing and implementing Industry 4.0 model in their industry.

References

1. Iqbal, R., et al.: Big Data analytics and Computational Intelligence for Cyber-Physical Systems: recent trends and state of the art applications. Future Gener. Comput. Syst. (2017). https://doi.org/10.1016/j.future.2017.10.021
2. Monostori, L., et al.: Cyber-physical systems in manufacturing. CIRP Ann. Manuf. Technol. **65**, 621–641 (2016). https://doi.org/10.1016/j.cirp.2016.06.005
3. Thramboulidis, K., Christoulakis, F.: UML4IoT—a UML-based approach to exploit IoT in cyber-physical manufacturing systems. Comput. Ind. **82**, 259–272 (2016). https://doi.org/10.1016/j.compind.2016.05.010
4. Song, Z., et al.: Data quality management for service-oriented manufacturing cyber-physical systems. Comput. Electr. Eng. **64**, 34–44 (2017). https://doi.org/10.1016/j.compeleceng.2016.08.010
5. Babiceanu, R., Seker, R.: Big Data and virtualization for manufacturing cyber-physical systems: a survey of the current status and future outlook. Comput. Ind. **81**, 128–137 (2016). https://doi.org/10.1016/j.compind.2016.02.004
6. Adamson, G., et al.: Feature-based control and information framework for adaptive and distributed manufacturing in cyber physical systems. J. Manuf. Syst. **43**, 305–315 (2017). https://doi.org/10.1016/j.jmsy.2016.12.003. Part 2
7. Wu, D., et al.: A fog computing-based framework for process monitoring and prognosis in cyber-manufacturing. J. Manuf. Syst. **43**, 25–34 (2017). https://doi.org/10.1016/j.jmsy.2017.02.011
8. Liu, X., et al.: Cyber-physical manufacturing cloud: architecture, virtualization, communication, and testbed. J. Manuf. Syst. **43**, 352–364 (2017). https://doi.org/10.1016/j.jmsy.2017.04.004
9. Valls, M., et al.: Adjusting middleware knobs to assess scalability limits of distributed cyber-physical systems. Comput. Stand. Interfaces **51**, 95–103 (2017). https://doi.org/10.1016/j.csi.2016.11.003
10. Colledani, M., et al.: A cyber-physical system for quality-oriented assembly of automotive electric motors. CIRP J. Manuf. Sci. Technol. **20**, 12–22 (2018). https://doi.org/10.1016/j.cirpj.2017.09.001
11. Chen, Y.: Integrated and intelligent manufacturing: perspectives and enablers. Engineering **3**, 588–595 (2017). https://doi.org/10.1016/J.ENG.2017.04.009
12. Zhong, R., et al.: Intelligent manufacturing in the context of Industry 4.0: a review. Engineering **3**, 616–630 (2017). https://doi.org/10.1016/J.ENG.2017.05.015
13. Liu, C., Lijang, P.: A cyber-physical system architecture in shop floor for intelligent manufacturing. Proc. CIRP **56**, 372–377 (2016). https://doi.org/10.1016/j.procir.2016.10.059

14. Majstorovic, V., et al.: Cyber-physical manufacturing metrology model (CPM^3) for sculptured surfaces – turbine blade application. Proc. CIRP **63**, 658–663 (2017)
15. Vanzca, J., Monostori, L.: Cyber-physical manufacturing in the light of Professor Kanji Ueda's legacy. Proc. CIRP **63**, 631–638 (2017)
16. Lee, J., et al.: Introduction to cyber manufacturing. Manuf. Lett. **8**, 11–15 (2016). http://dx.doi.org/016/j.mfglet.2016.05.002
17. Chai, J., et al.: Sensor data and information fusion to construct digital-twins virtual machine tools for cyber-physical manufacturing. Proc. Manuf. **10**, 1031–1042 (2017)
18. Soocheol, Y.: Manufacturing information bus from the perspective of cyber physical manufacturing system. IFAC-PapersOnLine **49–31**, 103–108 (2016)
19. Lu, Y., Ju, F.: Smart manufacturing systems based on Cyber-physical Manufacturing Services (CPMS). IFAC PapersOnLine **50**(1), 15883–15889 (2017)
20. Klitou, D., et al.: Germany: Industrie 4.0, Brussels. https://ec.europa.eu/growth/tools-databases/dem/
21. European Platform of National Initiatives on Digitizing Industry, Brussels (2018). https://ec.europa.eu/digital-single-market/en/digitising-european-industry-digital-day
22. A Holistic Strategy and a Strong Partnership in a New Industrial Age, Brussels (2018). https://ec.europa.eu/digital-single-market/en/digitising-european-industry
23. Digitizing European Industry: Working Group 2 – Digital Industrial Platforms, European Union, Brussels (2017)
24. European Commission: Digital Economy and Society Index (DESI) (2017). https://ec.europa.eu/digital-single-market/en/scoreboard/sweden. Accessed Feb 2018
25. Digital Capability Center (DCC) Aachen. https://www.dcc-aachen.com/. Accessed Feb 2018
26. UK Manufacturing and the 4th Industrial Revolution – 4IR Program (2017). https://www.eef.org.uk/campaigning/campaigns-and-issues/current-campaigns/industry-four/4th-industrial-revolution-facts-and-figures. Accessed Feb 2018
27. http://www.keidanren.or.jp/en/. Accessed Feb 2018
28. China Manufacturing 2025 – Putting Industrial Policy Ahead (2017). https://www.made-in-china.com/. Accessed Feb 2018
29. Majstorovic, V., et al.: Program - Advanced Industrialization of Serbia and Industry Policy, horizon 2020/2030, Mechanical Engineering Faculty (2016)
30. Majstorovic, V., et al.: Project - Advanced Industrialization of Serbia – Industry 4.0 for Serbia, Mechanical Engineering Faculty (2016)

An Automated Laser Control Technique for Improving Powder Bed Temperature Uniformity in Selective Laser Sintering

T. Phillips[1], L. Zhang[2], S. Fish[1], and J. Beaman[1(✉)]

[1] Department of Mechanical Engineering, The University of Texas at Austin, Austin, TX, USA
jbeaman@mail.utexas.edu

[2] Department of Computer Science, The University of Texas at Austin, Austin, TX, USA

Abstract. Selective Laser Sintering (SLS) of polymers is a common technique for creating structural plastic components in additive manufacturing. As a result, the mechanical strength of SLS parts is often of concern. Testing has revealed a large variation in mechanical properties of components within the same build. It is believed that this variation is a result of poor temperature control during the sintering process. The proposed control method involves using an infrared camera to measure initial powder temperature before computing an optimal laser power profile to sinter the powder. By controlling the laser energy, it is expected that a more uniform post lasing temperature can be achieved and variation in mechanical properties can be reduced. The feed-forward control method is presented along with preliminary results.

Keywords: Additive manufacturing · Process control
Feedforward laser control

1 Introduction

Selective Laser Sintering (SLS) is a subset of Powder Bed Fusion (PBF) additive manufacturing (AM) that uses a high-powered laser to fuse polymer powder into arbitrary 3D geometries whose features are defined in a 3D CAD file. This is accomplished by depositing a thin layer of polymer, which is typically nylon-based, onto a build surface and scanning a laser across the surface to fuse the material in a shape resembling a 2D cross-section of the final 3D part.

Another thin layer of material is spread on top of the previous and the laser now scans the surface according to the next geometric cross-section, fusing this material to the previous layer. Layer-by-layer, the laser fuses the material until the component is completed, after which the un-sintered powder is removed and the 3D component is revealed. While many other AM techniques require supporting structures to ensure the components retain their shape, SLS is capable of creating free-standing components without the need for supports. This enables a user to pack a large amount of components into a single build and faster production cycle times.

One of the reasons nylon is a common material for SLS is due to its thermal properties that allow for building components without introducing significant thermal stresses.

J. Ni et al. (Eds.): AMP 2018, LNME, pp. 239–249, 2018.
https://doi.org/10.1007/978-3-319-89563-5_18

Differential Scanning Calorimetry (DSC) is used to identify a range of optimal processing temperatures at which the nylon powder should be maintained throughout the entirety of a build. DSC is used to identify critical processing temperatures by locating inconsistencies in the amount of energy required to change the material's temperature, indicating the material is undergoing some phase change. A DSC curve for nylon 12 is presented in Fig. 1, which shows that melting occurs at 181 °C and re-crystallization begins at 148 °C. The range between these temperatures is known as the processing window and it allows for molten and solid material to coexist at the same temperature. The theory behind SLS is that by keeping the bulk powder temperature within this range, SLS can use the laser to selectively melt certain regions of the powder and allow those regions to cool back to the bulk powder temperature while remaining liquid [1]. In this way, molten volumes of arbitrary shape can be formed and cooled in a controlled way that mitigates thermal stresses that typically result in part curl.

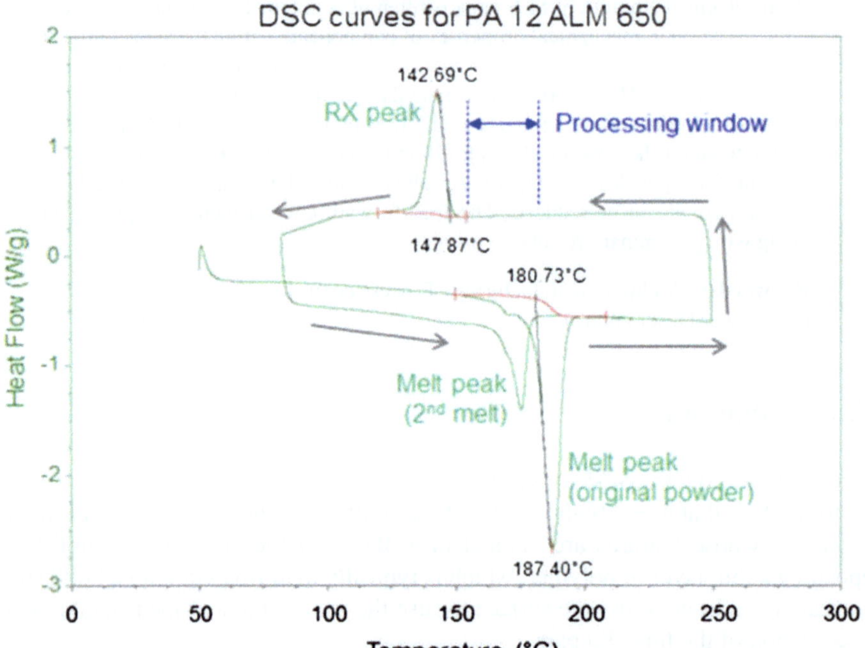

Fig. 1. DSC curve for Nylon [2]

In practice, the processing window shown on the DSC curve is a superset of the temperature range at which high quality parts can be produced. DSC itself is an idealized test and the numbers reported by it are dependent on heating and cooling rates, which are typically not controlled in AM [3]. Because of this, significant part curl can be seen when a temperature gradient much smaller than the DSC processing window is present on the build surface. In a machine that maintains a 10–15 °C gradient across its build surface, which is representative of many commercial machines, significant mechanical strength variation can be present. Previous work has shown that there is a correlation between the Ultimate Tensile Strength (UTS) of a SLS part and the temperature that component reaches during sintering [4, 5]. Therefore, in order to create parts with reduced UTS variations, improved thermal control is required. It is difficult to completely homogenize the temperatures of the build surface due to restrictions on heater size and quantity, so bulk heating alone is incapable of completely removing the thermal gradients that impact component strength. It is required, then, to use an alternative method for achieving consistent post-sintering temperatures with SLS. Other researchers that have worked on monitoring of laser fluence in additive manufacturing have typically been concerned with real-time data [6, 7]. This time constraint limits the controller options and requires significantly more expensive hardware.

This paper considers a feed-forward approach that allows for precomputation of optimal parameters. This reduces the cost of the system while still maintaining a high level of control over the processing temperatures.

2 Thermal Model

In order to improve the thermal control of SLS, a more complete understanding of what influences the post-sintering temperature is required. For this, a simple thermal model can be used to predict how powder temperatures change. For SLS, the simplified heat transfer model looks like [8]:

$$\rho C \frac{\delta T}{\delta t} = k\left(\frac{\delta^2 T}{\delta x^2} + \frac{\delta^2 T}{\delta y^2} + \frac{\delta^2 T}{\delta z^2}\right) + Q(x, y, z, t) \tag{1}$$

Equation 1: 3D Thermal Model

Having boundary conditions of radiation and convection at the top surface and adiabatic elsewhere. This equation is only valid when the material does not undergo phase change. If phase change occurs, as it does in SLS, the latent heat of fusion must also be considered. The laser heating term Q can be expanded for a moving, Gaussian beam to [9]:

$$Q(x, y, z, t) = (1 - R)\beta I_o \exp\left(-\frac{(x - v_x t)^2 + (y - v_y t)^2}{w^2} - \beta z\right) \tag{2}$$

Equation 2: Gaussian Laser Heating

Equation 1 can be simplified further by considering the order of magnitude each heat transfer mode contributes to the post-sintering temperature. If the energy lost by a voxel of powder due to convection, conduction, or radiation is significantly smaller than the energy deposited by the laser, it may be ignored with minimal impact and the maximum material temperature can be found with little error.

A finite differencing approach is taken to approximate nylon temperatures post-sintering with and without heat losses due to convection, conduction, and radiation.

Using values typical of SLS, it is seen that the maximum temperature achieved post-sintering varies by less than 1% when all heat loss is neglected. This being the case, convection, conduction, and radiation can be neglected in the model without significant impact on the maximum post-sintering temperature. The simplified thermal model for finding the maximum powder temperature is then reduced to:

$$T(t) = T_0 + \frac{Q}{\rho C}t \tag{3}$$

Equation 3: Simplified model for laser heating

Equation 3 shows that the temperature increase $(T - T_0)$ is independent of the initial powder temperature T_0 and is solely dependent on the laser power and time, assuming the material properties are constant within this temperature range. This implies that using a fixed laser power and speed will result in a constant temperature increase and by controlling the laser energy being deposited, the post-sintering temperature can be controlled. While this equation does not take into account the latent heat of fusion, the conclusion still holds if it is assumed that all powder fully melts during laser heating.

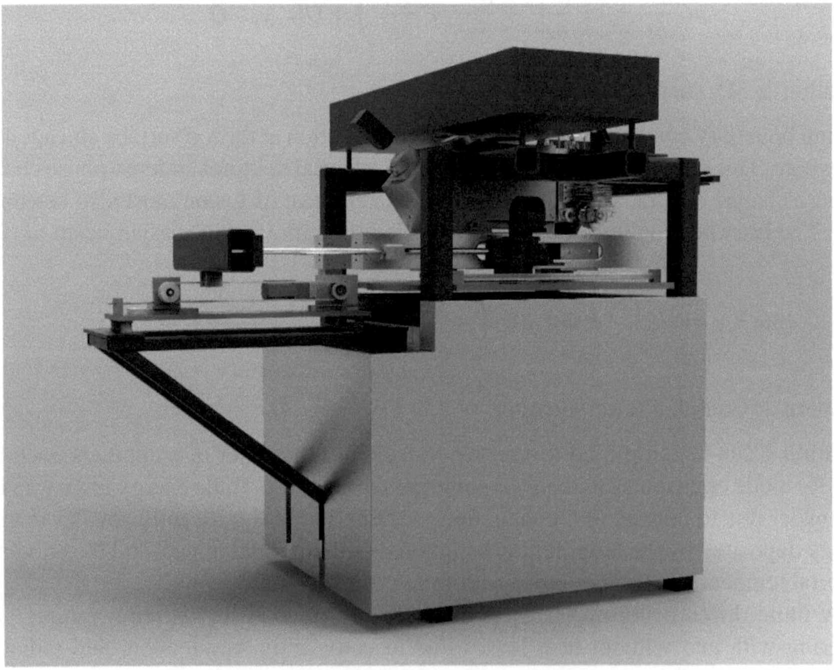

Fig. 2. Laser additive manufacturing pilot system

3 Experimental Setup

Thermal testing was accomplished in the Laser Additive Manufacturing Pilot System (LAMPS), an open hardware and software SLS machine developed at The University of Texas at Austin for SLS process monitoring and control. Renderings of the machine and its cross-section are provided in Figs. 2 and 3. LAMPS uses a 2-axis galvanometer set to scan a CO_2 laser over its 202 mm × 202 mm build surface. Powder recoating is accomplished via a thermally isolated hopper and counter-rotating roller mechanism. The majority of powder preheating is accomplished by heating the entire build chamber with an array of independently controlled strip heaters, while quartz lamps are used for the fine temperature control of the build surface. The LAMPS machine is controlled with in-house software written in National Instrument LabVIEW that communicates via a compact RIO system.

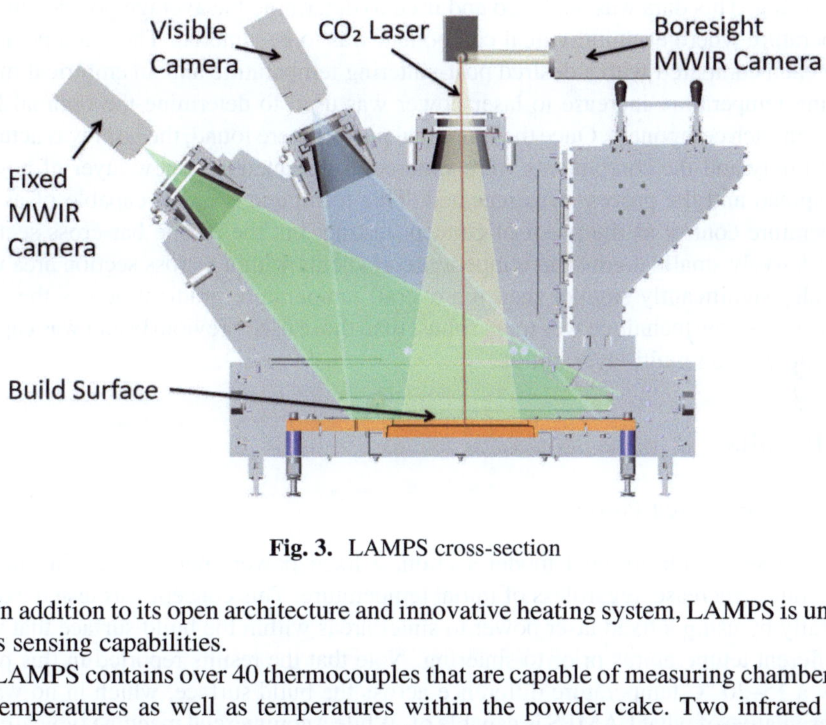

Fig. 3. LAMPS cross-section

In addition to its open architecture and innovative heating system, LAMPS is unique in its sensing capabilities.

LAMPS contains over 40 thermocouples that are capable of measuring chamber and air temperatures as well as temperatures within the powder cake. Two infrared (IR) cameras are also used for temperature monitoring. One camera is mounted in a boresight configuration, meaning a dichroic mirror is used to make the IR camera's field of view coaxial with the laser beam. This allows it to traverse the build surface with the laser and capture the laser-material interaction at high speeds. The other IR camera is mounted in a fixed position and is used as a feedback sensor for the quartz lamp control system.

Proof-of-concept testing for this algorithm was accomplished previously and detailed in [10]. The tests showed that modulating laser power throughout the length of a vector could control the post-sintering temperature. These previous tests focused on controlling temperature of individual lines and required significant user input. To extend that into a method that was capable of being fully automated and viable for sintering of larger areas, a slightly different approach was taken. At this time, LAMPS does not receive any real-time feedback for galvanometer position, so determining the position of the boresighted IR camera on the build surface is difficult. For sintering single lines, it is possible to interpolate between known points to get an approximation of the camera position but for sintering overlapping lines the error would be too great. For this reason, the controls technique was adapted to use thermal data taken with the fixed IR camera.

The goal for testing in this phase was to sinter a series of rectangles, similar in size to cross-sections of tensile bars oriented in the z direction, while using thermal data to control laser power to achieve consistent post-sintering temperatures.

The fixed IR camera was used to provide thermal feedback since no additional hardware was needed to translate between camera and build chamber coordinate systems. The controls technique began by capturing a thermal image of the powder surface prior to sintering. This data was analyzed and used to determine the average powder surface temperature where each individual component was to be sintered. These temperatures were each compared with a desired post-sintering temperature and an empirical model relating temperature increase to laser power was used to determine the optimal laser power in each component. Once these optimal powers were found, the laser was actuated accordingly and the components were sintered. Subsequently, a new layer of powder was spread and the process was repeated. This technique was not capable of as fine temperature control as the proof-of-concept testing, but the tensile bar cross sections had relatively small lengths and temperature gradients within a cross section area were typically significantly smaller than the overall temperature gradient across the build surface. The new technique was more robust than those used previously and was capable of being used on multi-layer builds.

4 Results

4.1 Baseline Fixed Power

As discussed in the thermal model section, a fixed power should result in uniform temperature increase, regardless of initial temperature. This concept was tested experimentally by using a fixed laser power to sinter areas within the build surface that were at different temperatures prior to sintering. Note that the results reported in this paper show a 15–20 °C temperature difference across the build surface, which in no way is representative of what LAMPS is capable of. While a commercial machine typically has 10–15 °C temperature variation, LAMPS typically has close to 5 °C temperature difference across its build surface. To highlight the capabilities of the proposed controls technique, a suboptimal heater configuration was purposely used to increase the baseline temperature variation profile in LAMPS.

The fixed IR camera was used to record the build surface temperature at 30 frames per second. After sintering was completed, the images were used to create a single composite image that tracks the maximum value each camera pixel saw during building of that layer. This composite image was treated as the maximum temperature reached post-sintering. While the frame rate was too slow to capture the laser interacting with every voxel of powder, the vectors being sintered were long enough such that the laser spot could be seen on most vectors and this method of creating the composite image provided good temperature data.

This composite image is presented in Fig. 3. This figure shows 3 rows of 10 rectangles sintered with a fixed power, a typical layer during a tensile bar build oriented in the z-direction.

While it is obvious in Fig. 3 that the post-sintering temperatures are not consistent, the extent of which is somewhat lost in the image. In order to display this more clearly, a single row of 10 rectangles is isolated and plotted as a 3D plot, with the z-axis representing temperature, as seen in Fig. 4. In this test, the pre-sintering powder temperature, which can be seen between the peaks representing the rectangular sintered regions, had a gradient of approximately 20°. The laser scan region to sinter the rectangles is shown by the peaks. On the surface, this data seems to not fit with the previous hypothesis that a fixed laser power produces a constant temperature increase, as the first 4 sintered regions appear to have larger temperature increase than the others. However, the pre-sintering temperature gradient was so great in this test that after laser scanning, the regions in the colder portion of the powder bed did not fully melt. Since melting consumes energy to overcome the latent heat of fusion, it is reasonable to expect powder that does not fully melt to experience a greater temperature increase. As the pre-sintering temperature increases (moving right in Fig. 4), the powder undergoes more melting, thus decreasing temperature rise until the powder within the sintered region fully melts and the temperature rise is constant.

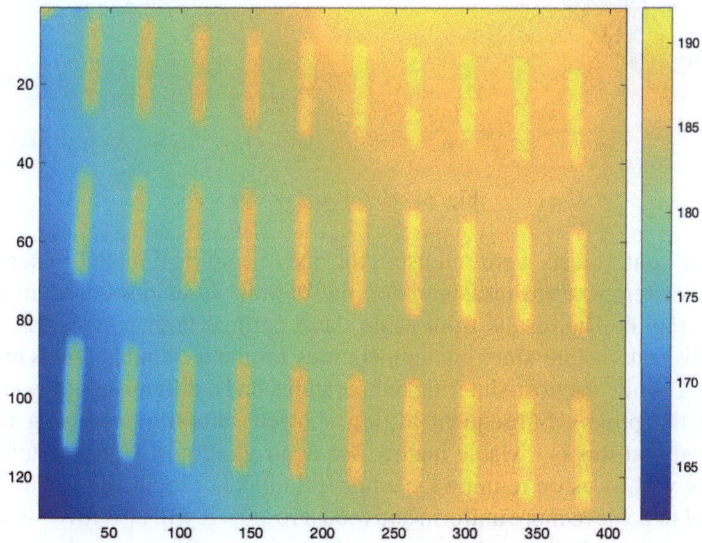

Fig. 4. Composite thermal image

4.2 Feed-Forward Laser Control

The laser control technique using the fixed IR camera as described above was used in order to homogenize the post-sintering temperatures. In order to determine the optimal laser powers, the laser power to temperature increase model needs to be calibrated. To do this, the baseline fixed power test is repeated with different fixed powers ranging from 20% to 100%. The thermal data is analyzed to extract how much each rectangular region's temperature raised for each laser power. The results for all 30 rectangular locations are shown in Fig. 5. Rather than averaging all locations together to form one overall transfer function, each location is considered separate to account for non-uniformities in the laser window transmittance.

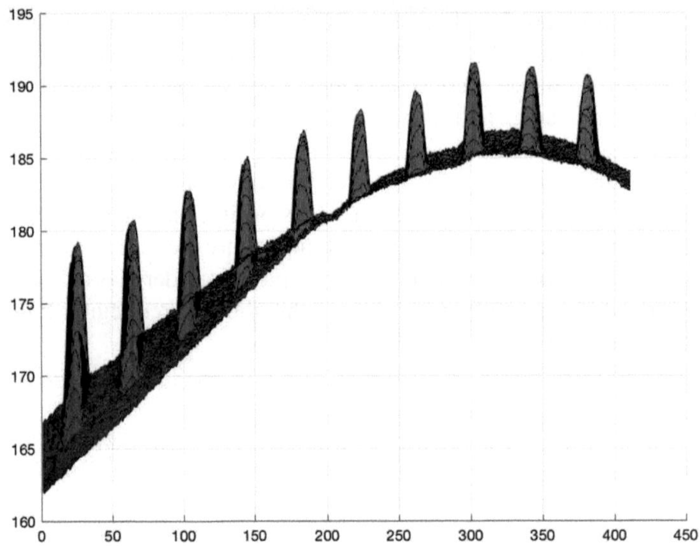

Fig. 5. 60% Fixed power

The laser control tests were run using the above empirical model. After taking the initial powder temperature measurement, the optimal laser powers were defined as depicted in Fig. 6. The results from using these optimal laser powers can be seen in Fig. 7. In this test, the pre-sintering temperatures for the rectangular areas had a difference of 14 °C while the post-sintering temperatures had a difference of fewer than 2 °C (shown by the peaks). Subsequent testing showed similarly promising results and demonstrated that this is a viable method for controlling post-sintering temperatures. Future work will focus on refining the empirical model or replacing it with a physics-based model to achieve more uniform and robust results. It will also involve completing full tensile specimen builds and comparing the variance of these specimens with those from standard builds (Fig. 8).

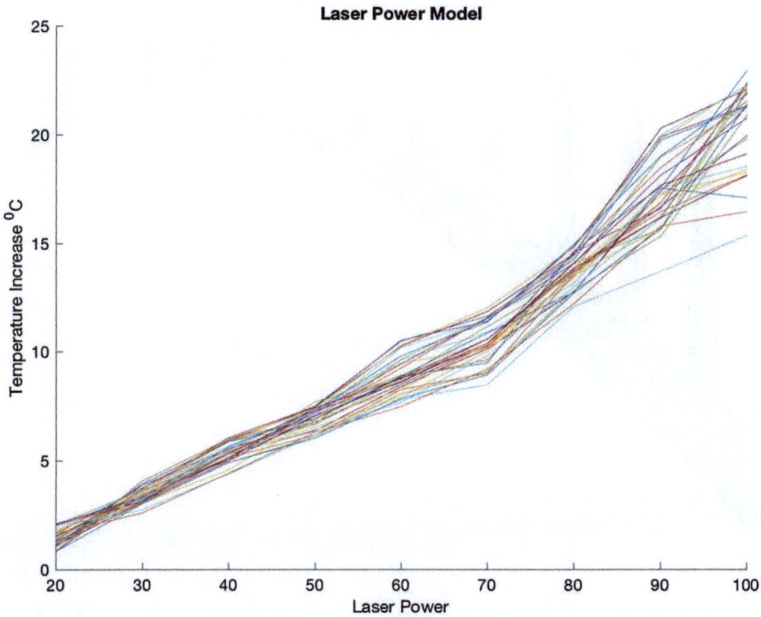

Fig. 6. Laser power empirical model

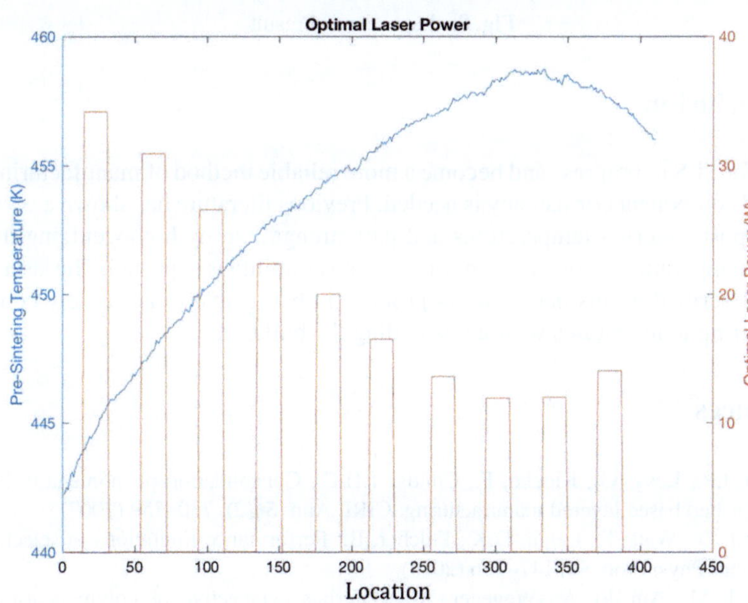

Fig. 7. Optimal laser power profile

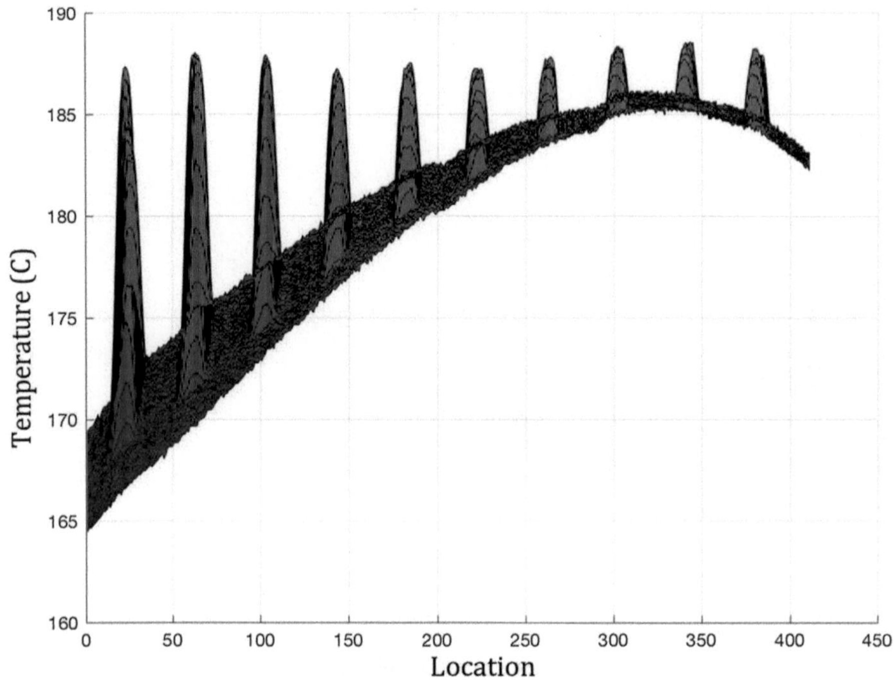

Fig. 8. Laser control result

5 Conclusion

In order for SLS to progress and become a more reliable method of manufacturing parts, improved component consistency is needed. Previous literature has shown a correlation between post-sintering temperatures and part strength, so by homogenizing the post-sintering temperatures, a decrease in component variability is expected. The laser control method described in this paper shows promise in being able to accurately control the post-sintering temperatures without extending the build time.

References

1. Kruth, J.P., Levy, G., Klocke, F., Childs, T.H.C.: Consolidation phenomena in laser and powder-bed based layered manufacturing. CIRL Ann. **56**(2), 730–759 (2007)
2. Bourell, D., Watt, T., Leigh, D.K., Fulcher, B.: Performance limitations in selective laser sintering. Phys. Proc. **56**, 147–156 (2014)
3. Schmid, M., Amado, A., Wegener, K.: Materials perspective of polymers for additive manufacturing with selective laser sintering. J. Mater. Res. **29**(17), 1824–1832 (2014)
4. Wroe, W.W.: Improvements and Effects of Thermal History on Mechanical Properties for Polymer Selective Laser Sintering (SLS). University of Texas, Austin, Austin (2015)
5. Benda, J.A.: Temperature-controlled selective laser sintering. In: Solid Freeform Fabrication Symposium, Austin (1994)

6. Clijsters, S., Craeghs, T., Buls, S., Kempen, K., Kruth, J.P.: In situ quality control of the selective laser melting process using a high-speed, real-time melt pool monitoring system. Int. J. Adv. Manuf. Technol. **75** (2014)
7. Craeghs, T., Stijn, C., Yasa, E., Kruth, J.-P.: Online quality control of selective laser melting. In: Solid Freeform Fabrication Symposium, Austin (2011)
8. Zeng, K., Pal, D., Stucker, B.: A review of thermal analysis methods in laser sintering and selective laser melting. In: Solid Freeform Fabrication Symposium, Austin (2012)
9. Sun, M.-S.M., Beaman, J.J.: A three dimensional model for selective laser sintering. In: Solid Freeform Fabrication Symposium, Austin (1991)
10. Phillips, T.: In-situ laser control method for polymer selective laser sintering. In: Solid Freeform Fabrication Conference, Austin (2016)

Gaussian Process Regression for Virtual Metrology of Microchip Quality and the Resulting Selective Sampling Scheme

Tyler Darwin[1], Roman Garnett[2], and Dragan Djurdjanovic[1(✉)]

[1] Department of Mechanical Engineering, University of Texas at Austin, Austin, TX, USA
dragand@me.utexas.edu

[2] Department of Computer Science and Engineering, Washington University in St. Louis, St. Louis, MO, USA

Abstract. Manufacturing of integrated circuits involves many sequential processes executed to nanoscale tolerances, and the yield depends on the often-unmeasured quality of intermediate steps. Taking physical quality measurements in this high-throughput industry can be expensive and time-consuming. Instead, we seek to predict product quality characteristics using readily available sensor readings of the tool environment during processing of each wafer and employ Gaussian Process Regression (GPR) paradigm to realize this Virtual Metrology (VM) concept. Convergence of the GPR based VM estimation of product quality is hastened through an active sampling scheme, whereby the predictive uncertainty of the GPR model informs which wafer's quality to measure next in order to obtain maximal additional information for the VM model. We evaluate these methods using a large dataset collected from a plasma-enhanced chemical vapor deposition (PECVD) process, with relevant tool sensor readings and the corresponding physical measurements of mean film thicknesses for 32,000 wafers. By selecting which wafers to physically measure for VM model updates, the GPR based VM method achieves ~10% greater accuracy on average than the partial least squares based method.

Keywords: Virtual metrology · Gaussian process regression
Plasma-enhanced chemical vapor deposition · Semiconductor manufacturing
Virtual metrology · Process drift

1 Introduction

To manufacture modern integrated circuits (IC), hundreds of sequential processes pattern, stack, and connect layers of electrical components that demand nanoscale tolerances across hundreds of millimeters of a wafer substrate. The dimensions necessary to pack more processing power or memory onto a single chip have shrunk for decades, pushing tolerances to limits of what may be measured, often necessitating the use of costly measurement equipment and complex, lengthy measurement procedures. By the nature of a high-throughput industry, semiconductor manufacturers cannot afford to

J. Ni et al. (Eds.): AMP 2018, LNME, pp. 250–264, 2018.
https://doi.org/10.1007/978-3-319-89563-5_19

physically measure all critical quality characteristics of these intermediate steps for every chip.

Instead, in this paper, we seek to predict such metrics of product quality through the concept of virtual metrology (VM), in which readily available sensor readings from the corresponding tool environment during processing of the workpiece are used to predict outgoing quality characteristics of that workpiece [1–3]. Due to the sheer complexity of the physical phenomena involved, mathematical models based on first principles, particularly those of plasma dynamics, typically struggle to provide reliable estimates of nanoscale deviations of wafer quality within time-scales comparable to production cycle times, which is necessary to utilize such results for subsequent process control. Instead, models in the context of VM utilize various data-driven regression models to approximate the dynamics that govern the wafer quality characteristic of interest using sensor traces that describe the corresponding production environment within the relevant manufacturing tool [3–7].

We propose a probabilistic approach to regress the metric of quality using Gaussian processes (GPs) [8], which can model the nonlinear dynamic dependencies typical of processes in semiconductor manufacturing. Such Gaussian process regression (GPR) predicts a Gaussian distribution for the target metric of quality given the signatures extracted from the sensor signals emitted during manufacturing of the wafer, thus enabling process engineers to elegantly assess the uncertainty of the VM predictions. Furthermore, this predictive uncertainty can be leveraged to inform which wafer's quality to measure next in order to maximize the information gained by the GPR-based VM model.

In theory, GPR can approximate arbitrarily complex functions, or exhibit universal consistency, assuming *stationary* Gaussian noise corrupts the observed outputs [8–10]. Artificial neural networks (ANNs), another powerful regression approach, offer similar appeal for their ability to model complex dependencies and discern patterns despite the presence of highly correlated input features, or those simply irrelevant to output [11]. However, despite the popularity of ANNs in the VM literature, the trained parameters yielded by such expressive yet complex architectures are typically not interpretable, which hampers their use for process control and decision-making.

As such, researchers often default to a linear model, such as partial least squares (PLS) [12], which can also handle highly correlated input features. PLS frames regression in manner akin to proposed GPR framework, by projecting the inputs to a latent space based on relevance to the output. Although simple and computationally efficient, such a linear approximation may not be adequate for VM purposes given the nonlinear dynamic dependencies that govern critical quality characteristics, especially as an operator may tweak the process during data collection.

Therefore, various approaches to adapt PLS for piecewise regression, or otherwise localize VM models have arisen. For example, Hirai and Kano [4] localize PLS by weighting examples based on proximity to the query point, after scaling the input dimensions with a global PLS model. However, this method requires one to set a free parameter that scales the dispersion of the relevant neighborhood of points, or the concentration of weights about the query. Bleakie and Djurdjanovic recommend a more strictly piecewise approach in [5], by partitioning the input space through unsupervised clustering, then fitting a local linear model within each cluster. As the model updates, various heuristics govern how to move the centers of the assigned clusters to regions of high modeling error and if necessary, create new clusters. Others have opted to localize models exclusively in terms of time, by windowing the data to regress on a set of previous examples before any given query point. For example, Lynn et al. [6] compare PLS, ANNs, and GPR for windows of various lengths, finding GPR to be about 10% more accurate for predicting plasma etch depth. However, windowing only marginally improved the GPR's accuracy compared to a global version, while the constant refitting greatly slowed execution.

Instead of complicating methods with heuristics, such as those to localize models, we confront a prevalent assumption in the VM literature: a fixed (perhaps random) scheme for measuring product quality. Such an assumption limits the VM model to passively learning from a given block of data. Recent work by Bleakie & Djurdjanovic recommends measuring at unscheduled moments to update the trained model, if the raw inputs are "unusually" distant from the known data [5]. The proposed, more principled approach lets the model decide which wafer's quality to measure based on expected information gain, in order to hasten convergence when starting from no initial data.

To thoroughly evaluate this modeling approach, we rely on an extensive dataset for which average film thickness across the wafer has been measured as the quality characteristic for each of the roughly 32,000 wafers produced over several months. Section 2 outlines the Bayesian framework for GPR-based VM. Section 3 describes the results of applying the GPR based VM approach to a large-scale dataset obtained from an industrial PECVD process, as well as comparisons of accuracy and convergence characteristics of the newly proposed VM approach against the traditional PLS-based VM method. Finally, Section 4 offers concluding thoughts and outlines possible directions for future work.

2 Bayesian Modeling Approach

2.1 Conceptual Background

Bayesian inference offers a consistent way to update prior beliefs by conditioning them on the data. By explicitly acknowledging uncertainty, especially when starting with no initial observations of the output and little to no prior knowledge of the input–output relationship, the process of "learning" from the data becomes more principled. In the context of the VM task, a Bayesian model would naturally express uncertainty in the predicted metric of product quality. Not only does this clarify whether the model is credible, but can also inform which output, given the inputs for a candidate set of examples, is expected to yield the greatest information gain by being observed or measured.

To match the complexity of the dynamics which govern the target metric of quality, we assumed a flexible, Gaussian process (GP) prior and a Gaussian likelihood for the measurement noise, following the technique for regression in. This special case yields exact inference of the predictive, or posterior distribution through Bayes' rule. Since a GP evaluated at any finite set of (input) points takes the form of a multivariate Gaussian distribution, GPs exhibit the same elegant properties as the Gaussian distribution itself, such as closure under affine transformations or under convolution with another Gaussian distribution. In effect, GPR models predict a Gaussian distribution for each product's metric of quality, as expressed by the predictive mean and variance. The mean corresponds to the Bayesian estimate that is optimal in the mean-squared sense, while two standard deviations above and below the mean bound a 95% credibility interval. The algebraic operations to update the GP prior take a similar form to how one conditions a Gaussian distribution on observed outputs given the associated inputs.

2.2 Mathematical Foundation

Recall that the quality characteristic of interest may be insensitive to some of the features extracted from sensor readings, which then act as conceptual noise to the model and hamper standard GP inference. The GP model proposed here learns the lower-dimensional space in which the dynamics operate, approximated by a linear embedding of the original features, while simultaneously learning the latent function mapping equipment signatures (inputs) to product quality (outputs). During the learning process described in the algorithm shown in Fig. 1, the embedding may be highly uncertain, which we seek to express in the predictive uncertainty.

```
Require: m, M; kernel κ, mean function μ; prior p(R); X ← ∅; Y ← ∅; ŷ ← ∅
  //Partition 32k wafers into 1280 lots of 25 wafers
  lot ← random integer from [1, 1280]
  lotsPerRun ← 100; finalLot ← lotsPerRun + lot
  initTrain ← 10; initTrainEnd ← initTrainEnd + lot
  while lot < finalLot
      if lot ≥ initTrainEnd
          ŷ ← [ŷ; q(f(X_lot))]
              //predict output for upcoming lot
      repeat on candidate wafers from given lot
          q(R) ← LaplaceApproximation (p(R | X, Y, κ, μ))
              //approximate posterior on embedding R
          q(f) ← ApproximateMarginal (p(f | R), q(R))
              //approximate marginal on function f
          x*   ← OptimizeUtility(q(f), q(R))
              //find approximate optimal evaluation point x*
          y*   ← Observe (f(x*))
          X    ← [X; x*]; Y ← [Y, y*]
      until budget depleted
      lot ← lot + 1
  return ŷ
  //Concludes one run of experiment
```

Fig. 1. Simultaneous active learning of function and linear embedding in the context of GPR based VM for the PECVD process (adapted from Garnett et al. [9]).

2.2.1 Incorporation of Prior Knowledge

Assuming the target metric of quality tends to a set point (i.e. constant mean value, μ), the GP's covariance function encapsulates our prior knowledge of the structure of the latent function of the underlying dynamics that maps input features to the outputs. Let x and x′ refer to inputs of dimensionality M for two examples, and assume a linear embedding, $R \in R^{m\times M}$ with m << M, such that the covariance between the examples may be expressed as

$$K(x, x') = \gamma^2 \exp\left(-\frac{1}{2}(x - x')R^T R(x - x')^T\right) \quad (1)$$

In other words, we modify the well-known squared-exponential kernel to project the model inputs into a lower-dimensional space through a linear mapping, R. Note that the resulting GP can still regress nonlinear interactions between these linearly embedded features that constitute a given input. The rows of R correspond to the directions of the most rapid change in the function, and the Euclidean length of each projection corresponds to its relevance, or inverse length scale. For the purposes of inference, examples more than a few input length scales apart along a given direction in R can be perceived as uncorrelated, or their observed outputs seen as irrelevant to each other.

2.2.2 Tuning of Parameters

As the model actively selects which wafers to physically measure so as to inform R and learn the latent input–output relationship, we seek to maximize the likelihood of the model with respect to the entries of R and other parameters of the GP prior. Let all these parameters[1] be denoted by a vector, θ. Then let $y \in R^{(N\times 1)}$ be the noisy measurements of the metric of product quality, and let the associated inputs be $X \in R^{(N\times M)}$, such that the marginal likelihood of the model for the given parameter values may be expressed by

$$\log p(y|X, \theta) = -\frac{(y-\mu)^T V^{-1}(y-\mu)}{2} - \frac{\log \det V}{2} - \frac{N \log 2\pi}{2} \quad (2)$$

in which σ^2 is the variance of the measurement noise, and $V = K(X, X) + \sigma^2 I$ corresponds to the covariance of y for inputs X. The first term of Eq. (2) penalizes poor fit to the data, whereas the second term penalizes the complexity of the model. In effect, this process of tuning the model's parameters by maximizing (2) balances its accuracy and simplicity.

2.2.3 Prediction Under Uncertain Embedding

To acknowledge uncertainty in R, let us assume a Laplace approximation on p(R) given the data, D:

[1] In deriving a GP, the underlying parametric model's weights are integrated out, so the tunable parameters of the GP prior are referred to as *hyperparameters* to emphasize the non-parametric nature of this approach.

$$p(R|D) = \mathrm{N}\left(R; \hat{R}, \textstyle\sum_R\right)$$
$$\textstyle\sum_R = \left(-\nabla\nabla p(R)|_{R=\hat{R}}\right)^{-1} \tag{3}$$

in which $\hat{R}$ refers to the maximum likelihood estimator (MLE) and Σ_R represents the inverse Hessian about the MLE. The GP then regresses the target metric of quality on $u = xR^T$, or a linearly transformed Gaussian expressed as $p(u) = N\left(xR^T; x^{\hat{R}^T}, x \sum_R x^T\right)$. The perceived inputs, u are uncertain, but correlated and Gaussian-distributed through R.

When estimating the predictive distribution for a test point, or $f_* = f(x_*)$, we wish to integrate out this uncertainty in the parameters θ, of which entries of R constitute a majority:

$$p(f_*|D) = \int p(f_*|D, \theta)\, p(\theta|D)\, d\theta \tag{4}$$

However, such an integral cannot be directly evaluated due to its intractability arising from the nonlinear dependence of f on R. In order to approximate (4), we rely on the method devised by Garnett et al. [9], first linearly approximate the dependence of $p(f_*|D,\theta)$ on θ, then match moments to those of the exact integral[2]. This technique inflates the predictive variance based on the uncertainty in θ, as expressed by

$$\begin{aligned} V_{f|D}(x_*) = \frac{4}{3}\hat{V}(x_*) &+ \left(\frac{\partial \hat{m}(x_*)}{\partial \theta}\right)^T \sum \left(\frac{\partial \hat{m}(x_*)}{\partial \theta}\right) \\ &+ \frac{1}{3\hat{V}(x_*)}\left(\frac{\partial \hat{V}(x_*)}{\partial \theta}\right)^T \sum \left(\frac{\partial \hat{V}(x_*)}{\partial \theta}\right) \end{aligned} \tag{5}$$

in which $p(\theta|D) = N\left(\theta; \hat{\theta}, \sum\right)$ and $p(f_*|D,\theta) = N\left(f_*; \hat{m}(x_*), \hat{V}(x_*)\right)$. Note that though the predictive variance yielded by this marginal GP (mGP) has increased, the mean remains the same (i.e. the value at $\hat{\theta}$, the MLE).

2.2.4 Active Learning of Embedding

Let us leverage this predictive uncertainty to hasten the convergence of the model to a credible form. In particular, we want to take physical measurements of product quality for cases that are expected to be the most informative, not only about the dynamics' latent function but also of its linear embedding, which facilitates this GP framework to correlate examples and generalize to unseen ones.

To assess the perceived information gain, or utility of observing a product's quality, we employ *Bayesian active learning by disagreement* (BALD), which assigns utility based on the mutual information, I(F; Θ) between the latent function, F and the hyper–

---

[2] Since the dependence of the latent function, f on the parameters, θ tends to be well-concentrated, approximating this dependence as a line through the MLE, $\hat{\theta}$ seems reasonable.

parameters, Θ. Mutual information, or relative entropy, may be interpreted through Kullback-Leibler divergence [13], as seen in the utility function:

$$\begin{aligned} v(x) &= I(F;\Theta) = D_{KL}(p(F,\Theta)\|p(F)p(\Theta)) \\ &= E\Theta(DKL(p(F|\Theta)\|p(F))) \\ &= H(F|\Theta) - H(F) \end{aligned} \tag{6}$$

in which H refers to the entropy, or log-variance for Gaussian distributions. Essentially, the greatest information gain would be expected at the candidate point for which the standard predictive variance and inflated version, from p(F|Θ) and p(F) respectively, diverge or "disagree" the most[3].

Since the GP framework yields predictions in the form of Gaussian distributions, the perceived utility of candidate points to measure may be readily approximated by a ratio of predictive variances using the mGP:

$$v'(x) = \frac{V_{f|D}(x)}{V_{f|D,\hat{\theta}}(x)} \tag{7}$$

In [9], Garnett et al. contrast this active sampling scheme with a more myopic one called uncertainty sampling, which arises from a utility function that uses only the standard predictive variance: $\tilde{v}(x) = V_{f|D,\hat{\theta}}$. Basically, by observing outputs that are expected to inform about an embedding, the predictive variance tends to shrink for all points, not just at the observed one.

3 Results

3.1 Description of Dataset

We apply the framework for GPR detailed in the previous section to data collected from a major semiconductor fabrication plant, in order to predict the average thickness of films deposited onto a wafer by a plasma-enhanced chemical vapor deposition (PECVD) process. In chemical vapor deposition, carrier gases containing volatile precursors decompose on the wafer substrate to yield a thin film of insulating or conducting material [14]. To exclude contaminating particles, the process is enclosed in a vacuum chamber, which can be pumped down to low pressure, typically 0.1–10 Torr, in order to enhance the film's density and purity. Energy for the reaction and subsequent reaction temperature can be reduced in the plasma environment, a strategy that PECVD processes exploit (see Fig. 2).

[3] The invariance of mutual information to homeomorphic transformations make this learning process robust to scaling or rotating of the current embedding, or to centering and scaling of the inputs and outputs.

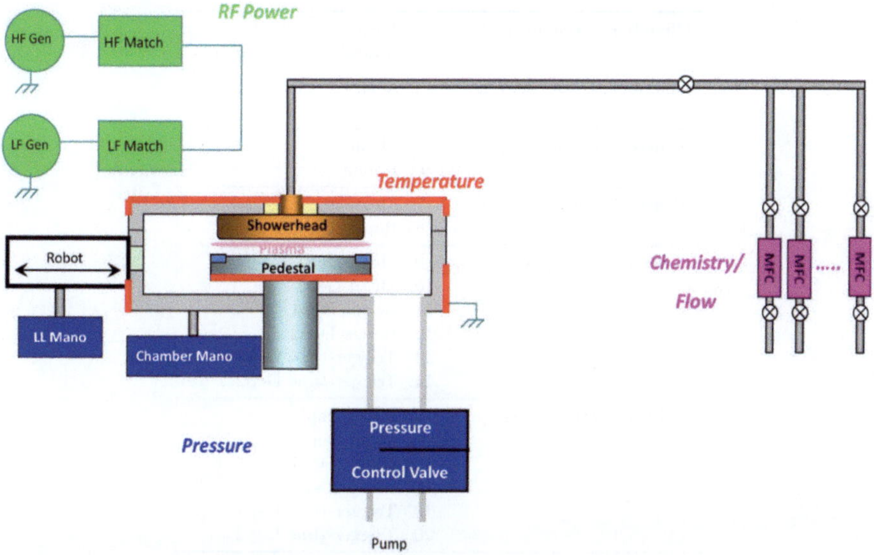

Fig. 2. Schematic diagram of a PECVD system.

The particular process analyzed here employs tetraethoxysilane (TEOS) that decomposes to form an insulating film of SiO_2 on the silicon substrate. The data gathered from the PECVD tool indirectly describe these dynamics through 11 sensor readings sampled at 10 Hz. The associated traces describe temperatures at the wafer pedestal and chamber walls, pressure, gas flows into the chamber, and the power and voltage associated with the radio-frequency (RF) system that energizes the plasma. Note that though the plasma supplies sufficient energy for the volatile precursors to decompose at room temperature, typically the wafer and walls of the chamber are heated to minimize the number of defects. These hot walls accumulate residue that can flake or otherwise disturb the plasma, prompting in-situ cleans to be conducted every 25–100 wafers.

The data includes physical measurements of mean wafer thickness (MWT) for each of the ~32,000 consecutively processed wafers, along with sensor readings collected from the tool at 10 Hz. Following [15], we condense the raw sensor traces into a set of features, describing each sensed parameter's steady-state and transient behavior, as well as transition times, yielding a total of 49 features (see list in Fig. 3). From this data, we could construct input–output pairs (x, y) in which a wafer's input vector, x, comprises this set of 49 features, and y denotes the corresponding MWT.

Chamber temperature 1	1. Mean 2. Range
Chamber temperature 2	3. Mean 4. Range
Pedestal temperature 1	5. Mean 6. Range
Pedestal temperature 2	7. Mean 8. Range
HF reflected power	9. Mean Dep1 10. Mean Dep2 11. Range Dep1 12. Range Dep2 13. Trigger-time Dep1 14. Trigger-time Dep2
LF reflected power	15. Mean Dep1 16. Mean Dep2 17. Range Dep1 18. Range Dep2 19. Trigger-time Dep1 20. Trigger-time Dep2
Flow rate of TEOS	21. Mean 22. Range 23. Over-shoot 24. Rise-time
Load capacitor voltage	25. Mean Dep1 26. Range Dep1 27. Max Dep2 28. Range Dep2
Chamber pressure	29. Rise-time of pump-up 30. Fall-time of pump-down 31. Peak 32. Mean Dep1 33. Range Dep1 34. Rise-time Dep2 35. Mean Dep2 36. Range Dep2 37. Min
Pendulum valve angle	38. Rise-time of pump-up 39. Max of pump-up 40. Mean of pump-up 41. Range of pump-up 42. Mean Dep2 43. Range Dep2 44. Max of pump-down 45. Mean post-process
Tune capacitor voltage	46. Mean Dep1 47. Range Dep1 48. Max Dep2 49. Range Dep2

Fig. 3. List of features derived from PECVD sensor traces, following procedures described in [15]. Dep1 and Dep2 refer to the two steps of the process.

Over the time-span of the data, the tool failed four diagnostic tests, signaling faulty behavior that required the tool to shut down for repair.

These faults correspond to bimodal thickness across the wafer, films exhibiting unacceptable range in thickness, particle contamination of the chamber, and Coulomb crystal formation in the chamber.

3.2 Description of Underlying Experiment

The following experiment mimics real-time prediction in a production environment, where model updates occur online, as measurements arrive. In particular, we assume only one wafer's quality characteristic can be "measured" or observed per process lot of 25 wafers, which is a typical constraint encountered in semiconductor manufacturing.

Consequently, the data was partitioned into lots of 25 wafers, which yielded approximately 1280 lots. In each run of the experiment, we predict MWT for the wafers in the next lot, select the wafer from that lot for which to measure the output MWT following the methodology described in Sect. 2.2.4, and then update the model accordingly. This procedure was conducted for 100 consecutive lots, and repeated from a random initial

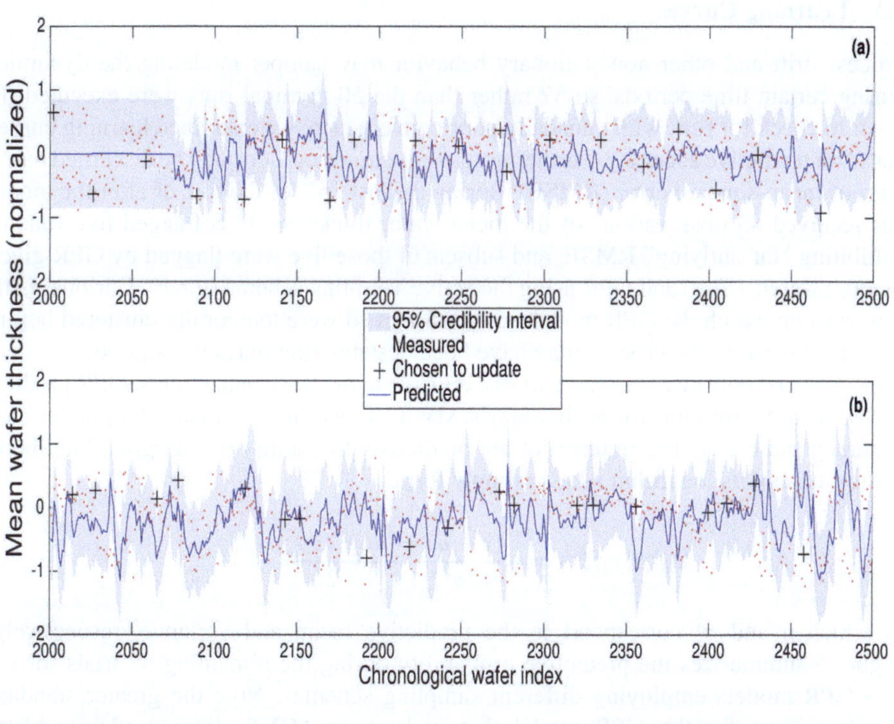

Fig. 4. Measured mean wafer thickness compared to that predicted by GPR using an active sampling scheme, during last 20% of two runs. To simulate online setting of a factory, GPR predicts one lot of 25 wafers ahead, selects the approximately optimal wafer to measure, updates, and repeats for 100 consecutive lots. Note that predictions in subplot (a) tend to be conservative (around a constant) before the embedding converges to a plausible form.

lot at least 50 times (see two examples of such runs in Fig. 4). Please note that we excluded the runs that would cross a fault or a 150-wafer buffer before the fault was flagged, which was done due to suspicions that process dynamics was disturbed close to faults and hence those lots were not included into the training dataset.

For each element of the linear embedding, we assigned a diffuse prior, following [9], while parameter updating followed procedures described in Sect. 2.2.2.

We compared the accuracy of GPR to that of PLS based regression, often used for VM purposes in industry. Note that such a pure regression model does not naturally depict information gains obtained by any given training sample and, therefore, we let the PLS regression based VM model update using a random wafer from each lot. We also maintained an auxiliary GPR model that updates on the same set of random wafers measured thicknesses.

In order to evaluate each model's convergence, we considered how the accuracy improves over time, as well as the marginal likelihood for the cases with GPR. Given these metrics represent predictive performance in an online scenario, we constructed each learning curve not by summarizing the improvement on a designated test set, but on all subsequent lots of wafers from a given point in the run.

3.3 Learning Curves

Process drift and other non-stationary behavior may hamper modeling the dynamics during certain time periods, so 57 rather than the 50 nominal runs were executed. In order to check for runs with unusually poor convergence, we examined learning curves based on predictive accuracy. Furthermore, we assumed outliers may be discerned based on root-mean-squared error (RMSE)[4] during the last 15% of the run, or after the model has received 85 observations of the mean wafer thickness. PLS flagged five runs as exhibiting "far outlying" RMSE, and subsets of those five were flagged by GPR given random wafers (four) and GPR using the active sampling scheme (three). Curiously, the four runs on which the GPR models and PLS agreed were temporally clustered before a fault, suggesting the process may have been destabilizing during that period.

Let us first consider learning curves exclusive to the Bayesian models. GPR predicts a Gaussian distribution for each wafer's MWT, so we may evaluate the predictions' accuracy and confidence in terms of the predictive probability (or marginal likelihood for test cases); its logarithm takes a simple form:

$$\log p\left(y_*|D\right) = -\frac{(y-\mu)^2}{2\sigma^2} - \frac{\log 2\pi\sigma^2}{2} \tag{8}$$

in which μ and σ^2 correspond to the predictive mean and variance, respectively. Figure 5 summarizes the predictive probability during the remaining 52 trials for the two GPR models employing different sampling schemes. Note the greater standard deviation bars for the GPR model that updates on MWTs from random wafers,

[4] RMSE more heavily penalizes large predictive errors than mean absolute error (MAE), and typically yields clearer outliers.

illustrating its struggle to extract sufficient information, as manifested in several runs with very low predictive probability. The tighter bars associated with GPR using BALD indicates a more robust strategy to learn from the seemingly repetitive dataset.

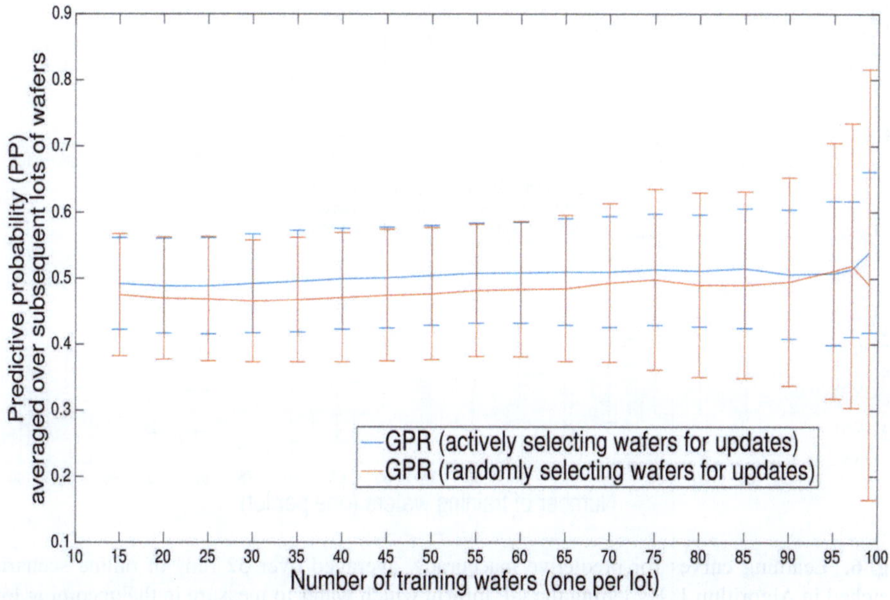

Fig. 5. Learning curves for predictive probabilities, averaged over 52 runs of online scenario sketched in Algorithm 1, illustrated in Fig. 1. The tighter standard deviation bars for the GP using an active sampling strategy suggest a more robust strategy for learning from the tightly controlled process' data.

As PLS only offers a point estimate, let us compare its performance to that of the GPR models on the basis of inaccuracy alone. Figure 6 summarizes their convergence in terms of RMSE and mean absolute error (MAE) over the 52 remaining trials of the experiment. Since dimensions beyond the tolerances incur disproportionately greater cost, RMSE would be a more appropriate metric for this task, as it more heavily penalizes large predictive errors than metrics of absolute error do. The MAE and RMSE in Fig. 6 share units of standard deviations, σ, of the TEOS thickness. As for the relative trends, when updating on MWTs from the same random wafers, GPR proves less accurate than the linear model. However, by letting the predictive uncertainty inform which wafer to measure next, GPR becomes about 10% more accurate on average than PLS, most likely due to its ability to model the nonlinear nature of the underlying dynamics.

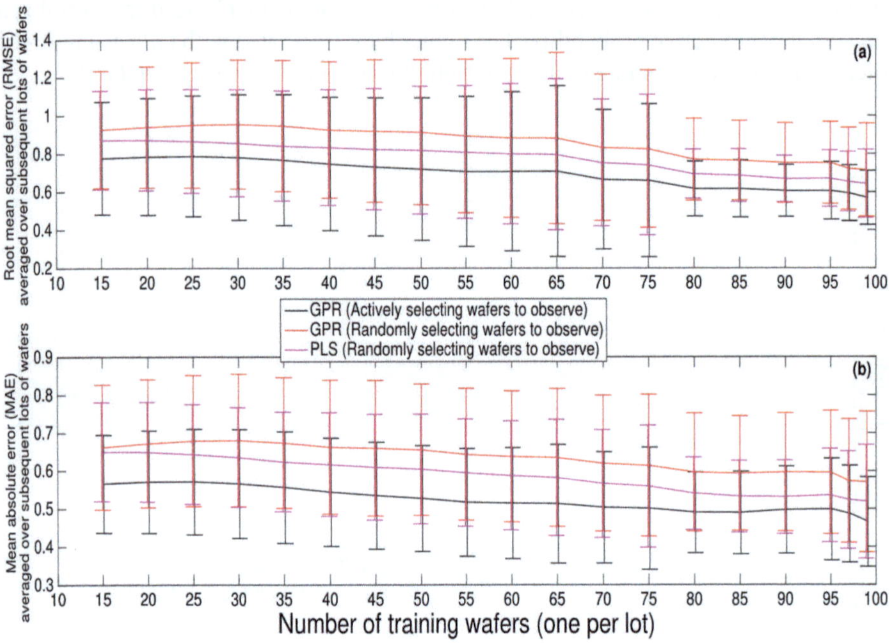

Fig. 6. Learning curves for predictive inaccuracy, averaged over 52 runs of online scenario sketched in Algorithm 1. By letting the GP inform which wafer to measure in the upcoming lot, it tends to obtain sufficient information to outperform the PLS based VM.

4 Conclusions and Future Work

The research presented in this paper indicates that GPR can approximate the nonlinear dynamics that govern metrics of product quality, such as mean film thickness in a PECVD process, by approximately optimizing the information extracted from a tightly controlled manufacturing process. This strategy for predicting microchip quality leverages the framework devised and implemented by Garnett *et al.* [9] in which a marginal GP actively learns the underlying dynamics as well as the embedded space in which the dynamics operate. This Bayesian method yields interpretable parameters, as well as clear ways to assess convergence and uncertainty in the prediction, thereby estimating the probability that a wafer's quality characteristics are within tolerances rather than just offering a "best guess" of their value.

In addition to providing greater predictive accuracy than a comparable linear model, the GP offers an integrated estimate of how relevant the features are to the target metric of quality. Such information can be utilized to inform measurement procedures which product (wafer) in a lot to physically measure in order to gain maximal amount of additional information that can be incorporated in to the VM model.

The GPR based VM approach proposed in this paper was evaluated in estimating mean film thicknesses in a large dataset obtained from an industrial PECVD process and compared against the industrial standard of PLS-based VM. It consistently outperformed the PLS-based VM in terms of accuracy, while in the same time naturally yielding uncertainty of those metrology estimates.

Future work in this domain needs to focus on expanding the GPR to incorporate non-Gaussian and non-stationary noise characteristics. Such traits can be pursued through consideration of regression models that include Gaussian mixtures rather than purely Gaussian distributions of output variables (product quality in the VM context). Furthermore, application of GPR-based or mixture GPR-based VM to more complex problems, such as estimation of critical dimensions in plasma-based etch processes also remains a challenge for the future. Finally, utilization of VM paradigm for process adaptations to remedy perceived disturbances in the underlying processes has not been done. Major problem in this direction is that all current VM models lack sufficient accuracy to facilitate such adaptations, which is less due to deficiency in models and more due to insufficient sensing of the underlying processes. Thus, improvements in sensing, coupled with development of better, more informative VM models should bring about future possibility of VM based automatic process control.

References

1. Ringwood, J.V., Lynn, S., Bacelli, G., Ma, B., Ragnioli, E., McLoon, S.: Estimation and control in semiconductor etch: practice and possibilities. IEEE Trans. Semicond. Manuf. **23**(1), 87–98 (2010)
2. An-Jhih, S., Jeng, J.-C., Huang, H.-P., Cheng-Ching, Yu., Hung, S.-Y., Chao, C.-K.: Control relevant issues in semiconductor manufacturing: overview with some new results. Control Eng. Pract. **15**(10), 1268–1279 (2007)
3. Yang, Y., Wang, M., Kushner, M.J.: Progress, opportunities and challenges in modeling of plasma etching. In: Proceedings of the 2008 IEEE International Interconnect Technology Conference (IITC), Burlingame, CA, USA, 1–4 June 2008 (2008). https://doi.org/10.1109/iitc.2008.4546934
4. Hirai, T., Kano, M.: Adaptive virtual metrology design for semiconductor dry etching process through locally weighted partial least squares. IEEE Trans. Semicond. Manuf. **28**(2), 137–144 (2015)
5. Bleakie, A., Djurdjanovic, D.: Growing structure multiple model system for quality estimation in manufacturing processes. IEEE Trans. Semicond. Manuf. **29**(2), 79–97 (2016)
6. Lynn, S.A., Ringwood, J., MacGearailt, N.: Global and local virtual metrology models for a plasma etch process. IEEE Trans. Semicond. Manuf. **25**(1), 94–103 (2012)
7. Lee, S., Kang, P., Cho, S.: Probabilistic local reconstruction for k-NN regression and its application to virtual metrology in semiconductor manufacturing. Neurocomputing **131**(5), 427–439 (2014)
8. Rasmussen, C.E.: Gaussian Processes for Machine Learning. MIT Press, Cambridge (2006). ISBN 026218253X
9. Garnett, R., Osborne, M.A., Hennig, P.: Active learning of linear embeddings for gaussian processes. In: Proceedings of the 13th Conference on Uncertainty in Artificial Intelligence, Quebec City, Quebec, Canada, 23–27 July 2014, pp. 230–239 (2014)
10. Osborne, M.A., Garnett, R., Roberts, S.J.: Gaussian processes for global optimization. In: Proceedings of the 3rd International Conference on Learning and Intelligent Optimization, 14–18 January 2009, Trento, Italy (2009)

11. Neal, R.M.: Bayesian Learning for Neural Networks. Springer Science and Business Media, New York (1996)
12. Wold, S., Sjöström, M., Eriksson, L.: PLS-regression: a basic tool of chemometrics. Chemometr. Intell. Lab. Syst. **58**(2), 109–130 (2001)
13. Kullback, S., Leibler, R.A.: On information and sufficiency. Ann. Math. Stat. **22**(1), 79–86 (1951)
14. McInerney, E.J.: Chemical vapor deposition. In: Geng, H. (ed.) Semiconductor Manufacturing Handbook. McGraw-Hill Inc., New York (2005). Chapter 14
15. Bleakie, A., Djurdjanovic, D.: Feature extraction, condition monitoring and fault modeling in semiconductor manufacturing systems. Comput. Ind. **64**(3), 203–213 (2013)

Cloud Enabled CPS and Big Data in Manufacturing

Lihui Wang(✉) and Wei Ji

Department of Production Engineering, KTH Royal Institute of Technology, 100 44 Stockholm, Sweden
lihuiw@kth.se

Abstract. This paper presents a cloud enhanced cyber-physical system (cloud CPS) in manufacturing by combining CPS and cloud technologies. The cloud CPS is enhanced by using the combined strength of holons, agents and function blocks (FBs). Here, a holarchy of multiple holons is a sub-CPS within cloud CPS, and a logical part and a physical part are involved in each holon, and they mimic the cyber and physical entities of the CPS. They are able to be realised by agents and FBs for the manufacturing processes. In addition, to address the weakness in operation level, big data analytics (BDA) is applied to optimise machining jobs and to predict faults in scheduling. Within the processes, machining relevant factors, including workpiece, machining requirement, machine tools, cutting tool, cutting conditions, machining process and machining results, are represented by data, which is able to solve the many operational issues in cloud CPS.

Keywords: CPS · Cloud technology · Big data analytics
Machining optimisation · Scheduling

1 Introduction

Over the last decades, Industry 4.0, initiated in Germany, is to promote the effective use of the latest information and communications technologies in real industrial applications towards smart factories of the future [1]. Cyber-physical system (CPS), as the core technology of Industry 4.0, is usually connected through the Internet, and recently applied the concept of the cloud to form cyber manufacturing [2]. CPS as a multi-disciplinary methodology challenges this idea in a larger scale where the computational entities and the physical entities are connected through networks and dynamically interact in order to optimise the overall system performance. It has recently gained a lot of attention and has shown promising applications in different fields of healthcare, traffic control, etc. [3]. Different attempts have been made for modelling CPS [4] and the 5C architecture has recently been proposed for CPS in Industry 4.0 manufacturing systems [5]. In the field of manufacturing, different applications of CPS have also been studied [4]. Recently, cloud manufacturing has been proposed by inheriting many interesting features from cloud computing, e.g. multi-tenancy, shared resource pooling, geo-distribution and ubiquitous network access, service oriented, and dynamic resource provisioning. Looking at CPS, "*an integrated cyber-physical system that can provide on-demand manufacturing*

J. Ni et al. (Eds.): AMP 2018, LNME, pp. 265–292, 2018.
https://doi.org/10.1007/978-3-319-89563-5_20

services, digitally and physically, at the best utilisation of manufacturing resources" [6], is the first version of so-called cloud CPS in manufacturing.

Could CPS in manufacturing does not have the capacities in low-level operations or low-level relevant operations, due to the differences between computing and manufacturing. For example, on shop floor, the optimisation of machine/cutter/cutting parameter selections are performed, which is the most important process in manufacturing, and is not able to be performed and optimised in cloud manufacturing or CPS. In scheduling level, today's scheduling systems lack prediction capability with regard to errors or potential faults of planned or ongoing tasks on the shop floor, which is another big gap in cloud CPS.

Toward a cloud enhanced CPS and low-level operation enabled CPS, the remainder of this paper is organised as follows. A review of related work is provided in Sect. 2. Combining CPS and cloud technologies, a cloud CPS is proposed in Sect 3, followed by an implementation method in manufacturing area by combining holons, agents, and FBs in Sect. 4. Big data analytics (BDAs) is employed to enhance cloud CPS in manufacturing in Sect. 5. Finally, the contributions of this work are given in Sect. 6, together with future work of this research.

2 Literature Review

2.1 Cyber-Physical Systems

In the past decades, advancements in Web- and Internet-based systems and applications have opened up the possibility for industries to utilise the cyber workspace to conduct efficient and effective daily collaborations from anywhere in distributed manufacturing environments [2]. Recent advances in manufacturing industry have paved way for a systematic deployment of CPS, within which information from all related perspectives is closely monitored and synchronised between the physical factory floors and the cyber computational space. CPS are engineered systems that are built from and depend upon the seamless integration of computational algorithms and physical components [7].

The term *Cyber-Physical Systems* was first proposed in the US in 2006 [8]. With the wide applications and development of CPS, the definition of CPS is multiple, and not clear and unified. For example, CPS are integrations of computation and physical processes. Embedded computers and networks monitor and control the physical processes, usually with feedback loops where physical processes affect computations and vice versa [9]. In other words, CPS use computations and communication deeply embedded in and interacting with physical processes so as to add new capabilities to physical systems [10]. Unlike traditional embedded systems that are typically standalone, a full-fledged CPS is characterised by a network of interacting elements with physical input and output, resembling the structure of a sensor network. Tremendous progress has been made in advancing CPS technology over the last five years. Certainly, new smart CPS will drive innovation and competition in sectors as diverse as aerospace, automotive, chemical process, civil infrastructure, energy, healthcare, manufacturing, transportation, and so forth. One example of CPS is an intelligent manufacturing line, where a machine can perform a variety of processes by communicating with the components. Ongoing advancement in science and engineering will continue to enhance the link between computational and physical elements, dramatically increasing the adaptability, autonomy, efficiency, functionality, reliability, safety, and usability of CPS [7]. The final aim of CPS is to realise "intelligent monitoring" and "intelligent control" [11, 12]. These are the processes that need to realise real-time information extraction, data analysis, decision making and data transmission. CPS is an emerging discipline and has attracted and engaged many researchers and vendors. For example, many universities and institutes (such as UC Berkeley, Vanderbilt, Memphis, Michigan, Notre Dame, Maryland, and General Motors Research and Development Centre) have joined one research project (http://newsinfo.nd.edu/news/17248-nsf-funds-cyber-physical-systems-project/). The European Union (EU) and other countries, such as China and Korea, also realised the importance and significance of CPS research (http://www.artemis.eu/). In addition, the American Government named CPS as a new development strategy [12]. In conclusions, research and applications of CPS have been active in such areas like transportation, smart home, robotic surgery, aviation, defence, critical infrastructure, etc. [13].

2.2 Machining Optimisation

Machining process optimisation started with mathematical model based methods that were popular in the 1990s. Chua et al. [14] proposed a series of mathematical formulations to optimise the cutting conditions and to reduce the operation time. He [15] developed an objective function for cost minimisation, by which the most economical dimensions and tolerances were calculated. However, accuracy is a big concern, which make the models works within the ideal scenarios. Towards those issues, experimental methods were applied. Yang et al. [16] applied a Taguchi experiment method to optimise the cutting parameters regarding the groove difference and the average roughness. Chen et al. [17] proposed an experimental plan of a four-factor D-optimal design to obtain the optimal spindle speed, feed rate, cutting depth, and the status of lubrication concerning vibration and surface roughness in precision turning.

Both physics based and experiment based approaches are limited within a small range; therefore, many methods were taken into the machining area from other fields, e.g. computer science, where machining process issue is regarded as a search problem, and it was treated based on search algorithms in which tabu search (TS) is a popular one. A constraint-based TS approach was applied in optimisation of selecting machining resources, determining setup plans and sequencing operations of a prismatic part [18]. Many other search algorithms were applied as well, e.g. throughput profit [19], simulated annealing (SA) [20, 21], search heuristics based on SA and TS [22], branch and fathoming algorithms [23], iterative approach [24], evolutionary strategy based optimisation [25], and harmony search [26]. Moreover, to enhance the processes, a set of intelligent algorithms have been employed. Genetic algorithm (GA) [27, 28], the most basic evolutionary algorithm, was a popular method used for optimisation. On the other hand, a series of swarm intelligence (SI) based optimisation algorithms were applied to process planning, e.g. particle swarm optimisation (PSO) [29, 30], ant colony optimisation (ACO) [31] and honey bees mating optimisation [32]. In addition, the hybrid methods were applied to relax the limitation of one singe algorithm in the past decade, such as GA based [33, 34], combined GA [35–37], and PSO based [38, 39]. Multi-objective optimisation was employed due to many limitations of single-objective optimisation methods in the real machining process [40–42].

In parallel, knowledge based method was developed to transfer the human experience and knowledge into the manufacturing systems. Expert system was developed to utilise the machining knowledge [43–46]. Data are a powerful carrier for information and knowledge. Therefore, data classification methods were also applied to optimise machining process. Data classification methods, e.g. decision trees [47, 48] and artificial neural networks (ANN) [49, 50] were also applied.

2.3 Scheduling

The research on scheduling started with static scheduling useful for the development of shop-floor scheduling systems for mass production. A mathematical model was developed by Manne [51]. Laporte et al. [52] analysed two integer linear programming formulas to address job sequencing and tool switching problems, in which branch-and-cut and branch-and-bound algorithms were developed and compared. The concept of integration of both process planning and scheduling was developed by Khoshnevis and Chen [53–55]. The objectives of scheduling are extended a lot, for instance, makespan, cost, idle time and efficiency for the integration of process planning and scheduling [56], as well as optimisation dispatching rules for complex manufacturing systems [57], and energy-efficient scheduling [58].

However, the static scheduling does not have the capacity to handle the situation with growing products of both small batch and wide variety, in particular unexpected faults. Therefore, dynamic scheduling method was developed, where the decisions can be made with rapid response automatically. Real-time scheduling of a manufacturing system involves scheduling and revised scheduling [59]. Vieira et al. [60] proposed a rescheduling method based on a wide variety of experimental and practical approaches. In the method, two common strategies were introduced, dynamic scheduling and predictive–reactive scheduling. Ham et al. [61] proposed a three-stage flexible job-shop scheduling method to deal with unpredictable system disturbances. Iwamura et al. [62] introduced an estimation of future status based real-time scheduling approach for holonic manufacturing systems (HMS). In their method, the future status of an HMS is predicted by applying a neural network model based simulation model. In addition, an agent-based service-oriented architecture was presented for real-time distributed shop-floor scheduling [63]. Semi-Markov decision models were also applied to real-time scheduling by Yih and Thesen [64]. ACO was applied to two dynamic job scheduling by Zhou et al. [65]. Within real-time scheduling, real-time decision making is a key, and there are many algorithms. By constructing a decision tree, Metan et al. [66] proposed a new scheduling system for selecting dispatching rules in real-time. The proposed scheduling system was developed by combining the techniques of simulation, data mining, and statistical process control charts. Bayesian algorithm was used to discover priority dispatching rules from large amounts of structured or unstructured data for the single machine scheduling problem [67]. In addition, real-time monitoring methods were also developed. Through applying RFID on shop floors, the real-time information of objects including operators, machines and materials, can be automatically captured, bound and synchronised with manufacturing orders [68–71]. According to the literature, dynamism, flexibility and adaptability are the important features in modern scheduling, and a scheduling system should be able to perform task rearrangement in case of unexpected events.

3 Could CPS in Manufacturing

In manufacturing area, a CPS allows a machine perform many processes by communicating with tasks. Within the processes, real-time information extraction, data analysis, decision making and data transmission are enable to be activated. Many technologies, e.g. wireless sensor network (WSN), BDA, and cloud technology, can support CPS in manufacturing. A cloud CPS is proposed by combining cloud manufacturing and CPS. Cloud manufacturing implies an integrated CPS that can provide on-demand manufacturing services, digitally and physically, at the best utilisation of manufacturing resources [72, 73]. The architecture of cloud-based cyber manufacturing system, shown in Fig. 1, adopts a three-tier view-control-model (VCM) design pattern and a segmented (public vs. private) network structure to address the requirement of efficient and secure data communication between cyber systems and physical systems in the cloud environment.

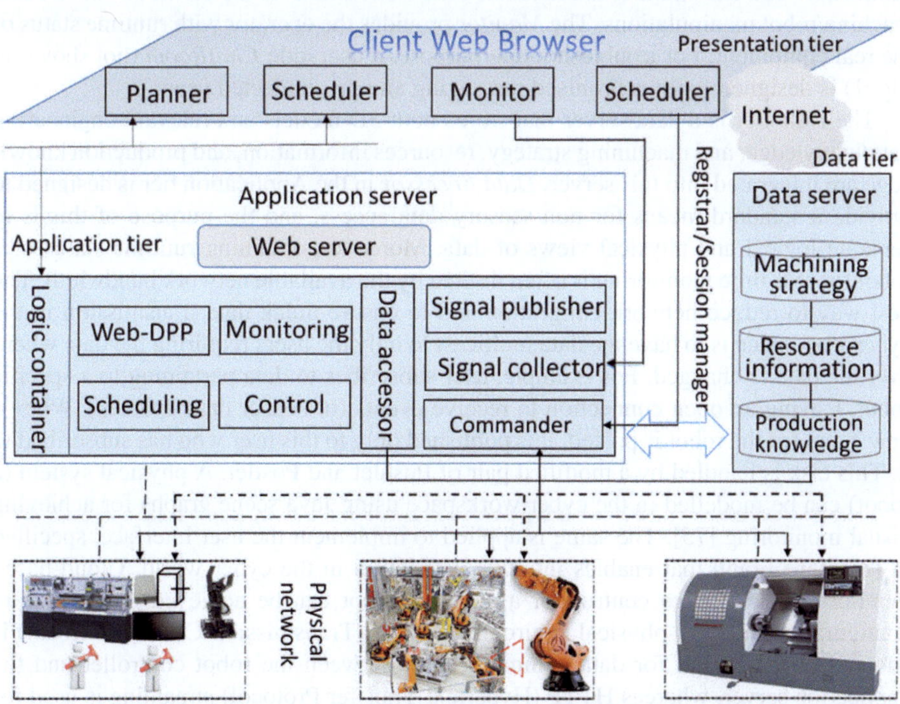

Fig. 1. Architecture of a cloud-based cyber manufacturing system

The Application tier is the application server in the cloud, which handles major security concerns; for example, session control, user registration, sensor data collection and distribution, and physical systems manipulation. This tier connects with the physical factory network, which contains the manufacturing resources and devices of factories, such as machines, and robots. *Signal Collector*, a server-side module, is responsible for collecting sensor data from networked physical machines or robots. After capturing the sensor data, *Signal Publisher* receives these data, and publishes and transmits the sensor data to the registered users, and uses the popular publish-subscribe design pattern to distribute the sensor data at the right time to the right users. A *Session Manager* is designed to search for issues such as user authentication, information synchronisation, session control and data logging. All the initial transactions must pass this module for access rights authorisation. The *Registrar* is designed to maintain a list of subscribers (users) together with the requested sensor data. For the sake of security, the *Commander* that functions like a proxy, through a distributed device network with tight security constraints, is used to control a physical machine or robot [74].

The Presentation tier is essentially a browser-based user interface where planner and scheduler can operate. Users still have the flexibility of monitoring machining and assembly operations from different perspectives, even though, the real-time sensor signals control the behaviours of the 3D models, e.g. selecting different 3D models and viewpoint control and so forth through the *Cyber Viewer*. An authorised user may submit a control command from anywhere to the cloud application server through the *Controller*. The *Commander* at the server-side then takes over the control rights for real machine/robot manipulations. The *Monitor* provides the operator with runtime status of the real equipment. For troubleshooting purpose, a user-side *ChatRoom* (not shown in Fig. 1) is designed for synchronised messaging among connected users.

The Data tier is a data server that stores both 3D models and relevant engineering data/knowledge, and machining strategy, resources information, and production knowledge are integrated into this server. *Data Accessor* in the Application tier is designed to provide a standard means for non-sensory data access, and the purpose of this is to separate logical and physical views of data. Moreover, obtaining runtime status of a robot for real-time monitoring is often limited by the available network bandwidth. The best way to reduce network congestion and to ensure quick data transmission in the cyber workspace is to have the data multicast to only the users requiring the data whenever the data is changed. For example, user subscribes to data pertaining to a specific robot, leaving an open connection to receive events (or sensor data updates). When a new event for the robot is posted, it is published only to this user who has subscribed to it. This task is handled by a modified pair of Pushlet and Postlet. A physical system (a robot) can be modelled in the cyber workspace using Java scene graphs for achieving visual monitoring [75]. The same is applied to implement the user interface, specified in the scene graph that enables intuitive navigation in the cyber world. Cloud-based monitoring and remote control for a physical robot can be achieved in the system configuration in cyber-physical environment. TCP (Transmission Control Protocol) is adopted in the design for data communication between the robot controller and the application server, whereas HTTP (Hypertext Transfer Protocol) streaming is used for data sharing from the server to the remote users. These control and transfer protocol can provide hardware and software protection for robots. Using this design, the CPS allows a remote user to monitor the motions of all joints and to control the robot for remote assembly operations. For the cloud-based remote monitoring and control mentioned above, an operator mainly focuses on what is going on at the robot side, once separated from a robot. This means that motion monitoring must be presented intuitively to guide the operator for remote control. This is then facilitated by condition monitoring in terms of vibration, force, temperature, and so forth, on how well it is doing. While camera-based approach is common and instrumental for motion monitoring, its bandwidth consumption can easily create a bottleneck for cloud-based real-time applications, which is the main concern of this application. To address this problem, a virtual 3D model entirely driven by real sensor data is used for cloud-based monitoring. The testing results of a mini-cell assembly case study reveal that a roundtrip latency of 30 ms is achieved using this approach, which is fast enough to be considered as real-time at the system level. This delay depends more on the network speed than the CPU speed.

4 A Method for Cloud CPS Implementation

In manufacturing area, holonic manufacturing and multi-agent systems are used for stable yet flexible manufacturing tasks, intelligent decision making and distributed manufacturing structures. Then FBs, as machine-level execution and control standards, are also considered a suitable approach to modelling distributed manufacturing systems and fit well with the concepts of holons and agents [76].

4.1 Concepts of Holons, Agents and Function Blocks

This section introduces the concepts of multi-agent systems (MAS), holonic manufacturing systems and FB, and they are as follows:

- Agents and multi-agent systems (MAS): A network of agents, will create an MAS of which members are not only responsible for satisfying their local objective, but also can interact and cooperate with each other to satisfy a global objective. MAS can be best characterised as a software technology that is able to model and implement individual and social behaviour in distributed systems [5].
- Holons and holonic manufacturing systems (HMS): Being closely related to the notion of multi-agent systems, an HMS is defined as a group of holons that integrate the entire range of manufacturing activities from order booking through design, production and marketing to realise the agile manufacturing [25].
- An FB is a control software unit that is embedded with necessary information and algorithms that are required for performing a task at the controller level.

4.2 A Cloud CPS Architecture

A CPS architecture combined with HMS holarchy, as shown in Fig. 2, includes mainly a cyber part and a physical part. CPS and HMS allow the integration of the physical world with the computational world where decisions are adaptively made according to the physical inputs, and are transferred to the physical entities in order to optimise the performance of the system. However, the holonic concept and representation must be realised in real-world applications. To implement this concept, agents and FBs are the key enablers. To address their specific characteristics, these two technologies are adopted at different levels of control. The cyber part of a CPS (or a holarchy) in most cases is non-real-time and can be implemented using the multi-agent approach. Once it comes to the physical part of the CPS, real-time behaviours must be respected and controlled. FBs are most suitable for the low-level device monitoring and control.

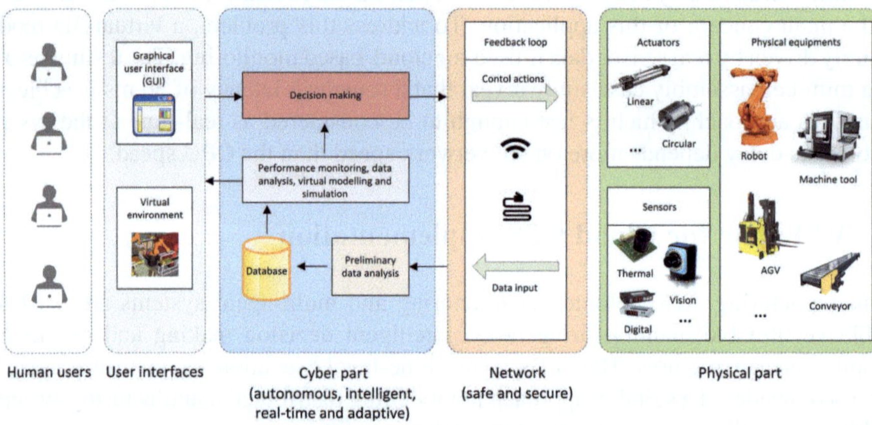

Fig. 2. CPS architecture combined with HMS holarchy

How MAS and FBs can be employed in different control levels of a holarchy representing a CPS is a key for combining holons, agents and FBs. MAS is used for soft real-time control, as the cyber part of a CPS or a holarchy to address the non-real-time issues at the high-level control. In this case, the tasks, e.g. planning or scheduling, contained in high level do not require immediate responses but can highly affect the productivity and profitability of the manufacturing system. As an example, failing to optimise the routing of machining tasks for a product is not considered a critical fault and does not necessarily stop the production flow, but can affect the number of system's throughput. Therefore, controlling the non-real-time tasks is necessary for optimising the performance and output of the system. Multi-agent systems have been very beneficial in this area especially due to their ability in making complex decisions using negotiation and cooperation techniques. Once a plan has been generated by MAS, the subtasks should be dispatched to the physical controllers for execution on the physical equipment. FBs can be used for encapsulation of tasks/processes and transferring them to the controller level of machines or other physical entities for execution.

At the shop-floor level, since many events are not predictable, a CPS should be therefore able to operate in the environment. A reaction to changes should occur according to specific timeframes and within strict deadlines. Therefore, a holon representing a physical equipment would require a hard real-time control system that can control the behaviours of the physical entity within a strict deadline. An FB, a reusable control unit, is applicable for real-time, distributed, intelligent and event-driven control, and execution of processes on physical equipment. The embedded algorithms in the FB can be controlled through its execution control chart, and the internal variables and algorithms can be tuned to match the environmental conditions. On the other hand, the process can be immediately paused at any moment and resumed when necessary via an FB control. Many benefits can be archived by the quick self-adjustment process, e.g. saving time, preventing system suspension. Therefore, FB is a good option for low-level control which is the closest control layer to the physical equipment. FB-enabled controllers are necessary for implementing this idea but few FB-enabled commercial controllers are currently available in the market. One solution would be to integrate an FB runtime environment hosted in a frontend computer with a commercial controller and consider the whole combination as an FB-enabled controller [42].

4.3 Implementation

To make a holonic CPS clear, the overall architecture of a robot holon that can be implemented by the multi-agent technology and FBs is presented shown in Fig. 3. It is composed of a layered structure from the bottom to the top, including the physical part layer, the control system layer, and the planning, scheduling and execution control layer.

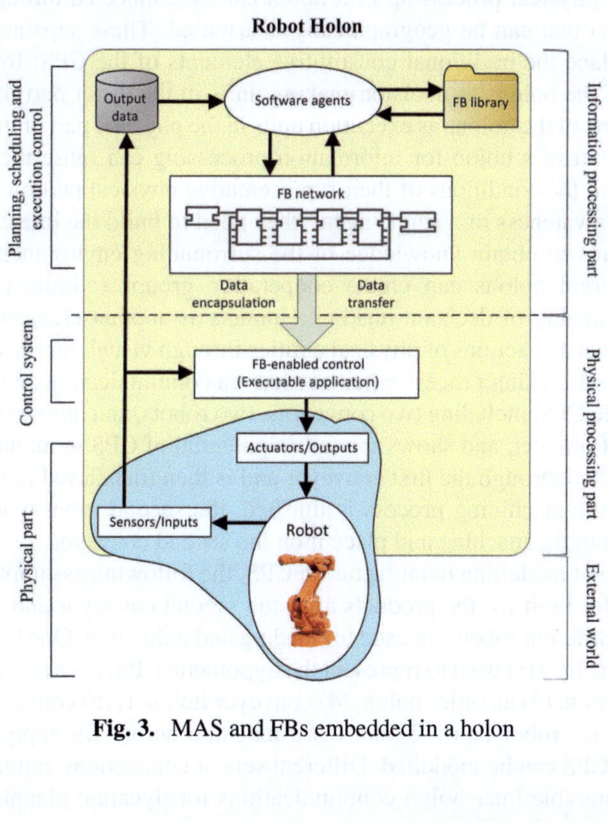

Fig. 3. MAS and FBs embedded in a holon

The physical part layer contains the actual physical equipment on the shop floor, such as sensors, robots, and actuators, and is connected to the control system layer. In the control system layer, FB-enabled control system completes the work of translating tasks to native machine control codes that are transmitted into the physical part layer, then, machines in the physical part layer execute operation instructs. The planning, scheduling and execution control layer is responsible for planning and scheduling, decision making, developing control instructions and transferring them to the control layer where these instructions can be executed as runtime codes. The software agents would be responsible for high-level decision-making and coordination, and FBs at the lower controller level would be responsible for execution of processes on the machine and quick fault recovery in case of disturbances on the shop floor. However, data will be processed by the software agents in complex cases because of the lack of the capacity of fault recovery for FBs. The decision-making agent, and communication/collaboration agent are integrated into the software agents at the information processing part of the holon, where the decision-making agent achieves communication and interaction with other decision-making agents embedded in other individual holons within a holarchy and humans through the communication agent. As mentioned before, a CPS considers two different layers of "cyber" and "physical". The layered structure of the robot holon matches the structure of CPS and allows for distributed control. The information processing and physical processing of a holon can be connected through a wide area network (WAN) that can be geographically distributed. These separated parts of the holons can replace the traditional constitutive elements of the CPS; for example, the logical parts of the holons as decision-making units in the cyber part of the CPS, and the physical parts of the holons as execution units in the physical part of the CPS. Agents in the logical part of a holon for information processing can sense their surrounding environment and the conditions of their representative physical part on the shop floor. Due to limited awareness of a single agent, they need to build the intercommunication with other agents to obtain knowledge of the surrounding environment. As a result, agents of different holons can create cooperative groups (within a holarchy) for improving the quality of decision making. Humans as another element of a CPS can monitor the status and actions of physical entities through visual interfaces and can also enter the decision-making process when needed via communication agents.

A hypothetic CPS, including two conveyors, two robots, and one machine, produces a single type of product, and shows the future potential of CPS in manufacturing. The raw material is fed through the first conveyor and is then transferred to the machine by a robot. Once the machining process is finished, the second robot would unload the finished part from the machine and place it on the second conveyor.

For the sake of modelling manufacturing CPS, the following assumptions are made: there is no buffer limit for the products after the second conveyor and before the first conveyor, and different robots are used for loading and unloading. One holarchy of eight defined holons (Fig. 4) is used to represent the hypothetic CPS: (1) a raw material holon, (2) a product holon, (3) an order holon, (4) conveyor holon-1, (5) conveyor holon-2, (6) robot holon-1, (7) robot holon-2, and (8) a machine holon. By applying the design concepts [77], CPS can be modelled. Different sets of connections among these holons represent the possible inter-holon communications for dynamic planning, scheduling

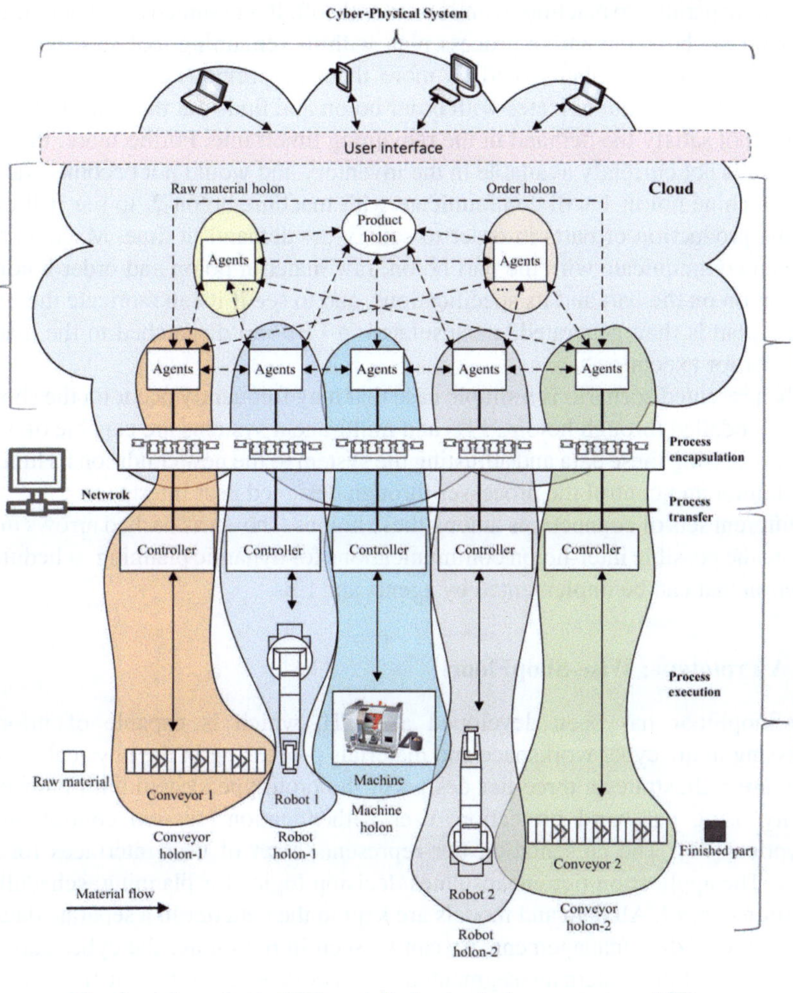

Fig. 4. A hypothetic CPS modelled using holons, agents and FBs

and execution that can be implemented by agents and FBs. In the case of tool breakage occurring while machine holon is in the middle of machining process, in order to stop the spread of damage, the hard real-time control system will halt the process in real time and retract the tool away from the part. Considering the current conditions, FBs will then adjust the process plan using the remaining available tools. The information of the new condition (i.e. tool breakage) is also sent to the information processing part of the machine holon. The information processing part will then communicate with the conveyor and robot holons, and updates them about the new condition so that they can make appropriate decisions such as delaying their processes or slowing down. To find an alternative process plan for the machine holon, its current machining state, geometry, etc., will be updated through building the communication between the machine holon and in-process part. For example, a second machine (machine holon-2) exists on the shopfloor in parallel to machine 1, and is currently off. It is assumed that machine holon-1 has found the best alternative process plan with its remaining tool sets; however, the cycle time using this plan would be more than the previous approach. As a result, machine holon-1 communicates with order holon and finds out that current production rate will not satisfy the demand in the remaining timeframe. Furthermore, the replacement tool is not currently available in the inventory and would not become available in time. Machine holon-1 will communicate with machine holon-2, to see if it can help with the production of parts in order to satisfy the demand in time. Machine holon-2 will then communicate with the part holon, raw material holon and order holon to get information on the part and its specifications, and to see if it can fabricate the part. The process plan is then generated, encapsulated in FBs and dispatched to the machine 2 controller for execution.

The presented scenario is a simple case that may frequently occur on the shop floor. A CPS modelled through holons, FBs and multi-agent systems are capable of automatically processing these data and adjusting the system to the new conditions. Humans can also monitor and control the processes through designed user interfaces.

Different sets of connections among these holons (shown as dashed arrows in Fig. 4) indicate the possible inter-holon communications for dynamic planning, scheduling and execution that can be implemented by agents and FBs.

4.4 A Prototype: Wise-ShopFloor

Wise-ShopFloor has been developed at KTH, which is capable of information processing in the cyber workspace and materials processing in the physical world.

Figure 5 illustrates a three-tier design of the prototype system with built-in cyber security using password protection, user authentication, session control, and data encryption [74]. The presentation tier represents a set of user interfaces for remote access. The application tier encapsulates decision logics for planning, scheduling and execution control. All data and models are kept in the data tier in a separate data server for the ease of data management. As can be seen in the figure, the cyber part and the physical part of the system are segmented by network accessibility via the Internet and/or a factory network. While agents are active in the cyber workspace, FBs are adopted for equipment control. However, in order to integrate other types of holons such as raw material/product holon, further research on the information modelling and interaction methods would be required.

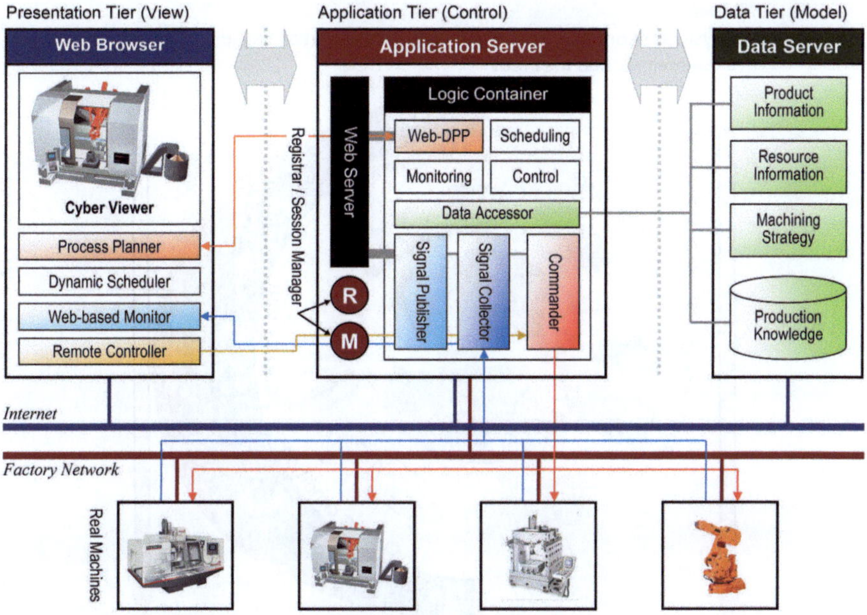

Fig. 5. A prototype CPS of Wise-ShopFloor

5 Big Data Analytics Enhanced Cloud CPS

BDA have shown big advantages in manufacturing, e.g. tool wears [78] and process planning [79]. Therefore, this section discusses how BDA can work in operations, machining and scheduling. BDA based on manufacturing data can provide an overall optimisation in planning stage for low-level operation, and also can improve the predictability of many events in low-level operations.

5.1 BDA Advantage in Manufacturing

Given the complexity of manufacturing operation processes, the theoretical and empirical research is to cover point-to-point issues, e.g. cutting parameter optimisation, coolant optimisation, cutting tool selection, and machine tool selection. Most of the processes are carried out separately. It is a major reason that it is impossible to solve the detailed machining related problems in the system level. However, BDA based methods are able to touch the machining process on the system side, to solve the machining problems at the system end, and to fill the differences between manufacturing and computing. The machining operation optimisation issues can be solved from big data perspective, as shown in Fig. 6, where machining process optimisation, from machine tool, cutting tool to cutting condition, are considered as a whole. All the machining resources are represented by data.

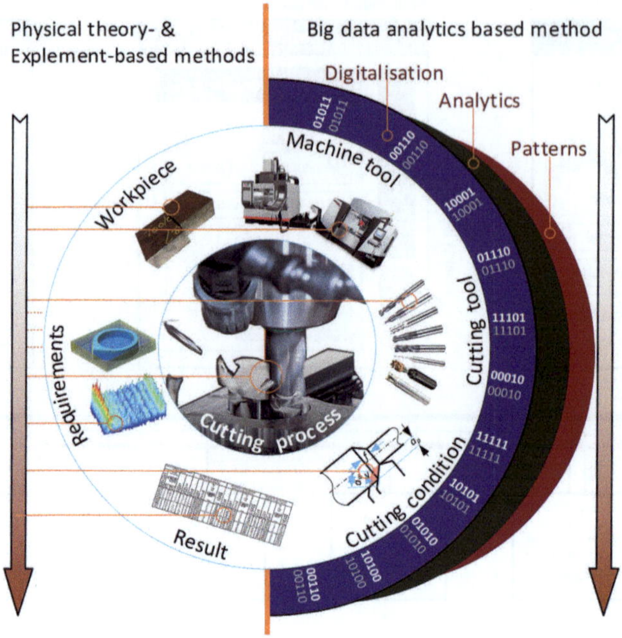

Fig. 6. Research focuses of BDA in manufacturing

5.2 BDA Enriched Distributed Process Planning (E-DPP)

The E-DPP, of which the architecture is shown in Fig. 7, is developed based on the original DPP which includes supervisory planning, execution control and operation planning as the three major system modules. In this design, the execution control module is placed in-between the supervisory planning and operation planning modules, and looks after jobs dispatching (in the unit of setups) based on up-to-date monitoring data, availability of machines and scheduling decisions [80, 81]. The distribution profile is a key feature. Within DPP, feature-based design and machining feature recognition are beyond the scope of this research. DPP assumes that machining features are already made available in the product data - they are either generated directly by using a feature-based design tool or recognised by a third-party machining feature recognition solution. Execution Control and Operation Planning consider machine tool selection, cutting tool selection and machining condition decision as a whole process driven by BDA. As the result of the optimisation, the machine resources are determined accordingly.

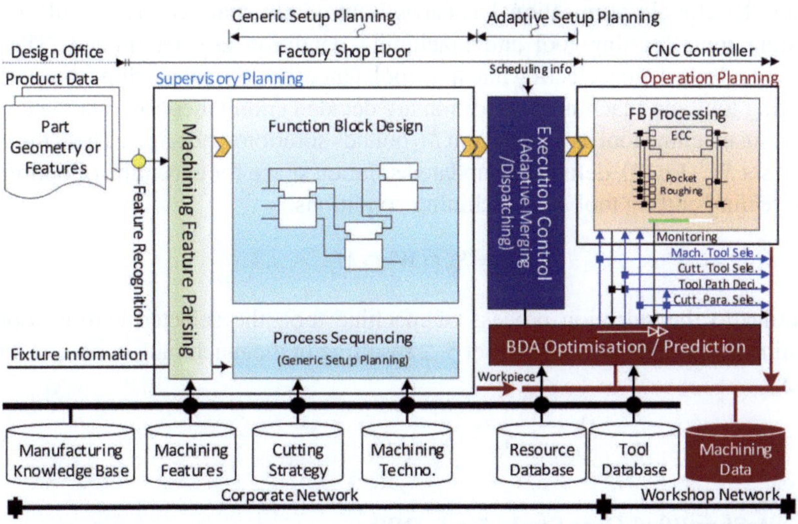

Fig. 7. Architecture of E-DPP

5.2.1 Problem Representation of E-DPP

The machining process is transformed into a statistic problem in the E-DPP. The available parameters of machine tool, cutting tool and machining condition are the original solution space S_0 with multi-dimensional data, as shown in Fig. 8. There are four steps in optimisation in the E-DPP, i.e., initialisation, optimisations 1, 2, and 3. Initialisation specifies the workpiece and machining requirements which are two constraints to the original solution space, i.e. workpiece constraint W_c and machining requirements M_r. Current solution space S_{mcm} can be calculated by

$$S_{mcm} = S_0 \cap W_c \cap M_r \tag{1}$$

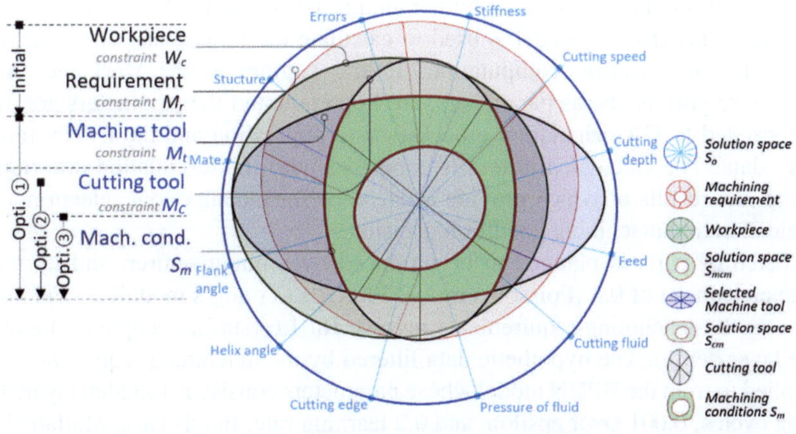

Fig. 8. E-DPP problem transformation [79]

Then the global optimisation 1 is carried out within space S_l, as a result of which machining tools, cutting tool and machining condition are determined. The result provides a reference for selecting machine tool. Once the optimal machine tool is chosen, the cutting tool and machining condition are decided simultaneously. On the contrary, a substitute machine tool as a constraint M_t bounds solution space S_{mcm}. Then, the current feasible set S_{cm} (Eq. 2), defines the updated solution space for execution of optimisation 2 for selecting cutting tool and machining conditions.

$$S_{mcm} = S_0 \cap W_c \cap M_r \tag{2}$$

Similar to the selection process of machine tool, the selected cutting tool, as a constraint M_c, bounds the feasible set S_{cm}, resulting in the machining condition solution space S_m,

$$S_m = S_{cm} \cap M_c \tag{3}$$

5.2.2 A Simplified Case Study for E-DPP

A simplified case study is chosen to show the process of the reported method as a proof of concept. A set of hypothetic data generated in-house are applied to train the BPNN model used to calculate the fitness of GA. The data used in the case are hypothetic data involving: (1) workpiece represented by machining feature geometrically and its material information expressed by Poisson's ratio (0.25, 0.45), Young's modulus (100, 250 GPa), and hardness (20, 70 HRC); (2) machining requirements including surface roughness *Ra* (0.4, 6.3) and white layer depth (0.3, 10 μm); (3) machine tool: errors of linear axis (1, 10 μm) and maximum spindle speed (3000, 20000 RPM); (4) machining conditions involving cutting parameters, i.e. cutting speed (30, 150 m/min) and feed rate (0.05, 0.5 mm/z); (5) cutting tool referring to rake angle (0, 30°) and flank angle (6, 20°); and (6) machining results associated with the machine requirements and used in training of BPNN model. The selected data are bounded within the limited feasible sets, and they include 10,000 row data generated randomly in Matlab®.

A detailed method is proposed by combining BPNN and GA, as shown in Fig. 8, where the trained BPNN model is used to calculate the fitness, and GA is applied to perform the optimisation computation. In the beginning, the workpiece and its machining requirements, as parameters, are imported, and the parameters are divided and proceeded to GA calculation processes as a population and the BPNN model as training data. The calculated fitness is compared with the termination criterion until finishing, the results of which provide a reference for machine tool selection, cutting tool selection and machining condition decision.

A batch of parts is planned to be produced by a manufacturer, and its material parameters consist of 0.3 (Poisson's ratio), 200 GPa (Young's modulus), and 40 HRC (hardness). Its machining requirements involve Ra 1.6 (surface roughness) and 7 μm (white layer depth). The hypothetic data filtered by the machining requirements filter, are applied to train the BPNN model whose parameters consist of 1 hidden layer, 10,000 training cycles, 0.001 error epsilon, and 0.2 learning rate. In this case, Matlab® is used for the computation, as a result of which the mean square error is stabilised around 0.375, when the number of iteration is reaching approximately 4×10^4 (due to that there is no pattern in the random data) (Fig. 9).

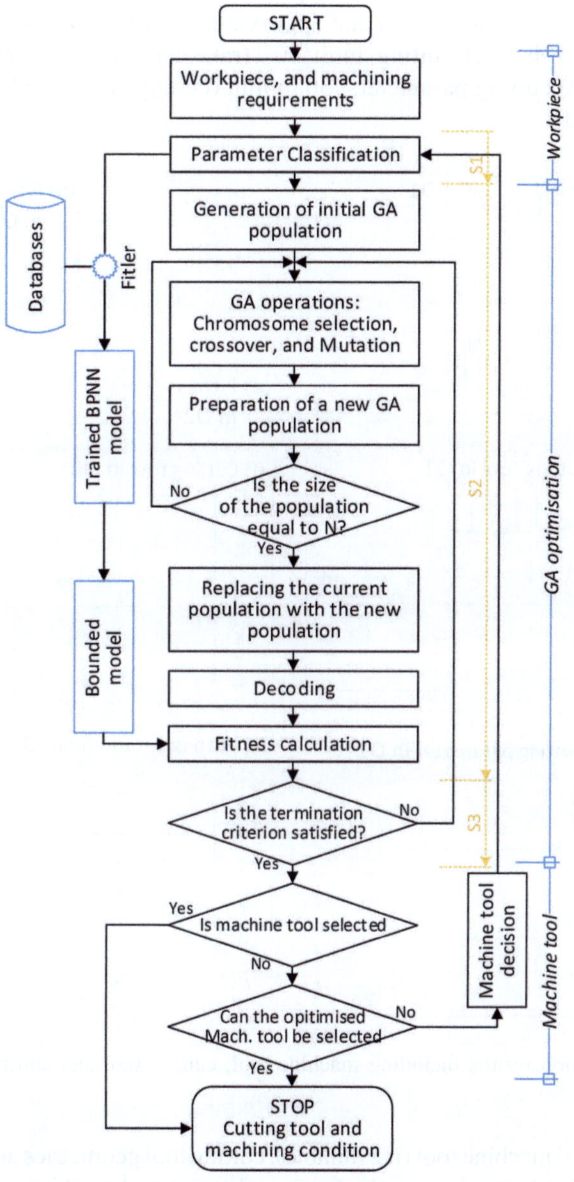

Fig. 9. A detailed algorithm combining GA and BPNN

Then the GA computation is performed to optimise machine tools, cutting tools and cutting parameters. The parameters of GA consist of (1) generations: 1000; (2) population size: 100; (3) crossover probability: 0.8; (4) mutation probability: 0.01; and

(5) computation accuracy: 0.001. The fitness of machining requirements is stable after approximately 650 generations, as shown in Fig. 10a, together with the parameters of machine tool, cutting tool and cutting parameters, detailed as follows: (1) machine tool: 10 μm (errors of linear axis) and approximately 4000 RPM (maximum spindle speed) (in Fig. 10b); (2) cutting tool: 30° (rake angle) and 6° (flank angle) (in Fig. 10c); and (3) cutting parameters: 30 m/min (cutting speed) and 0.05 mm/z (feed rate) (in Fig. 10d).

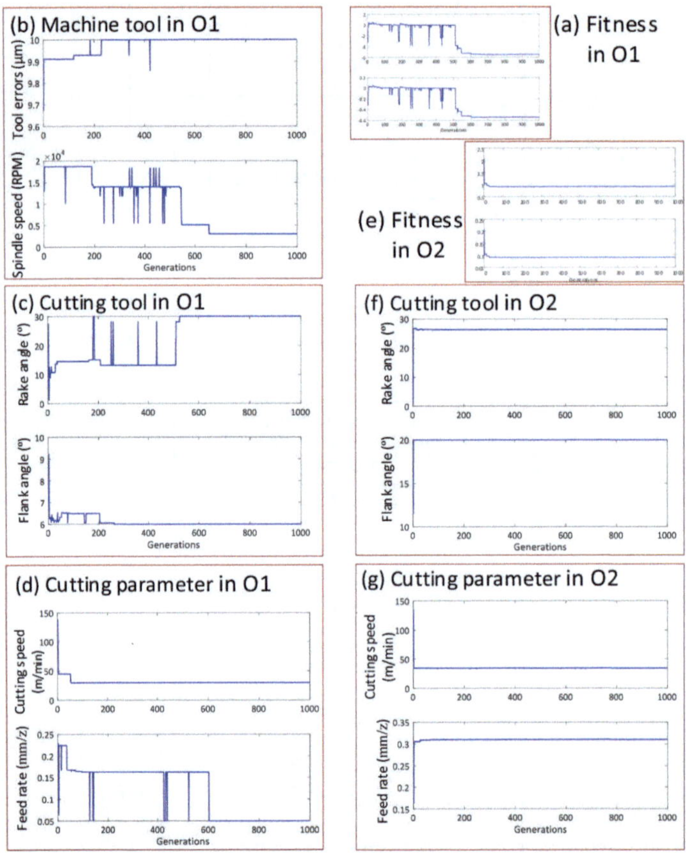

Fig. 10. Calculation results including machine tool, cutting tool and cutting parameters in 2 optimisations

The optimised machine tool (if available), cutting tool geometries and cutting parameters are determined, simultaneously. In case of no optimal machine tool, an alternative machine is selected, and its parameters are imported to GA. A machine tool, for instance, is equipped with a linear axis of 3 μm in error and a maximum spindle speed of 8000 RPM. Based on the selected machine tool and the initial workpiece parameters, the GA calculation is performed again. Figure 10e depicts the fitness of machining requirements after approximately 90 generations. The optimised results of cutting tools and cutting parameters consist of: (1) cutting tool: 28° (rake angle) and 20° (flank angle) (as shown Fig. 10f); and (2) cutting parameters: 30 m/min (cutting speed) and 0.3 mm/z (feed rate) (in Fig. 10g). The subsequent processes, i.e. cutting tool selection and machining condition optimisation, are the same in the statistics level, and are omitted in this chapter for the sake of page limit.

5.3 BDA Enhanced Scheduling

5.3.1 BDA for Fault Prediction in Scheduling

Between an order and the final delivery, the main elements on the manufacturing shop floor can be divided into several categories, according to their functions and properties; they are (1) physical operation level, (2) physical monitor and control level, (3) product level, and (4) cyber level: nets and database. Considering the complexity of machining processes, the current scheduling does not cover the machining process, and basically machining is the final point of scheduling. Therefore, fault prediction regarding machining is impossible due to the limitation of traditional methods. However, from the big data point of view, fault prediction can be archived. As shown in Fig. 11, tasks, represented by data attributes, are compared with the mined patterns based on relevant databases, resulting that the similarity is obtained [82]. Similarly, during the machining, real-time monitoring is used to collect data of ongoing tasks which are used to compare with the mined fault patterns. The two similarities are used for scheduling decision making.

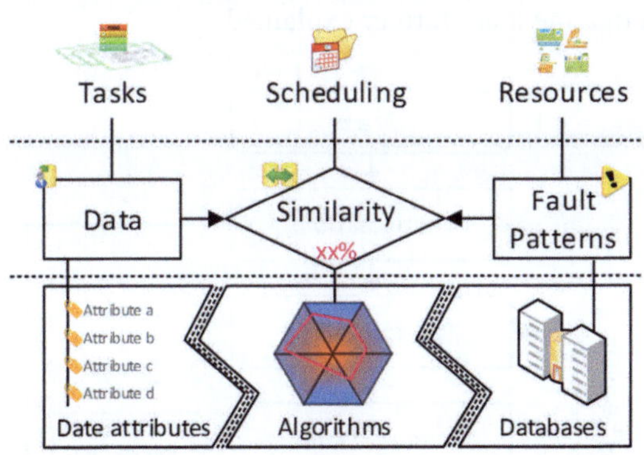

Fig. 11. Concept of fault prediction in scheduling

5.3.2 System Architecture

This section addresses two challenges: (1) to avoid any mismatch of machining tasks on machines before scheduling, and (2) to prevent faults from happening during machining via real-time monitoring and rescheduling. A BDA based fault prediction approach is therefore introduced for shop-floor scheduling. Figure 12 depicts the system architecture of three steps in horizontal direction and three levels in vertical direction. The *Analysis* step consists of order analysis and workpiece representation. The *Prediction* step contains pre-scheduling and fault prediction. These two steps are introduced

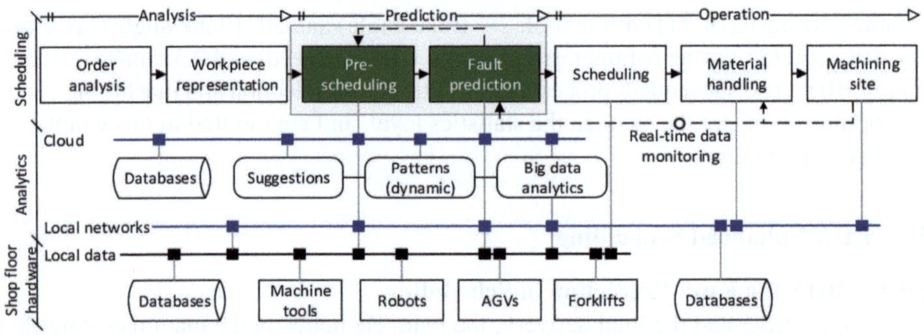

Fig. 12. Architecture of BDA in shop-floor scheduling

before scheduling in order to predict potential machining faults due to machine conditions and machining status via real-time monitoring. In the last *Operation* step, scheduling, material handling and machining take place. During machining, real-time data are collected and passed to the fault prediction module for fault predictions. Moreover, if there are still unpredicted errors during machining, the error message is sent to the pre-scheduling module, and the unfinished task is re-arranged for defect prevention.

In the analytics level, BDA is carried out resulting in the patterns used by fault prediction are discovered. Finally, suggestions if any are given to the shop floor operations. In the shop-floor hardware level, equipment such as machine tools, robots, AGVs and forklifts are connected by a local area network. For the sake of page limit, only fault prediction and data treatment are further explained.

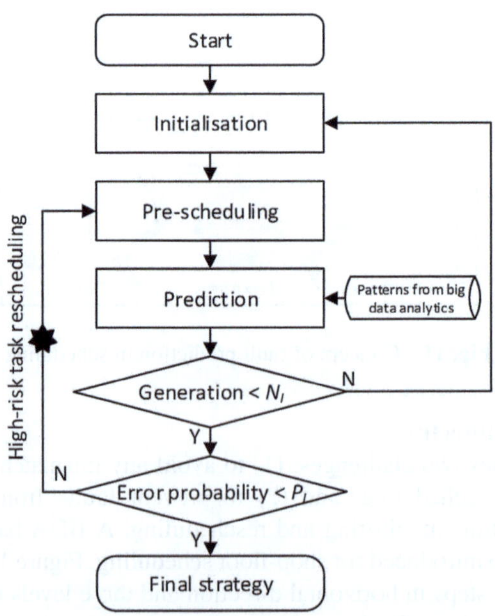

Fig. 13. Workflow of fault prediction

(1) Fault prediction

Figure 13 shows the workflow of fault prediction, including initialisation, pre-scheduling, prediction, termination criteria judgment, and error probability evaluation. Initial parameters are set up during initialisation, e.g. maximum number of calculation (generation) and the probability. Then the task plans are generated (by using the same procedures as in normal scheduling) in the pre-scheduling process, based on which the error probabilities in statistics level are calculated for the planned task in the prediction process. If the error probabilities are greater than the initial one P_I, the high-risk tasks will be rescheduled. However, the task plans may still be undecided after the number of generations reaches the initially-defined threshold value N_I. In such a case, the values of the initial parameters must be properly adjusted.

Among the processes, fault prediction is the key, where potential error patterns are mined by BDA based on the databases of the shop floor. Together, they provide a reference for scheduling decision making. The potential patterns from BDA are as follows:

- Potential faults of machine and workpiece: it refers to the fault probability of a class of machine tools or a single machine tool before task scheduling, and potential errors during machining regarding real-time data.
- Maintenance state of machine and workpiece: maintenance state of machine tool is related to its usage time, basically. However, if an unsuitable task is scheduled for the machine, the machine should be maintained in advance.
- Machining quality of workpiece and machine tool: machining quality is coupled with the state of the machine tool, e.g. stiffness, repeatability, and stability. Therefore, if the state of the machine tool is not suitable to machine a workpiece, the task should not be assigned to the machine.

(2) Data treatment

Considering the size of data, data should be divided and stored in different databases, i.e. local data, local network data and cloud data, as shown in Fig. 14. Local data are stored in distributed computers, including both historical data and real-time data from monitoring systems, and covering the information of machining features, machine tools, processes, operators, quality measurement and time. Local data analysis considers the patterns of each single machine tool with respect to the special characteristics of each machine tool. The local network data refer to the historical data covering machine tool class, parts, time and technical capability (one kind of evaluation based on process, operator and machining quality in the local data level). Local network data analysis focuses on the patterns of machine tool classes with respect to their performances. The cloud data also refer to the historical data involving time, shop floor level, manufacturing capacity and tasks. Cloud data analysis is to obtain the pattern of a shop floor, which is used to evaluate the manufacturing capacity of the shop floor.

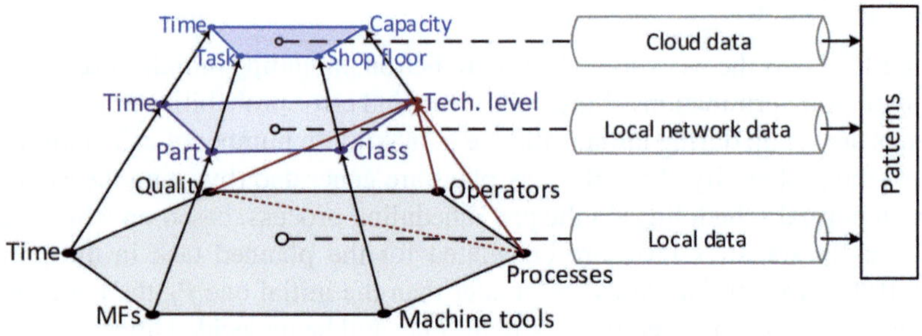

Fig. 14. Data stored and processed at three levels

5.3.3 A Simplified Case Study

A proof-of-concept case is given to demonstrate the proposed approach, where fault prediction for machine assignments are performed before scheduling decision making. For real-time monitoring of machining process, and machine maintenance state, fault prediction shares the same process with the case.

There are four candidate machine tools available, and their ID are M1–M4. Two tasks will be assigned to the shop floor at a certain date (Fig. 15a). Two machining features (MFs, one for each task) are selected together with material hardness, required tolerance and roughness, and delivery deadline (see Table 1).

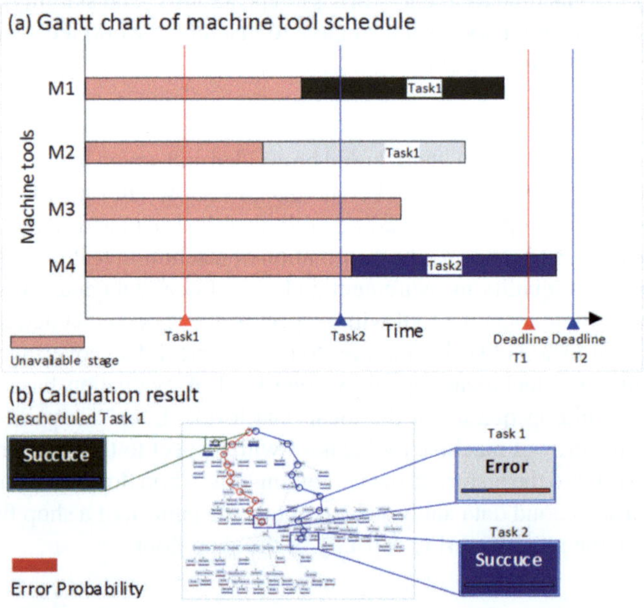

Fig. 15. Schedule and result of classification using decision tree

Table 1. Details of two machining tasks

Task	MF	Hardness (HRC)	Tolerance (mm)	Roughness (Ra)	Deadline
1	Side	31	0.005	1.2	T1
2	Slot	45	0.05	0.8	T2

Due to the fact that real data are not available, a hypothetic data sheet is generated randomly to test the system. The data are treated in a software, RapidMiner, where the decision tree model is used. Based on the trained results, as shown in Fig. 15b, the fault probability is bigger than 50% for assigning task 1 to M2, whereas it is almost 0 for arranging task 2 to M4. Therefore, machine tool carrying out Task 1 is thus rescheduled to M1, of which prediction result shows the error probability is changed into 0% (Fig. 15b).

6 Conclusions

In this paper, a cloud CPS is proposed by combining CPS and cloud technology consisting of holons, agents and FBs. Within the context, a CPS can be considered as a holarchy of multiple holons in a systematic way. Each holon corresponds to an entity in the CPS such as a piece of raw material, an order, a machine or a robot. The cyber part of the entity (the information processing part of the holon) is implemented using multi-agents whereas the physical part of the entity (or materials processing part of the holon) is controlled by FBs. To enable the system dealing with the operation issues in low level on shop floors, BDA is employed to optimise machining processes, e.g. machine tool selection, cutting tool selection and cutting condition optimisation, and to make the scheduling process predictable by comparing the tasks with mined error patterns.

The proposed method is still in the conceptual phase. Towards this issue and a fully functional cloud CPS, the future work of this research is therefore planned as follows: (1) development of Cloud CPS: the cyber security issue is considered, and then Wise-ShopFloor prototype will be enhanced to have the full functions of combined MAS, HMS, and FB, and (2) machining resource digitalisation: it is a key to archive a fully functional operation planning for cloud CPS. Finally, data collection will be carried out in the real world.

References

1. Wang, S., Wan, J., Zhang, D., et al.: Towards smart factory for industry 4.0: a self-organized multi-agent system with big data based feedback and coordination. Comput. Netw. **101**, 158–168 (2016). https://doi.org/10.1016/j.comnet.2015.12.017
2. Wang, L.: Cyber manufacturing: research and applications. In: TMCE, pp. 39–49 (2014)
3. Shi, J., Wan, J., Yan, H., Suo, H.: A survey of cyber-physical systems. In: 2011 International Conference on Wireless Communications and Signal Processing (WCSP), pp. 1–6 (2011)
4. Derler, P., Lee, E.A., Vincentelli, A.S.: Modeling cyber-physical systems. Proc. IEEE **100**, 13–28 (2012)

5. Lee, J., Bagheri, B., Kao, H.-A.: A cyber-physical systems architecture for Industry 4.0-based manufacturing systems. Manuf. Lett. **3**, 18–23 (2015). https://doi.org/10.1016/j.mfglet.2014.12.001
6. Wang, L., Gao, R., Ragai, I.: An Integrated Cyber-Physical System for Cloud Manufacturing. V001T04A029 (2014)
7. Chang, Q., Gao, R., Lei, Y., et al.: Cyber-physical systems in manufacturing and service systems. Math. Probl. Eng. **2015**, 2–4 (2015). https://doi.org/10.1155/2015/704213
8. Lee, E.A.: Cyber-Physical Systems - Are Computing Foundations Adequate? vol. 1, pp. 1–9, October 2006
9. Lee, E.A.: Cyber physical systems: design challenges. In: 2008 11th IEEE International Symposium on Object and Component-Oriented Real-Time Distributed Computing (ISORC), pp 363–369 (2008)
10. Wang, L., Törngren, M., Onori, M.: Current status and advancement of cyber-physical systems in manufacturing. J. Manuf. Syst. **37**, 517–527 (2015). https://doi.org/10.1016/j.jmsy.2015.04.008
11. Wan, J., Zhang, D., Zhao, S., et al.: Context-aware vehicular cyber-physical systems with cloud support: architecture, challenges, and solutions. IEEE Commun. Mag. **52**, 106–113 (2014). https://doi.org/10.1109/MCOM.2014.6871677
12. Wan, J., Yan, H., Suo, H., Li, F.: Advances in cyber-physical systems research. KSII Trans. Internet Inf. Syst. **5**, 1891–1908 (2011). https://doi.org/10.3837/tiis.2011.11.001
13. Wang, L., Törngren, M., Onori, M.: Current status and advancement of cyber-physical systems in manufacturing. J. Manuf. Syst. **37**(Part 2), 517–527 (2015). https://doi.org/10.1016/j.jmsy.2015.04.008
14. Chua, M.S., Loh, H.T., Wong, Y.S.: Optimization of cutting conditions for multi-pass turning operations using sequential quadratic programming. J. Mater. Process. Technol. **28**, 253–262 (1991). https://doi.org/10.1016/0924-0136(91)90224-3
15. He, J.R.: Tolerancing for manufacturing via cost minimization. Int. J. Mach. Tools Manuf. **31**, 455–470 (1991). https://doi.org/10.1016/0890-6955(91)90029-3
16. Yang, Y.K., Chuang, M.T., Lin, S.S.: Optimization of dry machining parameters for high-purity graphite in end milling process via design of experiments methods. J. Mater. Process. Technol. **209**, 4395–4400 (2009). https://doi.org/10.1016/j.jmatprotec.2008.11.021
17. Chen, C.-C., Chiang, K.-T., Chou, C.-C., Liao, Y.-C.: The use of D-optimal design for modeling and analyzing the vibration and surface roughness in the precision turning with a diamond cutting tool. Int. J. Adv. Manuf. Technol. **54**, 465–478 (2011). https://doi.org/10.1007/s00170-010-2964-0
18. Li, W., Ong, S.K., Nee, A.Y.C.: Optimization of process plans using a constraint-based tabu search approach. Int. J. Prod. Res. **42**, 1955–1985 (2004). https://doi.org/10.1080/00207540310001652897
19. Wei, S.Y., Lo, C.C., Chang, C.A.: Using throughput profit for selecting manufacturing process plan. Comput. Ind. Eng. **32**, 939–948 (1997). https://doi.org/10.1016/S0360-8352(97)00022-3
20. Brown, K.N., Cagan, J.: Optimized process planning by generative simulated annealing. AI Edam-Artif. Intell. Eng. Des Anal. Manuf. **11**, 219–235 (1997)
21. Ma, G.H., Zhang, Y.F., Nee, A.Y.C.: A simulated annealing-based optimization algorithm for process planning. Int. J. Prod. Res. **38**, 2671–2687 (2000). https://doi.org/10.1080/002075400411420
22. Lee, D.H., Kiritsis, D., Xirouchakis, P.: Search heuristics for operation sequencing in process planning. Int. J. Prod. Res. **39**, 3771–3788 (2001). https://doi.org/10.1080/00207540110061922

23. Lee, D.H., Kiritsis, D., Xirouchakis, P.: Branch and fathoming algorithms for operation sequencing in process planning. Int. J. Prod. Res. **39**, 1649–1669 (2001). https://doi.org/10.1080/00207540010028100
24. Lee, D.H., Kiritsis, D., Xirouchakis, P.: Iterative approach to operation selection and sequencing in process planning. Int. J. Prod. Res. **42**, 4745–4766 (2004). https://doi.org/10.1080/00207540410001720412
25. Wang, Y.C., Chen, T.: Modelling and optimization of machining conditions for the multi-pass dry turning process. Proc. Inst. Mech. Eng. Part B J. Eng. Manuf. **222**, 1387–1394 (2008). https://doi.org/10.1243/09544054JEM1162
26. Abhishek, K., Datta, S., Mahapatra, S.S.: Multi-objective optimization in drilling of CFRP (polyester) composites: application of a fuzzy embedded harmony search (HS) algorithm. Measurement **77**, 222–239 (2016). https://doi.org/10.1016/j.measurement.2015.09.015
27. Chen, M.C., Tseng, H.Y.: Machining parameters selection for stock removal turning in process plans using a float encoding genetic algorithm. J. Chin. Inst. Eng. **21**, 493–506 (1998). https://doi.org/10.1080/02533839.1998.9670412
28. Reddy, S.V.B., Shunmugam, M.S., Narendran, T.T.: Optimal sub-division of the depth of cut to achieve minimum production cost in multi-pass turning using a genetic algorithm. J. Mater. Process. Technol. **79**, 101–108 (1998)
29. Guo, Y.W., Mileham, A.R., Owen, G.W., et al.: Operation sequencing optimization for five-axis prismatic parts using a particle swarm optimization approach. Proc. Inst. Mech. Eng. Part B J. Eng. Manuf. **223**, 485–497 (2009). https://doi.org/10.1243/09544054JEM1224
30. Kafashi, S., Shakeri, M., Abedini, V.: Automated setup planning in CAPP: a modified particle swarm optimisation-based approach. Int. J. Prod. Res. **50**, 4127–4140 (2012). https://doi.org/10.1080/00207543.2011.592157
31. Abbas, A.T., Aly, M.F., Hamza, K.: Optimum drilling path planning for a rectangular matrix of holes using ant colony optimisation. Int. J. Prod. Res. **49**, 5877–5891 (2011). https://doi.org/10.1080/00207543.2010.507608
32. Wen, X.Y., Li, X.Y., Gao, L., Sang, H.Y.: Honey bees mating optimization algorithm for process planning problem. J. Intell. Manuf. **25**, 459–472 (2014). https://doi.org/10.1007/s10845-012-0696-8
33. Li, W.D., Ong, S.K., Nee, A.Y.C.: Hybrid genetic algorithm and simulated annealing approach for the optimization of process plans for prismatic parts. Int. J. Prod. Res. **40**, 1899–1922 (2002). https://doi.org/10.1080/00207540110119991
34. Venkatesan, D., Kannan, K., Saravanan, R.: A genetic algorithm-based artificial neural network model for the optimization of machining processes. Neural Comput. Appl. **18**, 135–140 (2009). https://doi.org/10.1007/s00521-007-0166-y
35. Wong, T.N., Chan, L.C.F., Lau, H.C.W.: Machining process sequencing with fuzzy expert system and genetic algorithms. Eng. Comput. **19**, 191–202 (2003). https://doi.org/10.1007/s00366-003-0260-4
36. Jayabal, S., Natarajan, U.: Optimization of thrust force, torque, and tool wear in drilling of coir fiber-reinforced composites using Nelder-Mead and genetic algorithm methods. Int. J. Adv. Manuf. Technol. **51**, 371–381 (2010). https://doi.org/10.1007/s00170-010-2605-7
37. Salehi, M., Bahreininejad, A.: Optimization process planning using hybrid genetic algorithm and intelligent search for job shop machining. J. Intell. Manuf. **22**, 643–652 (2011). https://doi.org/10.1007/s10845-010-0382-7
38. Tiwari, M.K., Dashora, Y., Kumar, S., Shankar, R.: Ant colony optimization to select the best process plan in an automated manufacturing environment. Proc. Inst. Mech. Eng. Part B J. Eng. Manuf. **220**, 1457–1472 (2006). https://doi.org/10.1243/09544054JEM449

39. Petrovic, M., Mitic, M., Vukovic, N., Miljkovic, Z.: Chaotic particle swarm optimization algorithm for flexible process planning. Int. J. Adv. Manuf. Technol. **85**, 2535–2555 (2016). https://doi.org/10.1007/s00170-015-7991-4
40. Sardinas, R.Q., Santana, M.R., Brindis, E.A.: Genetic algorithm-based multi-objective optimization of cutting parameters in turning processes. Eng. Appl. Artif. Intell. **19**, 127–133 (2006). https://doi.org/10.1016/j.engappai.2005.06.007
41. Solimanpur, M., Ranjdoostfard, F.: Optimisation of cutting parameters using a multi-objective genetic algorithm. Int. J. Prod. Res. **47**, 6019–6036 (2009). https://doi.org/10.1080/00207540802178109
42. Shin, K.S., Park, J.O., Kim, Y.K.: Multi-objective FMS process planning with various flexibilities using a symbiotic evolutionary algorithm. Comput. Oper. Res. **38**, 702–712 (2011). https://doi.org/10.1016/j.cor.2010.08.007
43. El-Sawy, A.A., Abdalla, H.S.: A hybrid approach for machining process optimization using multiple experts data. Comput. Ind. Eng. **37**, 445–448 (1999). https://doi.org/10.1016/S0360-8352(99)00114-X
44. Ribeiro, M.V., Coppini, N.L.: An applied database system for the optimization of cutting conditions and tool selection. J. Mater. Process. Technol. **93**, 371–374 (1999)
45. Arezoo, B., Ridgway, K., Al-Ahmari, A.M.A.: Selection of cutting tools and conditions of machining operations using an expect system. Comput. Ind. **42**, 43–58 (2000). https://doi.org/10.1016/S0166-3615(99)00051-2
46. Gupta, D.P., Gopalakrishnan, B., Chaudhari, S.A., Jalali, S.: Development of an integrated model for process planning and parameter selection for machining processes. Int. J. Prod. Res. **49**, 6301–6319 (2011). https://doi.org/10.1080/00207543.2010.523722
47. Sluga, A., Jermol, M., Zupanic, D., Mladenic, D.: Machine learning approach to machinability analysis. Comput. Ind. **37**, 185–196 (1998). https://doi.org/10.1016/S0166-3615(98)00098-0
48. Thimm, G., Britton, G.A., Whybrew, K., Fok, S.C.: Optimal process plans for manufacturing and tolerance charting. Proc. Inst. Mech. Eng. Part B J. Eng. Manuf. **215**, 1099–1105 (2001). https://doi.org/10.1243/0954405011519024
49. Wang, J.: Multiple-objective optimisation of machining operations based on neural networks. Int. J. Adv. Manuf. Technol. **8**, 235–243 (1993). https://doi.org/10.1007/BF01748633
50. Monostori, L., Viharos, Z.J., Markos, S.: Satisfying various requirements in different levels and stages of machining using one general ANN-based process model. J. Mater. Process. Technol. **107**, 228–235 (2000). https://doi.org/10.1016/S0924-0136(00)00698-1
51. Manne, A.S.: On the job-shop scheduling problem. Oper. Res. **8**, 219–223 (1960). https://doi.org/10.1287/opre.8.2.219
52. Laporte, G., Salazar-GonzÁLez, J.J., Semet, F.: Exact algorithms for the job sequencing and tool switching problem. IIE Trans. **36**, 37–45 (2004). https://doi.org/10.1080/07408170490257871
53. Khoshnevis, B., Chen, Q.M.: Integration of process planning and scheduling functions. J. Intell. Manuf. **2**, 165–175 (1991). https://doi.org/10.1007/bf01471363
54. Chen, Q., Khoshnevis, B.: Scheduling with flexible process plans. Prod. Plan. Control **4**, 333–343 (1993). https://doi.org/10.1080/09537289308919455
55. Tan, W., Khoshnevis, B.: Integration of process planning and scheduling—a review. J. Intell. Manuf. **11**, 51–63 (2000). https://doi.org/10.1023/a:1008952024606
56. Mohapatra, P., Nayak, A., Kumar, S.K., Tiwari, M.K.: Multi-objective process planning and scheduling using controlled elitist non-dominated sorting genetic algorithm. Int. J. Prod. Res. **53**, 1712–1735 (2014). https://doi.org/10.1080/00207543.2014.957872
57. Freitag, M., Hildebrandt, T.: Automatic design of scheduling rules for complex manufacturing systems by multi-objective simulation-based optimization. CIRP Ann. - Manuf. Technol. **65**, 433–436 (2016). https://doi.org/10.1016/j.cirp.2016.04.066

58. Gahm, C., Denz, F., Dirr, M., Tuma, A.: Energy-efficient scheduling in manufacturing companies: a review and research framework. Eur. J. Oper. Res. **248**, 744–757 (2016). https://doi.org/10.1016/j.ejor.2015.07.017
59. Khodke, P.M., Bhongade, A.S.: Real-time scheduling in manufacturing system with machining and assembly operations: a state of art. Int. J. Prod. Res. **51**, 4966–4978 (2013). https://doi.org/10.1080/00207543.2013.784414
60. Vieira, G.E., Herrmann, J.W., Lin, E.: Rescheduling manufacturing systems: a framework of strategies, policies, and methods. J. Sched. **6**, 39–62 (2003). https://doi.org/10.1023/a:1022235519958
61. Ham, M., Lee, Y.H., Kim, S.H.: Real-time scheduling of multi-stage flexible job shop floor. Int. J. Prod. Res. **49**, 3715–3730 (2011). https://doi.org/10.1080/00207543.2010.492797
62. Iwamura, K., Okubo, N., Tanimizu, Y., Sugimura, N.: Real-time scheduling for holonic manufacturing systems based on estimation of future status. Int. J. Prod. Res. **44**, 3657–3675 (2006). https://doi.org/10.1080/00207540600847129
63. Wang, C., Ghenniwa, H., Shen, W.: Real time distributed shop floor scheduling using an agent-based service-oriented architecture. Int. J. Prod. Res. **46**, 2433–2452 (2008). https://doi.org/10.1080/00207540701738052
64. Yih, Y., Thesen, A.: Semi-Markov decision models for real-time scheduling. Int. J. Prod. Res. **29**, 2331–2346 (1991). https://doi.org/10.1080/00207549108948086
65. Zhou, R., Lee, H.P., Nee, A.Y.C.: Applying Ant Colony Optimisation (ACO) algorithm to dynamic job shop scheduling problems. Int. J. Manuf. Res. **3**, 301–320 (2008). https://doi.org/10.1504/ijmr.2008.019212
66. Metan, G., Sabuncuoglu, I., Pierreval, H.: Real time selection of scheduling rules and knowledge extraction via dynamically controlled data mining. Int. J. Prod. Res. **48**, 6909–6938 (2010). https://doi.org/10.1080/00207540903307581
67. Jain, V.K., Premalatha, S., Baskar, N.: Implementation of supervised statistical data mining algorithm for single machine scheduling. J. Adv. Manag. Res. **9**, 170–177 (2012). https://doi.org/10.1108/09727981211271913
68. Kohn, W., Brayman, V., Littleton, J.: Repair-control of enterprise systems using RFID sensory data. IIE Trans. **37**, 281–290 (2005). https://doi.org/10.1080/07408170590516953
69. Wang, C., Jiang, P.: Manifold learning based rescheduling decision mechanism for recessive disturbances in RFID-driven job shops. J. Intell. Manuf. (2016). https://doi.org/10.1007/s10845-016-1194-1
70. Zhang, Y., Jiang, P., Huang, G.: RFID-based smart Kanbans for Just-In-Time manufacturing. Int. J Mater. Prod. Technol. **33**, 170–184 (2008). https://doi.org/10.1504/ijmpt.2008.01978
71. Qu, T., Yang, H.D., Huang, G.Q., et al.: A case of implementing RFID-based real-time shop-floor material management for household electrical appliance manufacturers. J. Intell. Manuf. **23**, 2343–2356 (2012). https://doi.org/10.1007/s10845-010-0476-2
72. Wang, L., Mohammed, A., Onori, M.: Remote robotic assembly guided by 3D models linking to a real robot. CIRP Ann. Manuf. Technol. **63**, 1–4 (2014). https://doi.org/10.1016/j.cirp.2014.03.013
73. Wang, L., Wang, X.V., Gao, L., Váncza, J.: A cloud-based approach for WEEE remanufacturing. CIRP Ann. **63**, 409–412 (2014). https://doi.org/10.1016/j.cirp.2014.03.114
74. Xu, Y., Song, R., Korba, L., et al.: Distributed device networks with security constraints. IEEE Trans. Ind. Inform. **1**, 217–225 (2005). https://doi.org/10.1109/TII.2005.843826
75. Wang, L., Wong, B., Shen, W., Lang, S.: A Java 3D-enabled cyber workspace. Commun. ACM **45**, 45–49 (2002). https://doi.org/10.1145/581571.581592
76. Monostori, L.: Cyber-physical production systems: roots, expectations and R&D challenges. Procedia CIRP **17**, 9–13 (2014). https://doi.org/10.1016/j.procir.2014.03.115

77. Wang, L., Haghighi, A.: Combined strength of holons, agents and function blocks in cyber-physical systems. J. Manuf. Syst. **40**, 25–34 (2016). https://doi.org/10.1016/j.jmsy.2016.05.002
78. Ji, W., Shi, J., Liu, X., et al.: A novel approach of tool wear evaluation. J. Manuf. Sci. Eng. **139**, 1–8 (2017). https://doi.org/10.1115/1.4037231
79. Ji, W., Wang, L.: Big data analytics based optimisation for enriched process planning: a methodology. Procedia CIRP **63**, 161–166 (2017). https://doi.org/10.1016/j.procir.2017.03.090
80. Wang, L., Shen, W.: DPP: An agent-based approch for distributed process planing. J. Intell. Manuf. **14**, 429–439 (2003)
81. Wang, L., Feng, H.-Y., Cai, N.: Architecture design for distributed process planning. J. Manuf. Syst. **22**, 99–115 (2003)
82. Ji, W., Wang, L.: Big data analytics based fault prediction for shop floor scheduling. J. Manuf. Syst. **43**, 187–194 (2017). https://doi.org/10.1016/j.jmsy.2017.03.008

Geometric Inspection Planning as a Key Element in Industry 4.0

Giovanni Moroni(✉) and Stefano Petrò

Department of Mechanical Engineering, Politecnico di Milano, Milan, Italy
{giovanni.moroni,stefano.petro}@polimi.it

Abstract. In the context of Industry 4.0, inspection is fundamental: if manufacturing opens the loop by converting digital parts into physical parts, inspection closes the loop by turning physical parts into information. The difference between an effective and a useless inspection is its planning. A well-planned inspection will provide the required data and information without wasting time and money.

In this work, we discuss the current role of geometric inspection planning, showing that not only it is a must for Industry 4.0, in order to guarantee a good link between the physical and digital world, but it can take advantage of this framework to improve itself. Methodologies for optimal inspection planning have been already conceptually proved and are just waiting for the required amount of data and information to be available, and Industry 4.0 will be the enabler to fill this gap.

1 What is Inspection Planning?

Manufactured parts must be inspected. Without inspection it is impossible to guarantee both product (which must conform to specifications) and process (which must be under control) quality. Therefore, inspection is among the main contributors to the value of a product [41].

In the field of mechanical engineering dimensional and geometric tolerances are the most relevant specifications. These tolerances define how and how much a real part can differ from its nominal geometry without affecting its functionality. The current solution for dimensional and geometric inspection is coordinate measurement, which consists of sampling a cloud of points on the surface of the part and then fit it to a perfect geometry for extracting all the dimensional and geometric measurements required. Coordinate Measuring Machines (CMM) were the first instruments for coordinate metrology, developed in the 60's. Today, many more Coordinate Measuring Systems (CMS) have been developed, but in general the structure of any CMS is constituted by:

- a base, carrying the mechanical structure and the probe;
- a mechanical structure, allowing the movement of the probe with the required degrees of freedom;
- a probe, capable of sampling the points on the surface of the part;
- a computer numerical control (CNC);

J. Ni et al. (Eds.): AMP 2018, LNME, pp. 293–310, 2018.
https://doi.org/10.1007/978-3-319-89563-5_21

- a computer which elaborates the gathered data and in most cases is the interface between the operator and the system;
- additional fixtures to fix the part to inspect on the machine.

Among the various CMSs, probably the most important distinction to consider is between contact and non-contact measuring systems [25]. The probe of contact CMSs requires a mechanical contact between the part and the probe itself. Contact CMSs are further distinguished in discrete-point measuring systems (which can sample only one point at once), and scanning measuring systems (in which the probe moves along the surface without loosing the contact, sampling points with a given spacing, so that points are acquired in profiles). In non-contacts CMS there is no mechanical contact, and the point is probed by e.g. optical triangulation, autofocus systems, or imaging probing systems.

CMSs are complex systems in which several components must work together, while the whole system is part of the overall manufacturing system. Therefore, measurement by CMSs must be planned carefully. Planning the measurement by CMSs means essentially:

- configuring the selected measuring system;
- defining an adequate sampling strategy.

To support the operator in these tasks the idea of a Computer Aided Inspection Planning (CAIP) software has been introduced [14,65] (Fig. 1). These software is analogue to the Computer Aided Process Planning for manufacturing. Although the concept of CAIP has been introduced in the 90's, no mature CAIP software is available on the market today: most software of this kind is limited to the possibility of simulating the probing operation, without a proper support to the planning.

The configuration of the measuring systems consists in defining the set of part orientations, fixtures and probes.

The fixture must be capable of stiffly holding the part to inspect while guaranteeing the accessibility to the surfaces to inspect, and avoiding any significant elastic deformation. Of course it is possible that no solution is found allowing the access to all surfaces to probe. In this case, more orientations of the part are required. To guarantee the maximum accuracy and minimize the inspection time the number of orientations should be minimized. These requirements are in general in contrast the one with the other, so a lot of care must be taken.

The configuration of the probe is a typical problem of CMM equipped with contact probes. It consists of the selection of a suitable stylus, stylus extensions, probe, and probe extensions [25]. Several issues must be considered, the first one being the accessibility of the surface. Considering for instance the measurement of holes, the stylus must be long enough to reach the bottom surface, the diameter of the probe tip must be smaller than the diameter of the hole, and a 90 angle joint is needed to measure holes perpendicular to the CMM ram. Moreover, it is preferable that the probe is either parallel or perpendicular to the surface to inspect. If a conventional 3-axis CMM is used, this is possible only in the case of simple geometric like planes and cylinders.

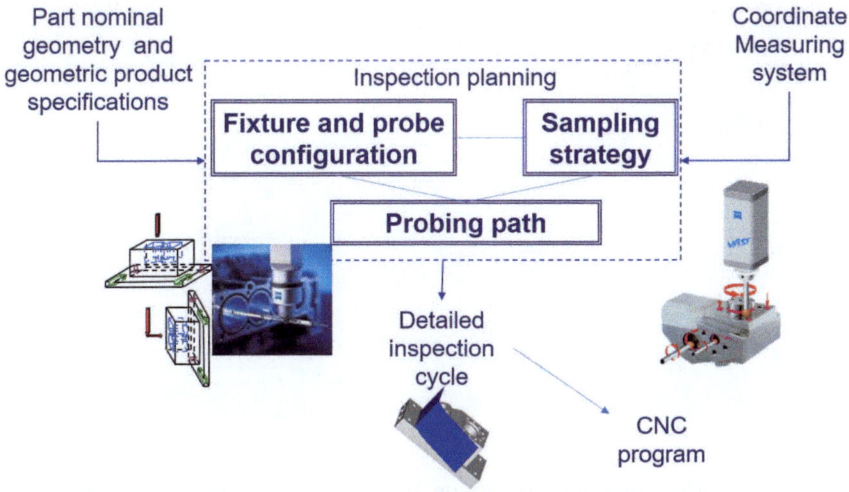

Fig. 1. Inspection planning in coordinate measurement.

The probe configuration problem is made easier by the use of articulating probing systems (5-axis CMM), which can change the orientation of the stylus. But in this case the probe configuration problem is substituted by the problem of the optimal stylus orientation, and path planning is more difficult as well.

Moreover, with contact probes the sampling strategy is among the main contributors to the measurement uncertainty. Probing a reduced number of points, or distributing them in an inefficient way, can lead to large measurement errors (Fig. 2). In general large sample size, guaranteeing a complete coverage of the surface, provides a higher accuracy. But increasing the sample size increases the measurement time as well. There can be constraints on the measurement time, in particular in the case of mass-production. And, in general, longer measurement time means higher inspection cost. This is particularly true for slow contact discrete-point probing CMMs. Modern systems equipped with scanning probing systems can acquire many points in a short time, but the problem still exists of ensuring an adequate coverage of the surface, as in this case points are concentrated in profiles.

Another relevant aspect of the sampling strategy is the definition of the probe path, i.e. the trajectory the probe follows to reach all the points to probe. The definition of the probe path is probably the main obstacle to the automation of the inspection: an incorrect path can lead to a collision between probe and part/fixture, and this will halt the measurement, and can even damage the delicate probe. Path planning is strictly linked to probe and fixture configuration: a well-designed probe and fixture makes path planning easier. Obviously, planning the path of a 5-axis CMM is more difficult than doing the same for a 3-axis CMM, even if the first solution is surely more flexible.

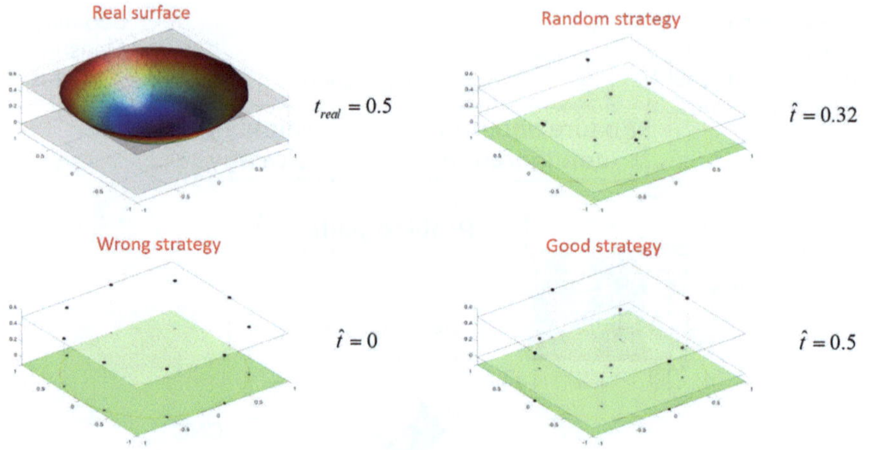

Fig. 2. Effect of the sampling strategy on the estimate of the peak-to-valley flatness deviation of an hollow surface.

In the case of non-contact probes the problem found in inspection planning can be different. When adopting discrete-point or profile scanning non-contact probing systems, the problems are similar to those encountered with contact probing systems. But when areal scanning systems are used, many points are sampled at once, in some cases without the need of moving the probe. In this case different problems arise, for instance:

- the selection of the probe optics, which affect the measurement resolution and scan area;
- the amount of overlap between adjacent scan patches, to allow their stitching;
- the partitioning of the point cloud, as often the sampling points are not associated with a specific geometric element.

In this paper the standard and advanced methodologies for planning the inspection will be reviewed. Then, the position of inspection planning within the Industry 4.0 framework will be highlighted.

2 Standard Practices in Inspection Planning

At most companies and calibration services, inspection planning is a human task. In this task, the operator is supported (or should be supported) by a series of "best practices", indicating which is the correct way of inspecting a part.

The best source for a best practice in inspection should be a series of international standards defining perfect verification operators [31]. Unfortunately, these standards are still missing. Just a few standards exist for form tolerances (cylindricity, roundness, straightness, flatness) [27–30], but they define only complete specification operators: in principle they define only the geometric specification, without any inspection link. For example, in the case of roundness [28] the complete specification operator defines:

> the transmission band for the roundness profile, together with an appropriate stylus tip geometry.

In general, this specification operator supposes that the geometry is defined along one or more profiles, sampled with suitable tip, and filtered with a (gaussian) filter. In the case of flatness and cylindricity, which are defined on surfaces rather than profiles, the standard adds that [26]:

> In practice, it is unrealistic to achieve comprehensive coverage of the flatness [the case of cylindricity is analogue] feature given by the theoretical minimum density of points [...] within an acceptable time span using current technology. Therefore, more limited extraction strategies are employed that give specific rather than general information concerning the deviations from flat form.

Actually, the specification operator is not so "complete". For example, it does not define:

- the wavelength of the filter(s);
- the fitting principle to be used to associate a substitute geometry to the part geometry [24];
- the diameter of the probing sphere.

Even if the specification operator only defines the tolerance, it is evident that it also influences the inspection: if the verification is not coherent with the specification, method uncertainty arises [31]. Unfortunately, the lack of completeness of the specification operator makes method uncertainty impossible to avoid. In addition, even if in principle the cited standards [27–30] should deal only with the specification operator, actually they give specific indication on the verification operator:

- given the wavelength cutoff of the filter, they require at least seven points per cutoff are sampled;
- in the case of flatness and cylindricity, a series of sampling strategies are suggested (Fig. 3), even if not considered mandatory.

These standards have been confirmed in 2016. However, even if still valid, they have been made obsolete by standards that have recently been reviewed: ISO 1101 [35] and ISO 14405 [34]. These standards cover chain links A, B, and C of the GPS matrix [33], that is, they cover all and only the specification aspects of the geometric and dimensional tolerances. However, they have a significant impact on inspection as well. In fact, they not only give a generic definition of the tolerance zone, but define an annotation scheme allowing to specify a lot of information on the tolerance, here including the intended filter to apply to the profile/surfaces, the association principle (least squares, minimum zone, two-points measurement,...),

Fig. 3. Sampling strategies suggested in ISO 12180-2 (redrawn from ISO 12180-2).

and so on. This method of annotation is also a Product Manufacturing Information (PMI) scheme, suitable for the implementation in a CAD model, thus allowing an digital management of this specification information in an Industry 4.0 optic. Moreover, as in the near future the new annotation scheme will not be applied widely, default specification operators have been defined, which will be applied in the case the annotation is not complete: this solves a long known gap in the standard, for which a standard method for association and filtering did not exist. In practice these new standards have significantly mitigated, and in some cases removed, the ambiguity of specification [31].

Even if the ISO 1101 and 14405 standards help the correct planning of the inspection, a lot of details are not defined by them, nor by any other standard, like e.g. the number and distribution of the sampling points on the profile/surface to inspect. In the common practice, these decisions are based on the experience of the operator, who can anyway take advantage of best practices that can be found in literature [19,20]. It is worth noting that most best practices are still linked to contact probing CMMs, considering problems like probe configuration and at most profile sampling. The modern non contact systems, requiring minimal configuration of the probe and allowing surface probing, are not treated in depth.

3 Advanced Methodologies for Sampling Strategy Design

Current practices for sampling strategy design in general see the operator as the sole actor. This is due to the limited circulation of the information about the part - quite often the operator just receives a drawing of the part with the tolerances to inspect. The availability of more information, both at system level and for the single machine during the measurement, can allow more effective solutions for the sampling strategy design. The challenge is the reduction of the measurement uncertainty without an increase of the inspection costs: reducing uncertainty is always possible by sampling more points, but in general this leads to higher costs. Intelligence is needed to optimize the sampling strategy.

The problem of defining the sampling strategy on the surface of a part changes depending on whether a discrete-point, profile scanning or areal scanning probing system is being adopted:

- for discrete point probing, one must define the number of points, their distribution on the part, and the path to travel from a point to next point;

- in the case of profile scanning systems, path planning is a similar problem, while point distribution and sample size are linked each other by the points being located in profiles: in this case the sampling speed and frequency must be defined, so that, once the profile(s) to scan have been defined, the sample size is defined as well;
- areal scanning systems sample many points at once, but with a point density and scan area depending on the scanning parameters (e.g. the optical magnification): again, this influences the number of points and their location.

3.1 Definition of the Point Distribution

In literature many ways of distributing sampling points have been proposed, but only for the case of discrete-point probing, so in the following we will refer to this case. These methodologies can be categorized in blind, manufacturing-based, and adaptive sampling strategies [53].

Blind point distributions include most of the operator designed distributions. In a blind point distribution the actual geometric deviation of the surface is not taken into account in the strategy design: the strategy is based only on the nominal geometry and specified tolerances. As such, points are usually spread across the surface in an almost uniform way, with just some "perturbation" aimed at avoiding an interaction between some periodic form deviation and a regular pattern of the points. Many strategies of this kind have been proposed in the past [9,39,43,63]. This can be considered the "Industry 3.0" way of designing sampling strategies: no knowledge on the manufacturing process is considered, neither in-line nor off-line.

A first a-priori information-based way of defining the point distribution is a manufacturing-based distribution. As well known, any manufacturing process leaves a "manufacturing signature" on the parts it produces: the geometric deviations will cause the part to differ from the nominal geometry, but the the geometric deviation themselves tend to be similar across all manufactured parts, as the process generates them the same way. Therefore, it is possible to develop point distributions specific for a particular manufacturing process: manufacturing-based distributions. In these distributions the sampling points are concentrated in those portions of the part which are more relevant in the definition of the global deviation (usually the parts deviating the most from the nominal geometry, Fig. 4).

Two approaches can be considered in manufacturing-based distributions. The first one is raw-data based [7,8,12,55,66]. In this kind of approach the algorithm defining the optimized sampling strategy is fed with just the sampling points from a series of parts characterized by the same manufacturing signature. The second approach is model-based [13,46,52,55,59,75], in which the algorithm is fed with a model of the manufacturing signature, which is itself usually developed based on some sampling points. It is clear that all these approaches require the management of a large amount of data from the manufacturing process,

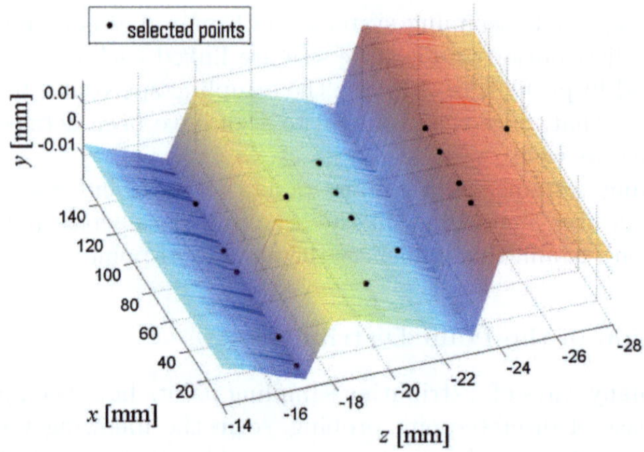

Fig. 4. Manufacturing-based point distribution for a contour milled surface. Peaks and valleys are generated by the different pass of the mill. Points are concentrated on the highest peaks and in the deepest valleys.

and also requires the complete traceability of these data. Recently, some authors has started trying to simulate the manufacturing signature generated by different processes [42].

The main drawback of manufacturing-based distribution is the deep knowledge of the manufacturing they need to be defined. When the lot size is small and the product variability is high, defining manufacturing-based sampling strategy is difficult. Manufacturing simulation can in part mitigate this problem. Another possible way is the continuous study of the signature left by the various manufacturing processes, in order to develop databases of signatures, and be able to experimentally forecast the signature. But also different solutions are possible.

Adaptive sampling is the best solution when little or no information is available on the manufacturing process, but one aims at an higher accuracy than blind strategies. In an adaptive sampling strategy there is no predefined point distribution, except for an initial sample (the definition of which can be seen as a seeding problem). The next sampling point is chosen based on points that have already been probed. A lot of different approaches to adaptive sampling have been proposed in literature. In some approaches [3–5, 22, 50] the next sampling point is chosen according to some optimization algorithm like taboo search, but in most approaches [2, 15, 42, 64, 67, 76] a model is fitted on the local deviations already measured on the part, and the next sampling point is selected based on this fitting (Fig. 5). The main limitation of adaptive sampling is their need of reconfiguring the probe path at each iteration of the algorithm. This can lead to collision risk. Recent researches on this subject, that will be introduced in Sect. 3.3, are trying to mitigate this problem.

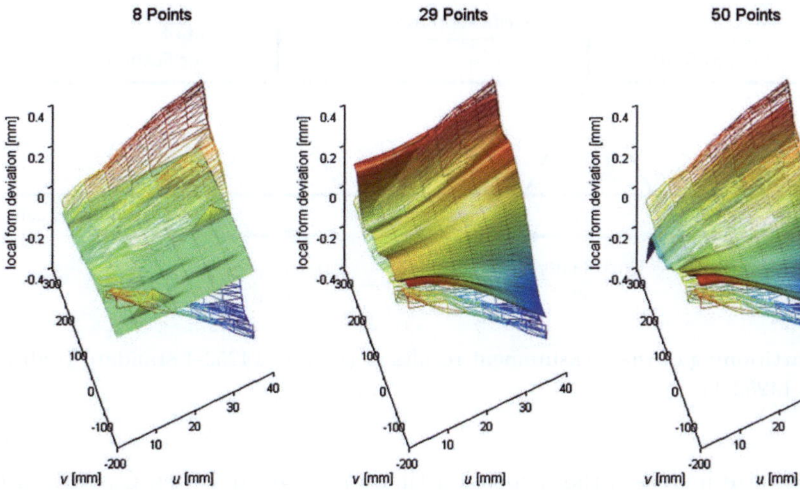

Fig. 5. Adaptive sampling. Modification of the fitted surface (mesh surface in the image) as the sample size increases. The wire-frame represent the real geometric deviation. The points are sequentially selected according to the criterion proposed by Ascione *et al.* [2].

3.2 Choice of the Sample Size: Economic Aspects

Blind, manufacturing-based and adaptive point distributions solve the problem of where to put sampling points, but still the number of sampling points has to be defined.

In general, the number of sampling points is considered the main factor affecting the inspection costs. As the sample size increases, the inspection time increases and as a consequence the inspection cost increases. This is not completely correct: the sample size also affects the measurement accuracy. If the measurement is not accurate, then one incurs in the risk of inspection errors: stating non-conforming a conforming part (type I error) or stating conforming a non-conforming part (type II error). Both these kind of error generate a cost for the company, and as such rules for reducing the impact of measurement errors have been developed and coded in the ISO 14253-1 standard [32]. This standard requires part verification considers the measurement uncertainty [36] by defining an "uncertainty range" (Fig. 6). Parts whose measurement result falls in this range are stated conforming if the aim is to prove non-conformance, and vice versa. In practice the rule tries to minimize the probability of either type I or II error.

In general the cost C_I of an inspection can be calculated as:

$$C_I = C_M + C_E \tag{1}$$

where C_M is the cost of performing the measurement, and C_E is the cost related to type I and type II inspection errors. As it has already been said, when

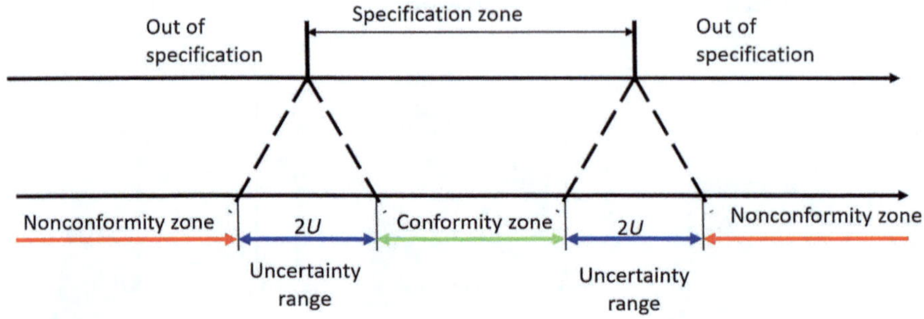

Fig. 6. Partitioning of the measurement results in the ISO 14253-1 standard (redrawn from ISO 14253-1).

the sample size increases the inspection time increases and then C_M increases. But at the same time if the sample size increases the measurement uncertainty decreases (measurement accuracy increases), unless some very odd distribution of the points is chosen. This in general leads to a reduction of inspection errors, both directly (more accurate measurements) and indirectly (the uncertainty range reduces). As such, in general when the sample size increases, C_E decreases. The optimal C_I is then linked to the optimal sample size that is a trade-off between C_E and C_M. In recent years, many authors have proposed scenarios for the optimization of the inspection costs [6,17,18,51,54,55,57,68,69].

3.3 Path Planning and Probe Configuration

Once the sample size and the point distribution have been selected, the sampling strategy cannot be considered completely defined. The sequence in which the sampling points are probed must be defined. This can seem a trivial problem. One could think that solving a "traveling salesman" problem is sufficient, because this minimizes the distance to travel during the probing. However, this solution neglects the problem of probe collisions. Collisions are a serious problem for all probing systems, in particular when fixturing is taken into account.

Although modern CMM software allows probing simulation and can identify collisions, the problem of path optimization is not yet completely solved. In past years, a series of approaches have been proposed to develop the fastest, collision-free path for probing [21,58,62]. In the last decade 5-axis systems, capable of fast scanning complex surfaced by freely orienting the stylus, have spread in industry. This kind of system has reduced the need for probe configuration, thus allowing better automation. However this kind of system is prone to the generation of inertial forces which could reduce the probing accuracy. In the last years, many works have been proposed trying to deal with these issues [37,38,45,47,70–74,78].

Even if 5-axis system reduce the problem of probe configuration, the problem still exists for high accuracy system, for which the reduction of accuracy due to the 5-axis movement is not acceptable. Furthermore, 3-axis systems are still the most diffused systems in industry, due to their reliability. The problem to be faced is the definition of a probe which is able to access all the surfaces to be probed. This problem is strictly linked to the design of the part orientation and fixture, so they are usually treated together in literature. In general, the operator is in charge of solving the problem of configuring the probe and the fixture. To automate the probe and fixture definition, databases configurations are being developed for gathering knowledge, and then develop expert systems capable of supporting the probe design, making it (almost) automatic [1,16,23,48,49,60,61].

4 The Role of Inspection in Industry 4.0

From the discussion proposed so far, it is apparent that the advanced approaches to inspection planning are de facto Industry 4.0 concepts.

Adaptive sampling falls in the context of self-adaptive manufacturing systems. A self-adaptive manufacturing system can adapt to what is currently being manufactured, e.g. a machining center shall modify the tool trajectory in order to compensate the tool wear or inflection. Similarly, adaptive sampling strategies self-adapt to the part that is being measured in order to improve the measurement accuracy. Although proof-of-concepts have been proposed on how to choose next set of sampling points based on the points which have been already sampled, still there are challenges to face before adaptive sampling can spread in industry. Often, the computational time required to choose the next sampling point is not compatible with an on-line application of adaptive sampling. Moreover, the path changes part to part: although methodologies already exist capable defining the probing path, in order to have a real application of the adaptive sampling the path must be defined in real time.

The development of manufacturing-based sampling strategies through the study of the manufacturing signature falls in the Industry 4.0 framework. As the products and production systems change, new manufacturing signatures arise. This could seem an obstacle to the diffusion of manufacturing-based sampling strategies, but in an Industry 4.0 context it enters into a measure/learn/improve cycle. New knowledge is gathered on the product (which is useful also to optimize the manufacturing process), which is virtuously used to reduce the sample size and then the costs. This is an example of so-called adaptive process and inspection planning. The need arises for proper systems capable of conveying the information required to build the knowledge base on which to study the manufacturing signature.

More in general, one could think about the role of geometric inspection in Industry 4.0. Geometric inspection is the main source of information on parts that directly qualifies their functionality. In Industry 4.0 one tries to base the on-line decisions on machine-generated signals. In this sense, the spread of in-situ geometric measurement on the machine tools provides a precious source of information on the parts. But also the off-line measurement on CMMs is anyway needed, as all the methodologies need validation. Therefore, geometric inspection cannot be neglected in Industry 4.0. In conclusion, all this information can be the base to improve the single manufacturing process and the overall system quality.

4.1 Inspection of Additive Manufacturing Parts

Among the enabling technologies of Industry 4.0, there is additive manufacturing. Additive manufacturing is a "complexity for free" manufacturing process. In practice, its cost for manufacturing complex and personalized products does not significantly differ from the cost of similar-sized and mass-produced products. This new paradigm is a revolution in manufacturing that is changing all the chain, from design [77] to inspection.

Unfortunately, the freedom of additive manufacturing does not find many corresponding methodologies in inspection. Many works have been proposed on the inspection of additive manufacturing parts [10,11,44], in part focusing on non-destructive testing, and in part on geometric inspection. In general, the authors agree that, due to the usual presence of complex surfaces in additive manufacturing parts, non-contact measuring systems are preferable. This does not solve anyway the problem of measuring internal surfaces, which are among the strong points of additive manufacturing parts, and are not accessible to conventional non-contact systems, and lattice geometries, which, apart from the accessibility issues they pose, are also characterized by a high surface/volume ratio, and as such would require a long time for the measurement.

Today there is only one kind of measuring system guaranteeing a "complexity for free" approach to measurement: 3D X-Ray Computed Tomography (XCT) [40]. XCT is a reconstruction technique that, based on a series of radiographic projections of the part, reconstruct a volumetric representation of the part itself (Fig. 7). This volumetric (voxel) representation is similar to a 3D photograph of the X-Ray adsorption of the material. In general it is possible to extract surfaces from the voxel representation, that can be analyzed with the conventional coordinate metrology tools.

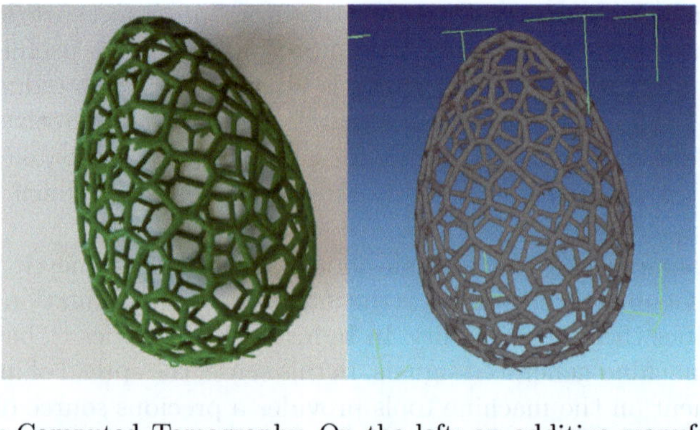

Fig. 7. X-Ray Computed Tomography. On the left: an additive manufacturing part; on the left: its XCT scan.

However, it is worth noting that extracting a surface from a voxel representation is not a straightforward process, and in general leads to an information loss. Moreover, it is possible to manufacture parts by additive manufacturing directly from a voxel representation of them [79]. As such, the authors of the present work think that in the future additive manufacturing parts will be designed, manufactured, and inspected using mainly, or even only, volumetric representations of them [56]. This would be a real change in the paradigm of manufacturing, that is anyway well coherent with Industry 4.0.

5 Conclusions

Industry 4.0 is a final shift of manufacturing from the physical to the digital world, so that products exist in the memory of computers before, in parallel, and after the existence of their physical counterparts. In this context inspection is fundamental, because, if manufacturing opens the loop by converting digital parts into physical parts, inspection closes the loop by turning physical parts into information. Geometric inspection, for mechanical components, is the perfect method for creating digital representations of real manufactured parts.

But obviously any paradigm can be applied effectively or badly. The difference between an effective and a useless inspection is its planning. A well planned inspection will provide the required information without wasting time and money and with the correct degree of accuracy.

We can conclude that geometric inspection and inspection planning is a core technology for Industry 4.0. As the main tool to verify the functionality of mechanical components, geometric inspection is the main source of information to close the physical part/digital part loop. And advanced inspection planning, by applying manufacturing-based and adaptive sampling strategies together with an automated definition of probe path and probe/fixture configuration, is a perfect example of self-adaptive system, the basis of Industry 4.0. As it is necessary and coherent, we can say that geometric inspection planning is a key element in Industry 4.0.

References

1. Anagnostakis, D., Ritchie, J., Lim, T., Sivanathan, A., Dewar, R., Sung, R., Bosché, F., Carozza, L.: Knowledge capture in CMM inspection planning: barriers and challenges. Procedia CIRP **52**, 216–221 (2016). https://doi.org/10.1016/j.procir.2016.07.045
2. Ascione, R., Moroni, G., Petrò, S., Romano, D.: Adaptive inspection in coordinate metrology based on kriging models. Precis. Eng. **37**(1), 44–60 (2013). https://doi.org/10.1016/j.precisioneng.2012.06.006
3. Badar, M.A., Raman, S., Pulat, P.S.: Intelligent search-based selection of sample points for straightness and flatness estimation. J. Manuf. Sci. Eng. **125**(2), 263–271 (2003). https://doi.org/10.1115/1.1556859

4. Badar, M.A., Raman, S., Pulat, P.S.: Experimental verification of manufacturing error pattern and its utilization in form tolerance sampling. Int. J. Mach. Tool Manufact. **45**(1), 63–73 (2005). https://doi.org/10.1016/j.ijmachtools.2004.06.017
5. Badar, M.A., Raman, S., Pulat, P.S., Shehab, R.L.: Experimental analysis of search-based selection of sample points for straightness and flatness estimation. J. Manuf. Sci. Eng. **127**(1), 96–103 (2005). https://doi.org/10.1115/1.1828051
6. Bosch-Mauchand, M., Siadat, A., Perry, N., Bernard, A.: VCS: value chains simulator, a tool for value analysis of manufacturing enterprise processes (a value-based decision support tool). J. Intell. Manuf. **23**(4), 1389–1402 (2012). https://doi.org/10.1007/s10845-010-0452-x
7. Capello, E., Semeraro, Q.: The harmonic fitting method for the assessment of the substitute geometry estimate error. Part I: 2D and 3D theory. Int. J. Mach. Tool Manufact. **41**(8), 1071–1102 (2001). https://doi.org/10.1016/S0890-6955(01)00020-7
8. Capello, E., Semeraro, Q.: The harmonic fitting method for the assessment of the substitute geometry estimate error. Part II: statistical approach, machining process analysis and inspection plan optimisation. Int. J. Mach. Tool Manufact. **41**(8), 1103–1129 (2001). https://doi.org/10.1016/S0890-6955(01)00019-0
9. Chan, F.M.M., King, T.G., Stout, K.J.: The influence of sampling strategy on a circular feature in coordinate measurements. Measurement **19**(2), 73–81 (1996). https://doi.org/10.1016/S0263-2241(96)00067-X
10. Chua, Z., Ahn, I., Moon, S.: Process monitoring and inspection systems in metal additive manufacturing: status and applications. Int. J. Precis. Eng. Manuf. Green Technol. **4**(2), 235–245 (2017). https://doi.org/10.1007/s40684-017-0029-7
11. Clark, D., Sharples, S., Wright, D.: Development of online inspection for additive manufacturing products. Insight Non Destr. Test Cond. Monit. **53**(11), 610–613 (2011). https://doi.org/10.1784/insi.2011.53.11.610
12. Colosimo, B.M., Gutierrez Moya, E., Moroni, G., Petrò, S.: Statistical sampling strategies for geometric tolerance inspection by CMM. Econ. Qual. Contr. **23**(1), 109–121 (2008). https://doi.org/10.1515/EQC.2008.109
13. Colosimo, B.M., Moroni, G., Petrò, S.: A tolerance interval based criterion for optimizing discrete point sampling strategies. Precis. Eng. **34**(4), 745–754 (2010). https://doi.org/10.1016/j.precisioneng.2010.04.004
14. Denti, P., Dondossola, G., Fiorentini, F., Moroni, G., Palezzato, P., Semeraro, Q., Tealdi, A.: CAIP: computer aided inspection planning. In: Proceedings of the 5th IMCC-International Manufacturing Conference in China, pp. 1–5 (1991)
15. Edgeworth, R., Wilhelm, R.G.: Adaptive sampling for coordinate metrology. Precis. Eng. **23**(3), 144–154 (1999). https://doi.org/10.1016/S0141-6359(99)00004-5
16. Emmer, C., Pfouga, A., Stjepandic, J., Tiringer, H.: Novel approach with 3D measurement data management for industry 4.0. Adv. Transdiscipl. Eng. **5**, 906–913 (2017). https://doi.org/10.3233/978-1-61499-779-5-906. Conference of 24th ISPE Inc. International Conference on Transdisciplinary Engineering, TE 2017; Conference Date: 10 July 2017 Through 14 July 2017; Conference Code:128611
17. Etienne, A., Dantan, J.Y., Qureshi, J., Siadat, A.: Variation management by functional tolerance allocation and manufacturing process selection. Int. J. Interact. Des. Manuf. **2**(4), 207–218 (2008). https://doi.org/10.1007/s12008-008-0055-3
18. Etienne, A., Mirdamadi, S., Mohammadi, M., Babaeizadeh Malmiry, R., Antoine, J.F., Siadat, A., Dantan, J.Y., Tavakkoli, R., Martin, P.: Cost engineering for variation management during the product and process development. Int. J. Interact. Des. Manuf., 1–12 (2016, in press). https://doi.org/10.1007/s12008-016-0318-3

19. Flack, D.: CMM measurement strategies. Technical report PDB: 2771, National Physical Laboratory (2014)
20. Flack, D.: CMM probing. Technical Report PDB: 2773, National Physical Laboratory (2014)
21. Grieco, A., Moroni, G., Nucci, F., Polini, W., Rasella, M.: Measuring points visiting sequence in CMM path planning. In: Proceedings of the 2nd CIRP International Seminar on Intelligence Computation in Manufacturing Engineering, pp. 485–490 (2000)
22. He, G., Huang, X., Ma, W., Sang, Y., Yu, G.: CAD-based measurement planning strategy of complex surface for five axes on machine verification. Int. J. Adv. Manuf. Technol. **91**(5–8), 2101–2111 (2017). https://doi.org/10.1007/s00170-016-9932-2
23. Hwang, C.Y., Tsai, C.Y., Chang, C.: Efficient inspection planning for coordinate measuring machines. Int. J. Adv. Manuf. Technol. **23**(9–10), 732–742 (2004)
24. International Organization for Standardization: ISO 14660-1: Geometrical product specifications (GPS) - geometrical features - part 1: General terms and definitions (1999)
25. International Organization for Standardization: ISO 10360-1: Geometrical Product Specifications (GPS) - Acceptance and reverification tests for coordinate measuring machines (CMM) - Part 1: Vocabulary (2000)
26. International Organization for Standardization: ISO/TS 12781-2: Geometrical Product Specifications (GPS) - Flatness - Part 2: Specification operators (2003)
27. International Organization for Standardization: ISO 12180-2: Geometrical Product Specifications (GPS) - Cylindricity - Part 2: Specification operators (2011)
28. International Organization for Standardization: ISO 12181-2: Geometrical Product Specifications (GPS) - Roundness - Part 2: Specification operators (2011)
29. International Organization for Standardization: ISO 12780-2: Geometrical Product Specifications (GPS) - Straightness - Part 2: Specification operators (2011)
30. International Organization for Standardization: ISO 12781-2: Geometrical Product Specifications (GPS) - Flatness - Part 2: Specification operators (2011)
31. International Organization for Standardization: ISO 17450-2: Geometrical product specifications (GPS) - general concepts - part 2: Basic tenets, specifications, operators and uncertainties (2012)
32. International Organization for Standardization: ISO 14253-1: Geometrical product specifications (GPS) - inspection by measurement of workpieces and measuring equipment - part 1: Decision rules for proving conformance or nonconformance with specifications (2013)
33. International Organization for Standardization: ISO 14238: Geometrical product specifications (GPS) - Matrix model (2015)
34. International Organization for Standardization: ISO 14405-1: Geometrical product specifications (GPS) - Dimensional tolerancing - Part 1: Linear sizes (2016)
35. International Organization for Standardization: ISO 1101: Geometrical product specifications (GPS) - Geometrical tolerancing - Tolerances of form, orientation, location and run-out (2017)
36. ISO/IEC: ISO/IEC GUIDE 99:2007(E/F): International vocabulary of metrology - basic and general concepts and associated terms (VIM) (2007)
37. Jia, Z.Y., Ma, J.W., Wang, F.J., Ding, Y.M.: Investigation of a measurement scheme based on IGES. Measurement **47**(1), 658–668 (2014). https://doi.org/10.1016/j.measurement.2013.09.017

38. Kamrani, A., Abouel Nasr, E., Al-Ahmari, A., Abdulhameed, O., Mian, S.: Feature-based design approach for integrated CAD and computer-aided inspection planning. Int. J. Adv. Manuf. Technol. **76**(9–12), 2159–2183 (2014). https://doi.org/10.1007/s00170-014-6396-0
39. Kim, W.S., Raman, S.: On the selection of flatness measurement points in coordinate measuring machine inspection. Int. J. Mach. Tool Manufact. **40**(3), 427–443 (2000)
40. Kruth, J., Bartscher, M., Carmignato, S., Schmitt, R., De Chiffre, L., Weckenmann, A.: Computed tomography for dimensional metrology. CIRP Ann. - Manuf. Technol. **60**(2), 821–842 (2011). https://doi.org/10.1016/j.cirp.2011.05.006
41. Kunzmann, H., Pfeifer, T., Schmitt, R., Schwenke, H., Weckenmann, A.: Productive metrology - adding value to manufacture. CIRP Ann. - Manuf. Technol. **54**(2), 155–168 (2005). https://doi.org/10.1016/S0007-8506(07)60024-9
42. Lalehpour, A., Barari, A.: Developing skin model in coordinate metrology using a finite element method. Measurement **109**, 149–159 (2017). https://doi.org/10.1016/j.measurement.2017.05.056
43. Lee, G., Mou, J., Shen, Y.: Sampling strategy design for dimensional measurement of geometric features using coordinate measuring machine. Int. J. Mach. Tool Manufact. **37**(7), 917–934 (1997). https://doi.org/10.1016/S0890-6955(96)00096-X
44. Lu, Q., Wong, C.: Additive manufacturing process monitoring and control by non-destructive testing techniques: challenges and in-process monitoring. Virtual Phys. Prototyping **13**(2), 39–48 (2018). https://doi.org/10.1080/17452759.2017.1351201
45. Majstorovic, V., Stojadinovic, S., Sibalija, T.: Development of a knowledge base for the planning of prismatic parts inspection on CMM. Acta IMEKO **4**(2), 10–17 (2015)
46. Majstorovic, V., Stojadinovic, S., Zivkovic, S., Djurdjanovic, D., Jakovljevic, Z., Gligorijevic, N.: Cyber-physical manufacturing metrology model (CPM3) for sculptured surfaces - turbine blade application. Procedia CIRP **63**, 658–663 (2017). https://doi.org/10.1016/j.procir.2017.03.093
47. Mansour, G., Tsagaris, A., Mansour, M.: Intelligent interaction with CMM. Int. J. Mech. Mech. Eng. **15**(3), 53–58 (2015)
48. Martìnez-Pellitero, S., Barreiro, J., Cuesta, E., Fernández-Abia, A.: KBE rules oriented to resources management in coordinates inspection by contact. J. Manuf. Syst. **37**, 149–163 (2015). https://doi.org/10.1016/j.jmsy.2015.07.005
49. Martínez-Pellitero, S., Barreiro, J., Cuesta, E., Fernández-Abia, A.: Knowledge base model for automatic probe orientation and configuration planning with CMMs. Rob Comput. Integr. Manuf. **49**, 285–300 (2018). https://doi.org/10.1016/j.rcim.2017.08.012
50. Martins, T., Tsuzuki, M., Takimoto, R., Barari, A., Gallo, G., Garcia, M., Tiba, H.: Algorithmic iterative sampling in coordinate metrology plan for coordinate metrology using dynamic uncertainty analysis. In: Proceedings of the 12th IEEE International Conference on Industrial Informatics, pp. 316–319 (2014). https://doi.org/10.1109/INDIN.2014.6945531
51. Mohammadi, M., Siadat, A., Dantan, J.Y., Tavakkoli-Moghaddam, R.: Mathematical modelling of a robust inspection process plan: Taguchi and Monte Carlo methods. Int. J. Prod. Res. **53**(7), 2202–2224 (2015). https://doi.org/10.1080/00207543.2014.980460
52. Moroni, G., Petrò, S.: Virtual CMM based sampling strategy optimization. In: Giordano, M., Mathieu, L., Villeneuve, F. (eds.) Product Life-Cycle Management: Geometric Variations, chap. 22, pp. 385–404. Wiley-ISTE, London (2010)

53. Moroni, G., Petrò, S.: Coordinate measuring machine measurement planning. In: Colosimo, B.M., Senin, N. (eds.) Geometric Tolerances - Impact on Product Design, Quality Inspection and Statistical Process Monitoring, Chap. 4, pp. 111–158. Springer, London (2011). https://doi.org/10.1007/978-1-84996-311-4
54. Moroni, G., Petrò, S.: Inspection strategies and multiple geometric tolerances. In: Jiang, X., Mathieu, L., Weckenmann, A. (eds.) Procedia CIRP, vol. 10, pp. 54–60. Huddersfield, UK (2013). https://doi.org/10.1016/j.procir.2013.08.012
55. Moroni, G., Petrò, S.: Optimal inspection strategy planning for geometric tolerance verification. Precis. Eng. **38**(1), 71–81 (2014). https://doi.org/10.1016/j.precisioneng.2013.07.006
56. Moroni, G., Petrò, S., Polini, W.: Geometrical product specification and verification in additive manufacturing. CIRP Ann. - Manuf. Technol. **66**(1), 157–160 (2017). https://doi.org/10.1016/j.cirp.2017.04.043
57. Moroni, G., Petrò, S., Tolio, T.: Early cost estimation for tolerance verification. CIRP Ann. - Manuf. Technol. **60**(1), 195–198 (2011). https://doi.org/10.1016/j.cirp.2011.03.010
58. Moroni, G., Polini, W., Rasella, M.: Feature based path planning for CMMs. In: Proceedings of the 31st CIRP International Seminar on Manufacturing Systems, pp. 161–167 (2001)
59. Moroni, G., Polini, W., Rasella, M.: Manufacturing signatures and CMM sampling strategies. In: Proceedings of the ASPE Summer Topical Meeting on Coordinate Measuring Machines, pp. 57–62 (2003)
60. Moroni, G., Polini, W., Semeraro, Q.: Knowledge based method for touch probe configuration in an automated inspection system. J. Mater. Process. Technol. **76**(1–3), 153–160 (1998)
61. Moroni, G., Polini, W., Semeraro, Q.: Part setup and probe design in CMM inspection. In: Proceedings of the 32nd CIRP International Seminar on Manufacturing Systems, pp. 63–71 (1999)
62. Moroni, G., Rasella, M.: An hybrid approach in cmm path planning. In: Proceedings of the 3rd CIRP International Seminar on Intelligent Computation in Manufacturing Engineering, pp. 473–478 (2002)
63. Namboothiri, V.N.N., Shunmugam, M.S.: On determination of sample size in form error evaluation using coordinate metrology. Int. J. Prod. Res. **37**(4), 793–804 (1999)
64. Pedone, P., Vicario, G., Romano, D.: Kriging-based sequential inspection plans for coordinate measuring machines. Appl. Stoch. Model Bus Ind. **25**(2), 133–149 (2009). https://doi.org/10.1002/asmb.746
65. Polini, W., Moroni, G.: A frame for a computer aided inspection planning system. Int. J. Eng. Technol. **4**(1), 125–138 (2015). https://doi.org/10.14419/ijet.v4i1.3937
66. Raghunandan, R., Rao, P.V.: Selection of an optimum sample size for flatness error estimation while using coordinate measuring machine. Int. J. Mach. Tool Manufact. **47**(3–4), 477–482 (2007). https://doi.org/10.1016/j.ijmachtools.2006.06.008
67. Rossi, A.: A form of deviation-based method for coordinate measuring machine sampling optimization in an assessment of roundness. Proc. IME B J. Eng. Manufact. **215**(11), 1505–1518 (2001). https://doi.org/10.1243/0954405011519411
68. Savio, E.: A methodology for the quantification of value-adding by manufacturing metrology. CIRP Ann. - Manuf. Technol. **61**(1), 503–506 (2012). https://doi.org/10.1016/j.cirp.2012.03.019
69. Savio, E., De Chiffre, L., Carmignato, S., Meinertz, J.: Economic benefits of metrology in manufacturing. CIRP Ann. - Manuf. Technol. **65**(1), 495–498 (2016). https://doi.org/10.1016/j.cirp.2016.04.020

70. Sibalija, T., Zivkovic, S., Fountas, N., Majstorovic, V., Macuzic, J., Vaxevanidis, N.: Virtual optimisation of CAI process parameters for the sculptured surface inspection. Procedia CIRP **57**, 574–579 (2016). https://doi.org/10.1016/j.procir.2016.11.099
71. Stojadinovic, S., Majstorovic, V., Durakbasa, N.: A feature -based path planning for inspection prismatic parts on CMM. In: Proceedings of the 21st IMEKO World Congress on Measurement in Research and Industry (2015)
72. Stojadinovic, S., Majstorovic, V., Durakbasa, N.: An advanced CAI model for inspection planning on CMM. LNME, pp. 57–65 (2017). https://doi.org/10.1007/978-3-319-56430-2_5
73. Stojadinovic, S., Majstorovic, V., Durakbasa, N., Sibalija, T.: Ants colony optimisation of a measuring path of prismatic parts on a CMM. Metrol Meas. Syst. **23**(1), 119–132 (2016). https://doi.org/10.1515/mms-2016-0011
74. Stojadinovic, S., Majstorovic, V., Durakbasa, N., Sibalija, T.: Towards an intelligent approach for cmm inspection planning of prismatic parts. Measurement **92**, 326–339 (2016). https://doi.org/10.1016/j.measurement.2016.06.037
75. Summerhays, K.D., Henke, R.P., Baldwin, J.M., Cassou, R.M., Brown, C.W.: Optimizing discrete point sample patterns and measurement data analysis on internal cylindrical surfaces with systematic form deviations. Precis. Eng. **26**(1), 105–121 (2002). https://doi.org/10.1016/S0141-6359(01)00106-4
76. Sun, L., Ren, M., Yin, Y.: Domain-specific gaussian process-based intelligent sampling for inspection planning of complex surfaces. Int. J. Prod. Res. **55**(19), 5564–5578 (2017). https://doi.org/10.1080/00207543.2017.1301688
77. Thompson, M., Moroni, G., Vaneker, T., Fadel, G., Campbell, R., Gibson, I., Bernard, A., Schulz, J., Graf, P., Ahuja, B., Martina, F.: Design for additive manufacturing: trends, opportunities, considerations, and constraints. CIRP Ann. - Manuf. Technol. **65**(2), 737–760 (2016). https://doi.org/10.1016/j.cirp.2016.05.004
78. Zhou, Z., Zhang, Y., Tang, K.: Sweep scan path planning for efficient freeform surface inspection on five-axis cmm. CAD Comput. Aided Des. **77**, 1–17 (2016). https://doi.org/10.1016/j.cad.2016.03.003
79. ZMorph: Voxelizer 3D (2016). www.voxelizer.com

MIX
Papier aus verantwortungsvollen Quellen
Paper from responsible sources
FSC® C105338

If you have any concerns about our products,
you can contact us on
ProductSafety@springernature.com

In case Publisher is established outside the EU,
the EU authorized representative is:
Springer Nature Customer Service Center GmbH
Europaplatz 3, 69115 Heidelberg, Germany

Printed by Libri Plureos GmbH
in Hamburg, Germany